Air Quality Research Using Remote Sensing

Air Quality Research Using Remote Sensing

Editors

Maria João Costa
Daniele Bortoli

MDPI • Basel • Beijing • Wuhan • Barcelona • Belgrade • Manchester • Tokyo • Cluj • Tianjin

Editors
Maria João Costa
Institute of Earth Sciences and
Earth Remote Sensing Lab
Portugal

Daniele Bortoli
Institute of Earth Sciences and
Earth Remote Sensing Lab
Portugal

Editorial Office
MDPI
St. Alban-Anlage 66
4052 Basel, Switzerland

This is a reprint of articles from the Special Issue published online in the open access journal *Remote Sensing* (ISSN 2072-4292) (available at: https://www.mdpi.com/journal/remotesensing/special_issues/AQ_RS).

For citation purposes, cite each article independently as indicated on the article page online and as indicated below:

LastName, A.A.; LastName, B.B.; LastName, C.C. Article Title. *Journal Name* **Year**, *Volume Number*, Page Range.

ISBN 978-3-0365-5893-6 (Hbk)
ISBN 978-3-0365-5894-3 (PDF)

Contents

About the Editors

Maria João Costa

Maria João Costa is Associate Professor at the Department of Physics, University of Évora, Portugal. She completed her habilitation in atmospheric and climate physics in 2020 and her PhD in physics, awarded by the University of Évora, in 2004. She is an integrated member of the Institute of Earth Sciences (ICT) and the coordinator of the Atmospheric Sciences, Water and Climate Group at ICT. She is also the director of the Earth and Space Sciences Doctoral Program and of the Earth Remote Sensing Laboratory (EaRSLab) at the University of Évora. Her main research interests concern cloud and aerosol physics, air and water quality, remote sensing, solar radiation, and atmospheric radiative transfer.

Daniele Bortoli

Daniele Bortoli is Assistant Professor at the Department of Physics, University of Évora, Portugal. He completed his PhD in physics in 2005 at the University of Évora. Since 2007, he has also worked as a non-stipendiary invited research fellow of the Italian Research Council. In 2007 (International Polar Year) he participated in the creation of the Portuguese Polar Program. He participated in scientific collaborations with the Antarctic Projects of Italy, India, Bulgaria, and New Zealand. His main research fields are the physics and chemistry of atmospheric compounds at high-/mid-latitudes, the ozone hole in the Arctic and Antarctica, air quality, the characterization of solar radiation, and the development of remote sensing instrumentation for the study of the atmospheric composition

remote sensing MDPI

Editorial

Editorial for the Special Issue "Air Quality Research Using Remote Sensing"

Maria João Costa [1,2,3,*] and Daniele Bortoli [1,2,3]

1 Institute of Earth Sciences (ICT), Institute of Research and Advanced Training, University of Évora, 7000-671 Évora, Portugal
2 Earth Remote Sensing Laboratory (EaRSLab), Institute of Research and Advanced Training, University of Évora, 7000-671 Évora, Portugal
3 Department of Physics, School of Sciences and Technology, University of Évora, 7000-671 Évora, Portugal
* Correspondence: mjcosta@uevora.pt

Air pollution is a worldwide environmental hazard with serious consequences for health and climate as well as for agriculture, ecosystems, and cultural heritage, among others. According to the WHO, there are 8 million premature deaths every year resulting from exposure to ambient air pollution. In addition, more than 90% of the world's population lives in places where air quality is poor, exceeding the recommended limits; most of these places are in low- or middle-income countries. Air pollution and climate influence each other through complex physicochemical interactions in the atmosphere, altering the Earth's energy balance, with implications for climate change and air quality.

It is vital to measure specific atmospheric parameters and pollutant concentrations, monitor their variations, and analyze different scenarios with the aim of assessing air pollution levels and developing early-warning and forecast systems; such developments provide a means of improving air quality and assuring public health in favor of a reduction in air pollution casualties and a mitigation of climate change phenomena. Eleven research papers were published in this Special Issue, comprising one communication paper [1], seven articles [2–8], two technical notes [9,10], and one letter [11]. The published research signals the potential of applying remote sensing data in air quality studies, including combination with in situ data [1,3,6,8], modeling approaches [2,9,11], and the synergy of different instrumentations and techniques [4,5,7,10]. Significant pollutants considered in the studies include aerosols—using $PM_{2.5}$ and aerosol optical depth (AOD) as quantification variables [1,2,4,5,9]—nitrogen dioxide (NO_2) [7,8,11], formaldehyde (HCHO) [3], and carbon monoxide (CO) [6,10], among others [10].

The influence of meteorology on seasonal $PM_{2.5}$ concentrations and AOD was analyzed, providing insight that may contribute to improving the retrievals of surface $PM_{2.5}$ from satellite AOD [2]. The mechanisms of $PM_{2.5}$ regional transport from biomass burning in Southeast Asia were examined for a case study during springtime, with an emphasis on the role of meteorology [9]. Furthermore, the influence of urban form on $PM_{2.5}$ surface concentrations was investigated, providing a seasonal analysis method which is relevant for urban planning strategies surrounding air quality improvement in populated areas [4]. New methods combining remote sensing data and additional ancillary datasets with machine learning algorithms were proposed, allowing us to retrieve surface $PM_{2.5}$ concentrations [1] and AOD [5]. Such prediction schemes can provide significant information for advances in air quality research.

The importance of drones for monitoring limited areas, often in areas of difficult access, is increasingly being recognized. An application of drones over a wastewater treatment plant, permitting the real-time monitoring of gaseous pollutants, was demonstrated in [10], and open challenges were identified.

An evaluation of satellite retrievals of HCHO, a recognized hazardous air pollutant, using ground-based data was carried out for a ten-year period [3]. Results suggest that

Citation: Costa, M.J.; Bortoli, D. Editorial for the Special Issue "Air Quality Research Using Remote Sensing". *Remote Sens.* **2022**, *14*, 5566. https://doi.org/10.3390/rs14215566

Received: 19 October 2022
Accepted: 31 October 2022
Published: 4 November 2022

satellite results are more prone to seasonal variations than ground-based measurements and show evidence of a latitude dependency with a seasonal bias. Studies of satellite retrievals in comparison with ground-based measurements are very pertinent considering the use of new Earth observation sensors for air quality monitoring. CO concentration variability was also assessed from both satellite and ground-based measurements [6]. The authors of [6] examined the horizontal and vertical variations in CO concentrations caused by the COVID-19 lockdown in 2020 and compared the contributions from different sources with results from 2019.

The distribution and trends of tropospheric NO_2 at a global scale were analyzed for a 13-year period using satellite retrievals [8]. Ground-based measurements were also used for comparison purposes. Hotspots of high concentrations of this air pollutant were identified, as well as regions of negative and positive trends during the period of study. The highest concentrations of tropospheric NO_2 were detected in recent years, indicating the importance of monitoring anthropogenic emissions and implementing further actions for their reduction. The authors of [11] used satellite data combined with air quality modelling to estimate the impact of the COVID-19 lockdown on tropospheric NO_2, while analyzing the role of meteorology and sampling variability in the process. Satellite data were used in combination with data from ground-based NO_2 concentration measurements, NO_x emissions, land uses, road networks, and population densities, in order to develop a regression model for determining surface NO_2 with a high spatial resolution [7]. The model was applied at a city scale, with the results highlighting the key role of Earth observation technologies in support of exposure assessments and policy development for air quality control.

The publications in this Special Issue highlight the importance and topicality of air quality studies and the potential of remote sensing, particularly from Earth observation platforms, in contributing to this topic.

Author Contributions: Conceptualization, M.J.C. and D.B.; methodology, M.J.C.; resources, M.J.C. and D.B.; writing—original draft preparation, M.J.C.; writing—review and editing, M.J.C. and D.B. All authors have read and agreed to the published version of the manuscript.

Funding: This research received no external funding.

Acknowledgments: The Guest Editors would like to thank all authors who contributed to this Special Issue for sharing their scientific findings in this forum. We would also like to thank the reviewers for their valuable work and the editorial team for all the support in the process.

Conflicts of Interest: The authors declare no conflict of interest.

References

1. Ibrahim, S.; Landa, M.; Pešek, O.; Brodský, L.; Halounová, L. Machine Learning-Based Approach Using Open Data to Estimate $PM_{2.5}$ over Europe. *Remote Sens.* **2022**, *14*, 3392. [CrossRef]
2. Qi, L.; Zheng, H.; Ding, D.; Ye, D.; Wang, S. Effects of Meteorology Changes on Inter-Annual Variations of Aerosol Optical Depth and Surface $PM_{2.5}$ in China—Implications for $PM_{2.5}$ Remote Sensing. *Remote Sens.* **2022**, *14*, 2762. [CrossRef]
3. Wang, P.; Holloway, T.; Bindl, M.; Harkey, M.; De Smedt, I. Ambient Formaldehyde over the United States from Ground-Based (AQS) and Satellite (OMI) Observations. *Remote Sens.* **2022**, *14*, 2191. [CrossRef]
4. Liu, Y.; He, L.; Qin, W.; Lin, A.; Yang, Y. The Effect of Urban Form on $PM_{2.5}$ Concentration: Evidence from China's 340 Prefecture-Level Cities. *Remote Sens.* **2022**, *14*, 7. [CrossRef]
5. Ibrahim, S.; Landa, M.; Pešek, O.; Pavelka, K.; Halounova, L. Space-Time Machine Learning Models to Analyze COVID-19 Pandemic Lockdown Effects on Aerosol Optical Depth over Europe. *Remote Sens.* **2021**, *13*, 3027. [CrossRef]
6. Zhou, M.; Jiang, J.; Langerock, B.; Dils, B.; Sha, M.K.; De Mazière, M. Change of CO Concentration Due to the COVID-19 Lockdown in China Observed by Surface and Satellite Observations. *Remote Sens.* **2021**, *13*, 1129. [CrossRef]
7. Zhang, L.; Yang, C.; Xiao, Q.; Geng, G.; Cai, J.; Chen, R.; Meng, X.; Kan, H. A Satellite-Based Land Use Regression Model of Ambient NO_2 with High Spatial Resolution in a Chinese City. *Remote Sens.* **2021**, *13*, 397. [CrossRef]
8. Jamali, S.; Klingmyr, D.; Tagesson, T. Global-Scale Patterns and Trends in Tropospheric NO_2 Concentrations, 2005–2018. *Remote Sens.* **2020**, *12*, 3526. [CrossRef]
9. Yang, Q.; Zhao, T.; Tian, Z.; Kumar, K.R.; Chang, J.; Hu, W.; Shu, Z.; Hu, J. The Cross-Border Transport of $PM_{2.5}$ from the Southeast Asian Biomass Burning Emissions and Its Impact on Air Pollution in Yunnan Plateau, Southwest China. *Remote Sens.* **2022**, *14*, 1886. [CrossRef]

10. Burgués, J.; Esclapez, M.D.; Doñate, S.; Pastor, L.; Marco, S. Aerial Mapping of Odorous Gases in a Wastewater Treatment Plant Using a Small Drone. *Remote Sens.* **2021**, *13*, 1757. [CrossRef]
11. Griffin, D.; McLinden, C.A.; Racine, J.; Moran, M.D.; Fioletov, V.; Pavlovic, R.; Mashayekhi, R.; Zhao, X.; Eskes, H. Assessing the Impact of Corona-Virus-19 on Nitrogen Dioxide Levels over Southern Ontario, Canada. *Remote Sens.* **2020**, *12*, 4112. [CrossRef]

remote sensing

Communication

Machine Learning-Based Approach Using Open Data to Estimate $PM_{2.5}$ over Europe

Saleem Ibrahim [1,*], Martin Landa [1], Ondřej Pešek [1], Lukáš Brodský [2] and Lena Halounová [1]

[1] Department of Geomatics, Faculty of Civil Engineering, Czech Technical University in Prague, 166 29 Prague, Czech Republic; martin.landa@fsv.cvut.cz (M.L.); ondrej.pesek@fsv.cvut.cz (O.P.); lena.halounova@fsv.cvut.cz (L.H.)
[2] Department of Applied Geoinformatics and Cartography, Faculty of Science, Charles University, 128 43 Prague, Czech Republic; lukas.brodsky@natur.cuni.cz
* Correspondence: saleem.ibrahim@fsv.cvut.cz

Abstract: Air pollution is currently considered one of the most serious problems facing humans. Fine particulate matter with a diameter smaller than 2.5 micrometres ($PM_{2.5}$) is a very harmful air pollutant that is linked with many diseases. In this study, we created a machine learning-based scheme to estimate $PM_{2.5}$ using various open data such as satellite remote sensing, meteorological data, and land variables to increase the limited spatial coverage provided by ground-monitors. A space-time extremely randomised trees model was used to estimate $PM_{2.5}$ concentrations over Europe, this model achieved good results with an out-of-sample cross-validated R^2 of 0.69, RMSE of 5 $\mu g/m^3$, and MAE of 3.3 $\mu g/m^3$. The outcome of this study is a daily full coverage $PM_{2.5}$ dataset with 1 km spatial resolution for the three-year period of 2018–2020. We found that air quality improved throughout the study period over all countries in Europe. In addition, we compared $PM_{2.5}$ levels during the COVID-19 lockdown during the months March–June with the average of the previous 4 months and the following 4 months. We found that this lockdown had a positive effect on air quality in most parts of the study area except for the United Kingdom, Ireland, north of France, and south of Italy. This is the first study that depends only on open data and covers the whole of Europe with high spatial and temporal resolutions. The reconstructed dataset will be published under free and open license and can be used in future air quality studies.

Keywords: $PM_{2.5}$; AOD; machine learning; Europe; open data

Citation: Ibrahim, S.; Landa, M.; Pešek, O.; Brodský, L.; Halounová, L. Machine Learning-Based Approach Using Open Data to Estimate $PM_{2.5}$ over Europe. *Remote Sens.* **2022**, *14*, 3392. https://doi.org/10.3390/rs14143392

Academic Editors: Maria João Costa and Daniele Bortoli

Received: 14 May 2022
Accepted: 12 July 2022
Published: 14 July 2022

Publisher's Note: MDPI stays neutral with regard to jurisdictional claims in published maps and institutional affiliations.

1. Introduction

Air quality monitoring is one of the most important fields when it comes to the individual's health due to the high risks related to its low quality. Fine particulate matter is an air pollutant that consists of liquid and solid molecules such as acid condensates, sulphates, and nitrates that have negative effects on human health [1]. The harmful effects of these particles vary depending on the concentrations, time exposure, and the particulate diameter. Risks are higher when the diameter gets smaller; $PM_{2.5}$ can penetrate deep into the lungs and may reach the blood circulation causing dangerous diseases such as cardiovascular problems, diabetes, prenatal disorder, and even mortality [2–5]. The effects are more notable in urban areas, where higher population density can be found, and more exposure will occur [6]. The form of the urban area plays an important role in the concentration of $PM_{2.5}$ [7].

The U.S. Environmental Protection Agency (EPA) has set an annual average standard of 12 $\mu g/m^3$ and a daily (24 h) of 35 $\mu g/m^3$ for $PM_{2.5}$ and when the amounts of these pollutants in the ambient air exceed these limits that could cause serious health issues [8]. The revised Directive 2008/50/EC of the European Parliament (EP) and of the Council on ambient air quality and cleaner air for Europe set limit values of annual $PM_{2.5}$ to 25 $\mu g/m^3$ since 1 January 2015 and not to exceed 20 $\mu g/m^3$ since 1 January 2020. $PM_{2.5}$

ground-based monitors are used to measure $PM_{2.5}$ with high accuracy. These stations are considered the backbone in almost all analyses related to these particles. However, the high cost of establishing these monitors limits the overall spatial coverage and the researchers who are focusing on air quality were seeking new methodologies to increase the spatial coverage so they have a better understanding on larger geographical scales. Numerous techniques were used to increase $PM_{2.5}$ spatial coverage, in other words, to estimate the pollutant concentrations in the areas where no monitors exist. Examples of that are interpolation techniques that count only on the ground stations [9,10]. The accuracy of these interpolations is highly related to the spatial distribution of the stations; although they can have good estimations in the areas that are surrounded by the network stations, they will probably fail to have good estimations where there is a lack of the stations [9]. Land use regression (LUR) models were also used to analyse pollution, particularly in densely populated areas [11,12].

Satellite remote sensing provides wide spatial coverage compared to the spatial coverage obtained from ground monitors. Aerosol optical depth (AOD) is an air quality indicator that can be observed from satellite remote sensing, and it is defined as the measure of the columnar atmospheric aerosol content. Numerous studies have found a positive correlation between satellite-based AOD and surface particulate matter [13,14]. Researchers have utilised satellite AOD to estimate $PM_{2.5}$ by developing different types of models such as physical models that were built based on the physical relationship between AOD and surface $PM_{2.5}$ [15]. Statistical methods which train the relationship between AOD and $PM_{2.5}$ using different statistical models [16,17] are suitable for the regions with a sufficient number of ground stations since they require a large amount of training data [18]. The generalised additive model (GAM) empowers the AOD–PM relationship by adding meteorological and land use information [19]. In the last few decades, artificial intelligence models have been applied to estimate $PM_{2.5}$ and were found to give a better description of the complex non-linear relationship between $PM_{2.5}$, AOD, and other independent variables than the previously mentioned methods [18] based on the usage of machine learning algorithms [20–22] or deep neural networks [23,24]. These algorithms utilise satellite observations, various modelled meteorological variables, population, land use, land cover, etc., to estimate $PM_{2.5}$. The importance of the inputs differs from one area to another, but generally, they can enhance $PM_{2.5}$ estimations since counting solely on AOD to estimate near-surface particulate matter values is not sufficient [25]. AOD without other variables was not enough to provide good $PM_{2.5}$ estimations over Europe [26]. In Great Britain, AOD was not among the 15 most important variables when predicting $PM_{2.5}$ levels [20]. Satellite AOD are more correlated with surface PM when the aerosols are well mixed within the planetary boundary layer height (PBLH) [9]. A global study found that 69% of the total AOD are within the PBLH [27], other studies have shown that temperature plays an important role in capturing AOD and understanding its vertical distribution that improve PM analysis [28]. Moreover, a higher humidity atmosphere is likely to have higher AOD without affecting the levels of $PM_{2.5}$ [9]. Other meteorological variables that affect $PM_{2.5}$ are the precipitation that showed a negative correlation in some areas [29] and a positive correlation in other parts of the world [30], and wind speed (WS) that also has different effects from one area to another [30,31].

In this study, we report the modelling of spatiotemporal heterogeneity of $PM_{2.5}$ using machine learning to generate daily estimations of $PM_{2.5}$ over the European Union member states, together with the United Kingdom, Iceland, Liechtenstein, Norway, Switzerland, Albania, Bosnia and Herzegovina, Kosovo, Montenegro, North Macedonia, and Serbia [32]. We will refer to the area of study as "Europe" located inside the coordinates box 26°W, 72°N, 42°E, and 36°S. The total study area covers 13,391,504 of 1 km grid cells; 5,450,009 of the total cell number are located over land. The study period covers the years 2018–2020 with full coverage of 1 km spatial resolution using various open data. In the following sections, we will introduce the study area and period and present the preliminary data that were tested while building the predicting model.

2. Primary Data

In this section, we will introduce the primary data we investigated while building the model. Not all these data were utilised while building the model. The chosen data can be found in Section 3.3.

2.1. $PM_{2.5}$ Measurements

$PM_{2.5}$ observations were collected from 848 stations across Europe represented in Figure 1. Data was downloaded from OpenAQ which is a non-profit organisation that collects air quality data from different governmental and research institutions and provides it to the users [33].

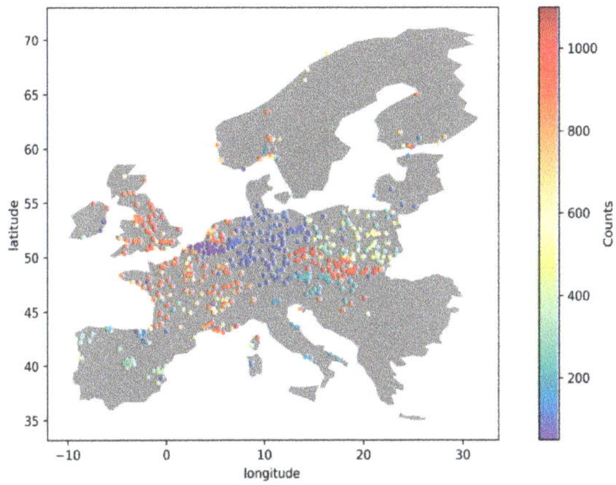

Figure 1. The location of $PM_{2.5}$ ground stations with the number of valid measurements used in this study.

For each station, data between 10 a.m. and 2 p.m. local time were averaged where there are at least 2 available observations to be consistent with MODIS satellites overpassing. We identified a skewed distribution for $PM_{2.5}$ as shown in Figure 2, we calculated the 25th percentile (Q1), the 75th percentile (Q2) of the dataset, and the inter-quartile range (IQR = Q3 − Q1). All $PM_{2.5}$ values that are higher than 2 × (Q3 + 3 × IQR which is refer as outer fence [34]) were removed, which counted less than 1% of the total data. The number of valid $PM_{2.5}$ observations was 123,248 in 2018, 143,048 in 2019, and 158,964 in 2020 totalling 425,260 observations throughout the study period.

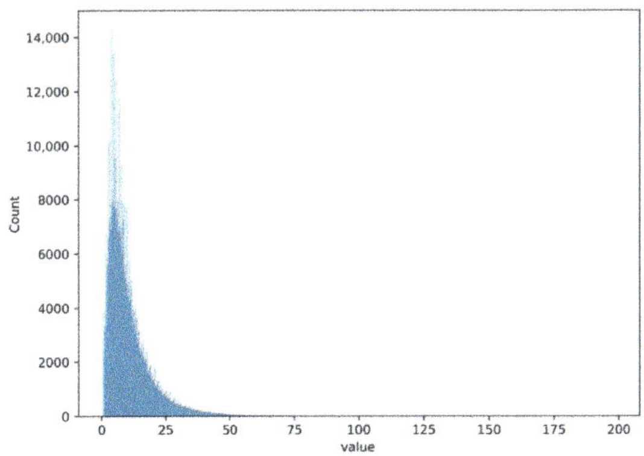

Figure 2. The distribution of the measured $PM_{2.5}$ used in this study.

2.2. AOD Data

AOD data were downloaded from GHADA, which is a Geo-Harmonized Atmospheric Dataset for Aerosol optical depth at 550 nm [35]. It contains daily estimations of AOD_{550} over Europe with 1 km spatial resolution. GHADA was built based on the MODIS MCD19A2 product [36] and modelled AOD data from Copernicus Atmosphere Monitoring Service (CAMS) [37] that were used to overcome the high percentage of gaps found in the MCD19A2 product. This dataset showed good results when validated with NASA's Aerosols Robotic Network (AERONET).

2.3. Meteorological Data

Meteorological data of the following variables wind component u, wind component v, PBLH, total column water vapour, total perception, evaporation, surface pressure, and temperature at 2 m (T2m) were collected from ERA5-Land which is a reanalysis dataset offering a consistent view of the development of land parameters over several decades with a spatial resolution of ~9 km. ERA5-Land was produced by replaying the land component of the European Centre for Medium-Range Weather Forecasts ERA5 climate reanalysis [38]. Relative humidity was collected from ERA5 with 0.25×0.25 horizontal resolution.

2.4. Digital Elevation Model

The Japan Aerospace Exploration Agency (JAXA) provides a worldwide digital surface model with a horizontal resolution of ~30 m by the Panchromatic Remote-sensing Instrument for Stereo Mapping (PRISM), which was carried on the Advanced Land Observing Satellite "ALOS" [39]. Data were accessed on the 8 March 2021 from https://www.eorc.jaxa.jp/ALOS/.

2.5. Normalised Difference Vegetation Index

MODIS Terra satellite provides a monthly normalised difference vegetation index (NDVI) product called MOD13A3 [40]. It has 1 km spatial resolution, and it quantifies vegetation presence with values ranging between −1 and 1. NDVI is commonly expressed as shown in Equation (1):

$$\text{NDVI} = \frac{\text{NIR} - \text{Red}}{\text{NIR} + \text{Red}} \quad (1)$$

where NIR and Red are spectral reflectance values in the near-infrared and red wavelengths.

2.6. Land Cover

Land cover data were extracted from the 2018 CORINE Land Cover (CLC) inventory that was built based on ortho-rectified satellite images having a spatial resolution ranging from 5 m to 60 m and were aggregated into 100 m. We grouped the original 44 CLC classes into seven level 1 classes defined as: agricultural areas, artificial areas, continues urban areas, discontinues urban areas, forests, industrial areas, and water surfaces. Then, we calculated the percentage of each class in every 1×1 km^2 grid cell.

2.7. Population Data

Population data was extracted from the Visible Infrared Imaging Radiometer Suite (VIIRS) night-time lights (NTL) data by averaging the monthly data of the year 2019.

3. Methodology

3.1. Data Pre-Processing

All data were reprojected to the European Terrestrial Reference System 1989 (EPSG:3035) that uses metres as measuring units. This system is used for statistical mapping and other purposes which requires a true area representation, using a 1 km grid cell with bilinear interpolation method for ECMWF data and the cubic convolution for the ALOS elevation model. In addition, we calculated WS based on the two wind U and V components.

A spatio-temporal dataset was created by extracting the information from all input data at the locations of $PM_{2.5}$ stations. The Julian day, month, and year were added as the temporal information; longitude and latitude were added as the spatial information. The generated dataset was used to train and test the model.

3.2. Model Development

We first analysed the linear relationship between the primary independent variables and $PM_{2.5}$ values. PBLH was negatively correlated to $PM_{2.5}$ with Pearson correlation of $r = -0.24$. Most of the meteorological variables were also negatively related to $PM_{2.5}$ with $r = -0.2$ for WS, $r = -0.15$ for T2m, $r = -0.13$ for RH, and $r = -0.1$ for TP. AOD and evaporation had the highest positive correlation with $PM_{2.5}$ $r = 0.14$. Based on this initial data exploratory analysis, we excluded some primary inputs that had high correlation with other inputs such as skin temperature, which was correlated to T2m with $r = 0.93$. We tested linear models to estimate $PM_{2.5}$. These models suffered from underfitting issues and failed to describe the relationship between the independent variables and $PM_{2.5}$. Therefore, we used a more complex algorithm called Extremely Randomised Trees (ET).

ET is a very similar decision tree-based ensemble method to the widely used Random Forest (RF). Both algorithms are composed of large number of trees, where the final decision is obtained from the prediction of every tree by majority vote in classification problems and arithmetic average in regression problems. Both algorithms have the same growing tree procedure and selecting the partition of each node. Additionally, both algorithms randomly choose a subset of input features.

ET, on the other hand, strongly randomises the selection of both attribute and cut point while splitting a tree node using the whole learning sample to grow the trees which adds randomisation, making it a more robust algorithm against overfitting. From computational point of view, the complexity of the tree growing procedure is on the order of N log N with respect to learning sample size [41]. The main parameters in the ET splitting process are the number of attributes that are randomly selected at each node and the minimum sample size for splitting a node. For further information on how the ET algorithm operates refer to Table 1 in [41]. In addition to accuracy, the ET algorithm has higher computational efficiency than the RF algorithm since it chooses the splits randomly and does not look for the optimum split as the latter one [41]. The number of estimators (number of trees in the forest), the maximum depth of the trees, the number of samples required to split an internal node, and the minimum number of samples required to be at a leaf node were the main parameters while tuning our model.

Table 1. The dependent and independent variables used to build the ET model.

Name of the Variable	Unit	Minimum	Maximum	Mean	STD
$PM_{2.5}$	$\mu g/m^3$	2	80	11.81	9.26
Aerosol optical depth	-	0.01	3.12	0.13	0.08
PBLH	m	73.90	3420.17	933.39	463.59
WS	m/s	0.23	18.12	3.88	2.13
T2m	K	249.86	314.15	287.03	8.17
Relative Humidity	%	0.04	110.82	68.53	22.93
Total precipitation	mm	0	8	0.1	0.3
Total Column Water Vapour	Kg/m^2	0.95	50.61	16.76	7.88
NDVI	-	−0.3	0.73	0.25	0.12
Evaporation	mm	−0.744	0.065	−0.164	0.109
Elevation	m	−3.88	914.26	151.66	156.01

To reduce model complexity due to the large number of independent variables we excluded the input variables based on the feature importance in the ET algorithm. Besides the spatio-temporal information, we used $PM_{2.5}$ with the independent variables that are shown in Table 1 to develop our model.

3.3. Model Validation

3.3.1. Sample-Based Cross Validation

Cross validation (CV) is a common method to analyse the model performance and detect potential overfitting problems where the model achieves high accuracy on the training set and performs badly on new data or the test set. We applied a 10-fold CV where all samples in the training dataset were randomly divided into 10 equal subsets. Then, in each round, 9 subsets were used to fit the model, and the remaining subset was used for testing the model performance [42]. This approach is used widely in PM studies [20,21,43–45].

3.3.2. Spatial and Temporal 10-Fold Cross Validation

In this validation, we divided the samples based on two factors. For the spatial 10-fold cross validation we splatted the data based on the location of the stations, the stations were divided randomly into 10 folds. In each fold, the model was trained on the samples from 90% of the stations and the samples from the remaining 10% for testing. For the temporal 10-fold cross validation, we divided the samples into 10 folds based on the Julian day and applied the cross validation in a similar way to the previously mentioned one.

4. Results

The results of sample-based, spatial, and temporal 10-fold cross validation are shown in Table 2. The density scatter plot for the sample-based cross validation is shown in Figure 3.

Table 2. R^2, RMSE, and MAE of the sample-based 10-CV, spatial 10-CV, and the temporal 10-CV.

10-CV	R^2	RMSE	MAE
Sample-based	0.69	5.0	3.3
Spatial	0.69	4.9	3.2
Temporal	0.53	6.1	4.1

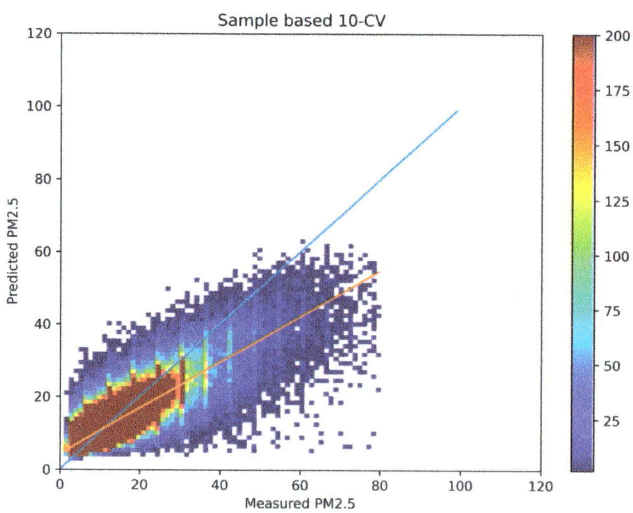

Figure 3. Density scatter plot of the sample-based 10-CV results of the model.

It must be noted that $PM_{2.5}$ levels in general are low in Europe when compared to more polluted areas and this is reflected by the low RMSE we obtained in our study when compared to some studies outside Europe with higher R^2 values [44,45]. Our model proved its efficiency in predicting $PM_{2.5}$ when our results (out-of-sample R^2 = 0.69, RMSE = 5 μg/m^3)

were comparable with results obtained from a recent study over a smaller geographic area in Europe (Great Britain; out-of-sample $R^2 = 0.77$, RMSE = 4 μg/m^3) [20]. It is also noted that the model underestimates high $PM_{2.5}$ values (>40 μg/m^3) since such values are not abundant over our study area.

To justify the difference in the model performance spatially and temporally, we applied site-based cross validation where we used samples from one station as the test set, and the samples from all remaining stations were used to train the model. We applied this method to analyse the model performance spatially, since the standard 10-CV may not be able to detect potential spatial overfitting [18].

The results are shown in Figure 4. The model performs well in most of the locations in Central Europe with an average R^2~0.7. A total of 63% of all stations in Europe have $R^2 > 0.6$. The accuracy of the model is lower in the northern and southern parts of Europe. However, the RMSE and MAE are relatively small even in the northern and southern parts.

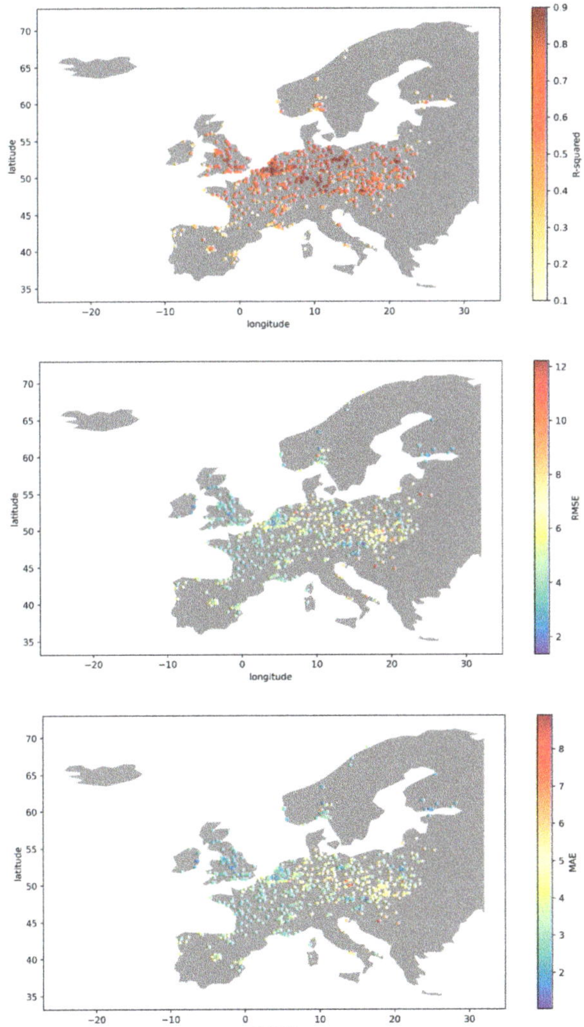

Figure 4. Spatial distribution of the site-based cross validation of coefficient of determination, the root mean square error, and the mean absolute error.

5. Creating $PM_{2.5}$ Maps

Daily $PM_{2.5}$ maps during MODIS satellite overpassing were created for the period 2018–2020 over Europe. Figure 5 shows the average $PM_{2.5}$ for the year 2018, 2019, and 2020.

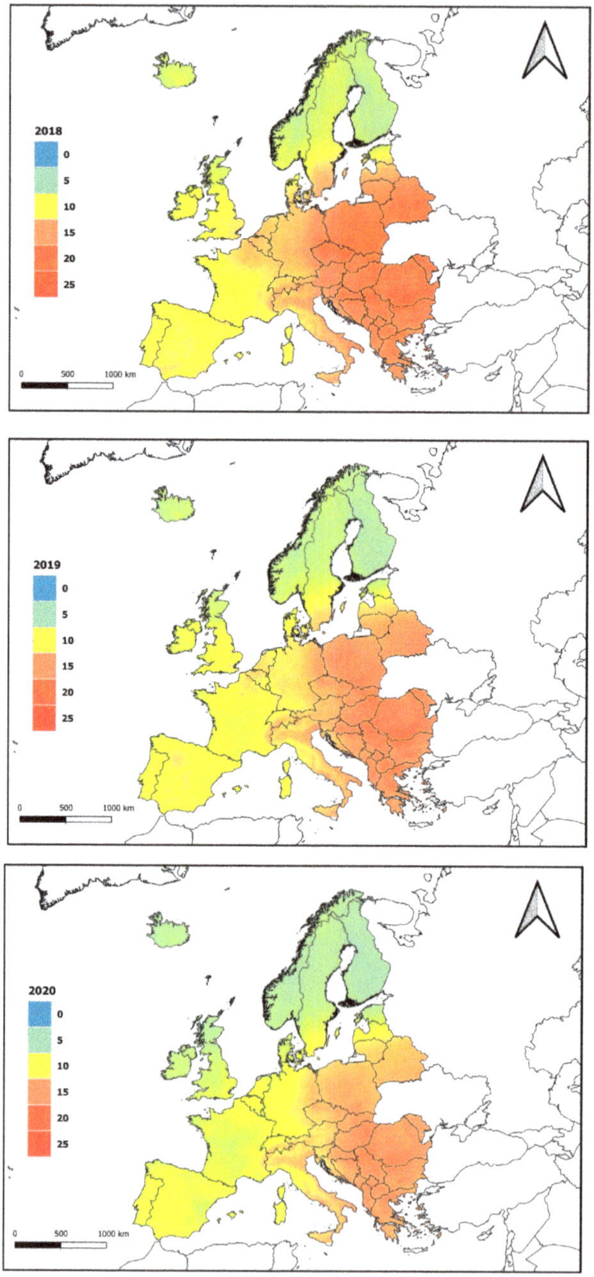

Figure 5. The average $PM_{2.5}$ for the years 2018, 2019, and 2020 over Europe.

A significant decline in $PM_{2.5}$ levels has occurred over Europe throughout the study period. Poland had the highest $PM_{2.5}$ average level in the year 2018 with an average level~19.5 μg/m^3, in 2019 Romania had the highest average~16.5 μg/m^3 whereas Serbia had the highest average in 2020 with an average~15.8 μg/m^3. Finland had the lowest $PM_{2.5}$ average level in all three years with 7.1 in 2018, 6.3 in 2019, and 5.8 in 2020. Comparing the results of the average $PM_{2.5}$ levels for the years 2018, 2019, and 2020 were highly compatible with the reports of the European Environment Agency (EEA). According to EEA the highest $PM_{2.5}$ concentrations were found in central and eastern Europe and northern Italy. For the central and western parts, the main reason for high $PM_{2.5}$ is the usage of solid fuels with older vehicle compared to other parts of Europe [46], besides using the solid fuels for heating as was found in Poland [47]. For the northern part of Italy, the high levels of $PM_{2.5}$ are due to the combination of a high density of anthropogenic emissions and meteorological conditions [46,48]. Furthermore, Milan, the largest city in the north of Italy previously reported levels of $PM_{2.5}$ exceeding the safety limit set by the EU [49].

As an application, we used the proposed machine learning-based prediction approach in $PM_{2.5}$ levels analysis to study the effect of the COVID-19 lockdown (March to June of the year 2020) on air quality over Europe. As an attempt to verify the influence of the lockdown on air quality, we compared the average $PM_{2.5}$ of the previous 4 months (November to December in 2019 and January to February 2020) and the following 4 months (July to October 2020) to the 4 months of the lockdown by calculating the relative percentage difference (RPD). By doing so, we masked the general improvement trend in air quality over Europe. RPD calculated using Equation (2).

$$\text{RPD} = \frac{\text{PM}_{2.5}\ (\text{lockdown}) - \text{PM}_{2.5}\ \text{avg}(\text{before lockdown, after lockdown})}{\text{PM}_{2.5}\ \text{avg}(\text{before lockdown, after lockdown})} \times 100 \quad (2)$$

We found a significant improvement in air quality over Europe except for UK, Ireland, north of France, and south of Italy as shown in Figure 6. Our results are in agreement with another study over Poland (Eastern Europe), where the air quality represented by $PM_{2.5}$ has significantly improved in the months of March to April in 2020 when the authors compared to the same months from the previous two years [50]. Interestingly, the unusual increase in $PM_{2.5}$ levels in the UK was consistent with what was reported in [51] as the authors justified such increase by unusual meteorological conditions. The latter conditions may also justify the increase in $PM_{2.5}$ over northern France. In Italy, where people were spending most of their time at home, the increased house heating during the lockdown period limited the decrease in $PM_{2.5}$ levels besides the effects of the agriculture sector that kept performing during the lockdown [52].

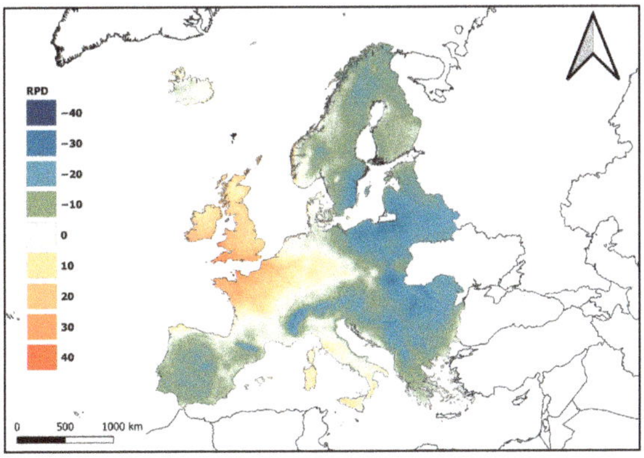

Figure 6. Relative percentage difference of $PM_{2.5}$ for the lock down period of the year 2020 with the average of the previous 4 months and the following 4 months.

6. Discussion

In this study, we proposed the first machine learning-based scheme to estimate $PM_{2.5}$ levels over Europe with high spatial resolution of 1 km. We trained an extra trees model using observed $PM_{2.5}$ from 848 stations as the target variable. AOD, different meteorological variables, land variables, and NDVI as the independent variables.

The sample-based 10-fold CV showed that our model underestimates high $PM_{2.5}$ values (>40 μg/m^3) which may limit the model ability to detect hazard situations. This underestimation occurred since high $PM_{2.5}$ values were not common over our study area as shown in Figure 2. The spatial cross validation showed that the model estimates $PM_{2.5}$ with a higher R^2 in the areas with high ground stations density the compared to the areas with a lower density. The occurred spatial overfitting is expected to happen due to spatially unbalanced data.

In Central Europe (Czech Republic, Poland, Slovakia, and surrounding areas), the model performed with a higher R^2 compared to the northern and southern parts of Europe. However, the RMSE in Central Europe was comparably higher than the ones in the previously mentioned parts. This is due to the fact that the average $PM_{2.5}$ value in Central Europe is higher and have more variations than the northern and southern parts. The highest RMSE in Central Europe can be found in three stations in the Czech Republic. These stations are located near mining areas with higher $PM_{2.5}$ values compared to other stations that are mostly located in urban areas. This issue can be potentially solved by including a detailed land cover data with an appropriate classification for each station which is usually difficult to achieve on a large scale such as in our study.

Having unbalanced spatial-temporal data made the modelling more complex than other studies which focused on smaller areas with well-balanced data and with similar instruments in measuring $PM_{2.5}$ values. However, by tuning the parameters in the model, we were able to achieve acceptable results for most parts of our study area. The effect of the chosen independent variables in estimating $PM_{2.5}$ differs across the study area. We analysed the spatial potential relationships of the independent variables in estimating $PM_{2.5}$ by calculating feature importance in four parts of Europe: north-west (latitude > 50 and longitude < 10), north-east (latitude > 50 and longitude > 10), south-west (latitude < 50 and longitude < 10), and south-east (latitude < 50 and longitude > 10). AOD and PBLH had the most feature importance in all parts of Europe with an average of 10.4% and 14.1%, respectively. WS and temperature had more effect in estimating $PM_{2.5}$ in the south of Europe compared to the north. Rh had more importance in estimating $PM_{2.5}$ in the western part of Europe compared to the eastern part.

Table 3 shows the effects of AOD and the most important meteorological variables on $PM_{2.5}$ estimates. We tried to train multiple models based on the area. However, this approach did not improve the overall performance over the whole study area.

Table 3. The effects (%) of AOD and the most important meteorological variables on $PM_{2.5}$ estimations in the four chosen parts of our study area.

Independent Variable	North-West	North-East	South-West	South-East
AOD	13.25	8.81	10.43	9.11
BLH	15.89	15.22	14.98	10.41
T2m	8.62	6.25	10.13	10.71
Rh	6.41	3.99	5.82	4.71
E	3.58	5.99	3.44	7.96
WS	5.18	4.25	7.32	5.82
TCWV	4.469	3.63	4.55	4.07

7. Conclusions

In this study, we developed a spatio-temporal machine learning model to estimate daily $PM_{2.5}$ levels for the years 2018–2020 with 1 km spatial resolution over Europe using open data from multiple sources such as remote sensing satellite-based products, meteorological reanalysis datasets, and other land variables.

The developed model was used to estimate $PM_{2.5}$ values over 5,450,009 land cells (1 km^2) for a 3-year period (1096 days) totalling more than 5.973 billion estimations with a good sample-based CV coefficient of 0.69, RMSE of 5 $\mu g/m^3$, and MAE of 3.3 $\mu g/m^3$.

We calculated the yearly average of $PM_{2.5}$ levels, and we found that $PM_{2.5}$ values have dropped in almost all parts of Europe during the study period.

The full coverage dataset of $PM_{2.5}$ that we produced can be used to investigate air quality over Europe with higher spatial resolution compared to the available products which may provide better understanding in time series analysis in this field.

Author Contributions: Conceptualization, S.I.; Data curation, S.I.; Formal analysis, S.I., M.L., O.P. and L.B.; Funding acquisition, M.L.; Investigation, S.I.; Methodology, S.I., L.B. and L.H.; Project administration, M.L. and L.H.; Resources, S.I. and L.H.; Software, S.I., M.L. and O.P.; Supervision, L.H.; Validation, S.I.; Visualization, S.I.; Writing—original draft, S.I.; Writing—review & editing, S.I., M.L., O.P., L.B. and L.H. All authors have read and agreed to the published version of the manuscript.

Funding: This work is co-financed under the Grant Agreement Connecting Europe Facility (CEF) Telecom project 2018-EU-IA-0095 by the European Union and by the Grant Agency of the Czech Technical University in Prague, grant No. SGS22/047/OHK1/1T/11.

Data Availability Statement: The data and data analysis methods are available upon request.

Acknowledgments: The authors sincerely thank the OpenAQ organisation for providing $PM_{2.5}$ observations, NASA EOSDIS for providing the daily MODIS MAIAC AOD product (MCD19A2) which was used to build GHADA and that is available from the Land Processes Distributed Active Archive Centre (LPDAAC), the European Centre for Medium-Range Weather Forecasts (ECMWF) for providing global reanalysis of atmospheric composition, and the Japan Aerospace Exploration Agency (JAXA) for providing the digital surface model used in this study.

Conflicts of Interest: The authors declare no conflict of interest.

References

1. Li, L.; Losser, T.; Yorke, C.; Piltner, R. Fast Inverse Distance Weighting-Based Spatiotemporal Interpolation: A Web-Based Application of Interpolating Daily Fine Particulate Matter $PM_{2.5}$ in the Contiguous U.S. Using Parallel Programming and k-d Tree. *Int. J. Environ. Res. Public Health* **2014**, *11*, 9101–9141. [CrossRef] [PubMed]
2. Crippa, M.; Janssens-Maenhout, G.; Guizzardi, D.; Van Dingenen, R.; Dentener, F. Contribution and uncertainty of sectorial and regional emissions to regional and global $PM_{2.5}$ health impacts. *Atmos. Chem. Phys.* **2019**, *19*, 5165–5186. [CrossRef]
3. Pascal, M.; Falq, G.; Wagner, V.; Chatignoux, E.; Corso, M.; Blanchard, M.; Host, S.; Pascal, L.; Larrieu, S. Short-term impacts of particulate matter (PM10, PM10–2.5, PM2.5) on mortality in nine French cities. *Atmos. Environ.* **2014**, *95*, 175–184. [CrossRef]
4. Liu, C.; Chen, R.; Sera, F.; Vicedo-Cabrera, A.M.; Guo, Y.; Tong, S.; Coelho, M.S.Z.S.; Saldiva, P.H.N.; Lavigne, E.; Matus, P.; et al. Ambient Particulate Air Pollution and Daily Mortality in 652 Cities. *N. Engl. J. Med.* **2019**, *381*, 705–715. [CrossRef]
5. Martins, N.R.; da Graça, G.C. Impact of $PM_{2.5}$ in indoor urban environments: A review. *Sustain. Cities Soc.* **2018**, *42*, 259–275. [CrossRef]
6. Baklanov, A.; Molina, L.T.; Gauss, M. Megacities, air quality and climate. *Atmos. Environ.* **2016**, *126*, 235–249. [CrossRef]
7. Mao, X.; Wang, L.; Pan, X.; Zhang, M.; Wu, X.; Zhang, W. A study on the dynamic spatial spillover effect of urban form on $PM_{2.5}$ concentration at county scale in China. *Atmos. Res.* **2022**, *269*, 106046. [CrossRef]
8. Environmental Protection Agency 40 CFR Part 50 Review of the National Ambient Air Quality Standards for Particulate Matter. Available online: https://cfpub.epa.gov/ncea/ (accessed on 19 December 2021).
9. Lee, H.J. Advancing Exposure Assessment of $PM_{2.5}$ Using Satellite Remote Sensing: A Review. *Asian J. Atmos. Environ.* **2020**, *14*, 319–334. [CrossRef]
10. Deng, L. Estimation of $PM_{2.5}$ spatial distribution based on kriging interpolation. In Proceedings of the First International Conference on Information Sciences, Machinery, Materials and Energy, Chongqing, China, 11–13 April 2015; Volume 126. [CrossRef]
11. Vienneau, D.; De Hoogh, K.; Beelen, R.; Fischer, P.; Hoek, G.; Briggs, D. Comparison of land-use regression models between Great Britain and the Netherlands. *Atmos. Environ.* **2010**, *44*, 688–696. [CrossRef]
12. Briggs, D.J. The use of GIS to evaluate traffic-related pollution. *Occup. Environ. Med.* **2006**, *64*, 1–2. [CrossRef]
13. You, W.; Zang, Z.; Pan, X.; Zhang, L.; Chen, D. Estimating $PM_{2.5}$ in Xi'an, China using aerosol optical depth: A comparison between the MODIS and MISR retrieval models. *Sci. Total Environ.* **2015**, *505*, 1156–1165. [CrossRef] [PubMed]

14. Yao, F.; Si, M.; Li, W.; Wu, J. A multidimensional comparison between MODIS and VIIRS AOD in estimating ground-level $PM_{2.5}$ concentrations over a heavily polluted region in China. *Sci. Total Environ.* **2017**, *618*, 819–828. [CrossRef] [PubMed]
15. Zhang, Y.; Li, Z. Remote sensing of atmospheric fine particulate matter (PM2.5) mass concentration near the ground from satellite observation. *Remote Sens. Environ.* **2015**, *160*, 252–262. [CrossRef]
16. Kanabkaew, T. Prediction of Hourly Particulate Matter Concentrations in Chiangmai, Thailand Using MODIS Aerosol Optical Depth and Ground-Based Meteorological Data. *EnvironmentAsia* **2013**, *6*, 65–70. [CrossRef]
17. Gupta, P.; Christopher, S.A. Particulate matter air quality assessment using integrated surface, satellite, and meteorological products: Multiple regression approach. *J. Geophys. Res. Earth Surf.* **2009**, *114*. [CrossRef]
18. Ma, Z.; Dey, S.; Christopher, S.; Liu, R.; Bi, J.; Balyan, P.; Liu, Y. A review of statistical methods used for developing large-scale and long-term $PM_{2.5}$ models from satellite data. *Remote Sens. Environ.* **2021**, *269*, 112827. [CrossRef]
19. Liu, Y.; Paciorek, C.J.; Koutrakis, P. Estimating Regional Spatial and Temporal Variability of $PM_{2.5}$ Concentrations Using Satellite Data, Meteorology, and Land Use Information. *Environ. Health Perspect.* **2009**, *117*, 886–892. [CrossRef] [PubMed]
20. Schneider, R.; Vicedo-Cabrera, A.M.; Sera, F.; Masselot, P.; Stafoggia, M.; de Hoogh, K.; Kloog, I.; Reis, S.; Vieno, M.; Gasparrini, A. A Satellite-Based Spatio-Temporal Machine Learning Model to Reconstruct Daily $PM_{2.5}$ Concentrations across Great Britain. *Remote Sens.* **2020**, *12*, 3803. [CrossRef]
21. Wei, J.; Li, Z.; Cribb, M.; Huang, W.; Xue, W.; Sun, L.; Guo, J.; Peng, Y.; Li, J.; Lyapustin, A.; et al. Improved 1 km resolution $PM_{2.5}$ estimates across China using enhanced space–time extremely randomized trees. *Atmos. Chem. Phys.* **2020**, *20*, 3273–3289. [CrossRef]
22. Chen, G.; Li, S.; Knibbs, L.D.; Hamm, N.A.S.; Cao, W.; Li, T.; Guo, J.; Ren, H.; Abramson, M.J.; Guo, Y. A machine learning method to estimate $PM_{2.5}$ concentrations across China with remote sensing, meteorological and land use information. *Sci. Total Environ.* **2018**, *636*, 52–60. [CrossRef]
23. Xiao, F.; Yang, M.; Fan, H.; Fan, G.; Al-Qaness, M.A.A. An improved deep learning model for predicting daily $PM_{2.5}$ concentration. *Sci. Rep.* **2020**, *10*, 20988. [CrossRef] [PubMed]
24. Li, L.; Girguis, M.; Lurmann, F.; Pavlovic, N.; McClure, C.; Franklin, M.; Wu, J.; Oman, L.D.; Breton, C.; Gilliland, F.; et al. Ensemble-based deep learning for estimating $PM_{2.5}$ over California with multisource big data including wildfire smoke. *Environ. Int.* **2020**, *145*, 106143. [CrossRef] [PubMed]
25. Van Donkelaar, A.; Martin, R.V.; Brauer, M.; Kahn, R.; Levy, R.; Verduzco, C.; Villeneuve, P.J. Global Estimates of Ambient Fine Particulate Matter Concentrations from Satellite-Based Aerosol Optical Depth: Development and Application. *Environ. Health Perspect.* **2010**, *118*, 847–855. [CrossRef]
26. Koelemeijer, R.; Homan, C.; Matthijsen, J. Comparison of spatial and temporal variations of aerosol optical thickness and particulate matter over Europe. *Atmos. Environ.* **2006**, *40*, 5304–5315. [CrossRef]
27. Bourgeois, Q.; Ekman, A.M.L.; Renard, J.-B.; Krejci, R.; Devasthale, A.; Bender, F.A.-M.; Riipinen, I.; Berthet, G.; Tackett, J.L. How much of the global aerosol optical depth is found in the boundary layer and free troposphere? *Atmos. Chem. Phys.* **2018**, *18*, 7709–7720. [CrossRef]
28. Liu, B.; Ma, X.; Ma, Y.; Li, H.; Jin, S.; Fan, R.; Gong, W. The relationship between atmospheric boundary layer and temperature inversion layer and their aerosol capture capabilities. *Atmos. Res.* **2022**, *271*, 106121. [CrossRef]
29. Li, X.; Feng, Y.J.; Liang, H.Y. The Impact of Meteorological Factors on $PM_{2.5}$ Variations in Hong Kong. *IOP Conf. Series Earth Environ. Sci.* **2017**, *78*, 012003. [CrossRef]
30. Wang, J.; Ogawa, S. Effects of Meteorological Conditions on $PM_{2.5}$ Concentrations in Nagasaki, Japan. *Int. J. Environ. Res. Public Health* **2015**, *12*, 9089–9101. [CrossRef]
31. Wang, S.; Gao, J.; Guo, L.; Nie, X.; Xiao, X. Meteorological Influences on Spatiotemporal Variation of $PM_{2.5}$ Concentrations in Atmospheric Pollution Transmission Channel Cities of the Beijing–Tianjin–Hebei Region, China. *Int. J. Environ. Res. Public Health* **2022**, *19*, 1607. [CrossRef]
32. Open Data Science Europe. *Geo-Harmonizer Project Implementation Plan 2020–2022*; Open Data Science Europe: Wageningen, The Netherlands, 2020.
33. OpenAQ. Available online: https://openaq.org/ (accessed on 8 May 2022).
34. Tukey, J.W. *Exploratory Data Analysis*; Addison-Wesley Publishing Company: Boston, MA, USA, 1977.
35. Ibrahim, S.; Landa, M.; Pešek, O.; Pavelka, K.; Halounova, L. Space-Time Machine Learning Models to Analyze COVID-19 Pandemic Lockdown Effects on Aerosol Optical Depth over Europe. *Remote Sens.* **2021**, *13*, 3027. [CrossRef]
36. Lyapustin, A.; Wang, Y.; Laszlo, I.; Kahn, R.; Korkin, S.; Remer, L.; Levy, R.; Reid, J.S. Multiangle implementation of atmospheric correction (MAIAC): Part 2. Aerosol algorithm. *J. Geophys. Res.* **2011**, *116*. [CrossRef]
37. Inness, A.; Ades, M.; Agustí-Panareda, A.; Barré, J.; Benedictow, A.; Blechschmidt, A.-M.; Dominguez, J.J.; Engelen, R.; Eskes, H.; Flemming, J.; et al. The CAMS reanalysis of atmospheric composition. *Atmos. Chem. Phys.* **2019**, *19*, 3515–3556. [CrossRef]
38. Muñoz-Sabater, J.; Dutra, E.; Agustí-Panareda, A.; Albergel, C.; Arduini, G.; Balsamo, G.; Boussetta, S.; Choulga, M.; Harrigan, S.; Hersbach, H.; et al. ERA5-Land: A state-of-the-art global reanalysis dataset for land applications. *Earth Syst. Sci. Data* **2021**, *13*, 4349–4383. [CrossRef]
39. Tadono, T.; Ishida, H.; Oda, F.; Naito, S.; Minakawa, K.; Iwamoto, H. Precise Global DEM Generation by ALOS PRISM. *ISPRS Ann. Photogramm. Remote Sens. Spat. Inf. Sci.* **2014**, *II-4*, 71–76. [CrossRef]

40. Didan, K. MOD13A3 MODIS/Terra Vegetation Indices Monthly L3 Global 1 km SIN Grid V006 [Dataset]. NASA EOSDIS Land Processes DAAC. 2015. Available online: https://doi.org/10.5067/modis/mod13a3.006 (accessed on 14 March 2021).
41. Geurts, P.; Ernst, D.; Wehenkel, L. Extremely randomized trees. *Mach. Learn.* **2006**, *63*, 3–42. [CrossRef]
42. Rodriguez, J.D.; Perez, A.; Lozano, J.A. Sensitivity Analysis of k-Fold Cross Validation in Prediction Error Estimation. *IEEE Trans. Pattern Anal. Mach. Intell.* **2009**, *32*, 569–575. [CrossRef] [PubMed]
43. Li, T.; Shen, H.; Zeng, C.; Yuan, Q.; Zhang, L. Point-surface fusion of station measurements and satellite observations for mapping $PM_{2.5}$ distribution in China: Methods and assessment. *Atmos. Environ.* **2017**, *152*, 477–489. [CrossRef]
44. He, Q.; Huang, B. Satellite-based mapping of daily high-resolution ground $PM_{2.5}$ in China via space-time regression modeling. *Remote Sens. Environ.* **2018**, *206*, 72–83. [CrossRef]
45. Wei, J.; Huang, W.; Li, Z.; Xue, W.; Peng, Y.; Sun, L.; Cribb, M. Estimating 1-km-resolution $PM_{2.5}$ concentrations across China using the space-time random forest approach. *Remote Sens. Environ.* **2019**, *231*, 111221. [CrossRef]
46. European Environment Agency. Available online: https://www.eea.europa.eu (accessed on 19 December 2021).
47. Baborska-Narozny, M.; Szulgowska-Zgrzywa, M.; Mokrzecka, M.; Chmielewska, A.; Fidorow-Kaprawy, N.; Stefanowicz, E.; Piechurski, K.; Laska, M. Climate justice: Air quality and transitions from solid fuel heating. *Build. Cities* **2020**, *1*, 120–140. [CrossRef]
48. Perrone, M.G.; Zhou, J.; Malandrino, M.; Sangiorgi, G.; Rizzi, C.; Ferrero, L.; Dommen, J.; Bolzacchini, E. PM chemical composition and oxidative potential of the soluble fraction of particles at two sites in the urban area of Milan, Northern Italy. *Atmos. Environ.* **2016**, *128*, 104–113. [CrossRef]
49. Perrone, M.; Larsen, B.; Ferrero, L.; Sangiorgi, G.; De Gennaro, G.; Udisti, R.; Zangrando, R.; Gambaro, A.; Bolzacchini, E. Sources of high $PM_{2.5}$ concentrations in Milan, Northern Italy: Molecular marker data and CMB modelling. *Sci. Total Environ.* **2012**, *414*, 343–355. [CrossRef] [PubMed]
50. Filonchyk, M.; Hurynovich, V.; Yan, H. Impact of Covid-19 lockdown on air quality in the Poland, Eastern Europe. *Environ. Res.* **2020**, *198*, 110454. [CrossRef] [PubMed]
51. Jenkins, N.; Parfitt, H.; Nicholls, M.; Beckett, P.; Wyche, K.; Smallbone, K.; Gregg, D.; Smith, M. *Estimation of Changes in Air Pollution Emissions, Concentrations and Exposure during the COVID-19 Outbreak in the UK*; Report for The Air Quality Expert Group, on Behalf of Defra: Analysis of Air Quality Changes Experienced in Sussex and Surrey since the COVID-19 Outbreak; UK Air, Department for Food and Rural Affairs: London, UK, 2020; 57p.
52. Pala, D.; Casella, V.; Larizza, C.; Malovini, A.; Bellazzi, R. Impact of COVID-19 lockdown on PM concentrations in an Italian Northern City: A year-by-year assessment. *PLoS ONE* **2022**, *17*, e0263265. [CrossRef]

remote sensing

MDPI

Article

Effects of Meteorology Changes on Inter-Annual Variations of Aerosol Optical Depth and Surface $PM_{2.5}$ in China—Implications for $PM_{2.5}$ Remote Sensing

Ling Qi [1], Haotian Zheng [2], Dian Ding [3], Dechao Ye [2] and Shuxiao Wang [2,4,*]

1 School of Energy and Environmental Engineering, University of Science and Technology Beijing, Beijing 100083, China; lingqi@ustb.edu.cn
2 State Key Joint Laboratory of Environment Simulation and Pollution Control, School of Environment, Tsinghua University, Beijing 100084, China; hzheng@mail.tsinghua.edu.cn (H.Z.); yedechao@mail.tsinghua.edu.cn (D.Y.)
3 Institute for Atmospheric and Earth System Research (INAR)/Physics, Faculty of Science, University of Helsinki, 00014 Helsinki, Finland; dian.ding@helsinki.fi
4 State Environmental Protection Key Laboratory of Sources and Control of Air Pollution Complex, Beijing 100084, China
* Correspondence: shxwang@tsinghua.edu.cn

Abstract: $PM_{2.5}$ retrieval from satellite-observed aerosol optical depth (AOD) is still challenging due to the strong impact of meteorology. We investigate influences of meteorology changes on the inter-annual variations of AOD and surface $PM_{2.5}$ in China between 2006 and 2017 using a nested 3D chemical transport model, GEOS-Chem, by fixing emissions at the 2006 level. We then identify major meteorological elements controlling the inter-annual variations of AOD and surface $PM_{2.5}$ using multiple linear regression. We find larger influences of meteorology changes on trends of AOD than that of surface $PM_{2.5}$. On the seasonal scale, meteorology changes are beneficial to AOD and surface $PM_{2.5}$ reduction in spring (1–50%) but show an adverse effect on aerosol reduction in summer. In addition, major meteorological elements influencing variations of AOD and $PM_{2.5}$ are similar between spring and fall. In winter, meteorology changes are favorable to AOD reduction (-0.007 yr^{-1}, -1.2% yr^{-1}; $p < 0.05$) but enhanced surface $PM_{2.5}$ between 2006 and 2017. The difference in winter is mainly attributed to the stable boundary layer that isolates surface $PM_{2.5}$ from aloft. The significant decrease in AOD over the years is related to the increase in meridional wind speed at 850 hPa in NCP ($p < 0.05$). The increase of surface $PM_{2.5}$ in NCP in winter is possibly related to the increased temperature inversion and more stable stratification in the boundary layer. This suggests that previous estimates of wintertime surface $PM_{2.5}$ using satellite measurements of AOD corrected by meteorological elements should be used with caution. Our findings provide potential meteorological elements that might improve the retrieval of surface $PM_{2.5}$ from satellite-observed AOD on the seasonal scale.

Keywords: meteorology; $PM_{2.5}$; AOD

Citation: Qi, L.; Zheng, H.; Ding, D.; Ye, D.; Wang, S. Effects of Meteorology Changes on Inter-Annual Variations of Aerosol Optical Depth and Surface $PM_{2.5}$ in China—Implications for $PM_{2.5}$ Remote Sensing. *Remote Sens.* **2022**, *14*, 2762. https://doi.org/10.3390/rs14122762

Academic Editors: Maria João Costa and Daniele Bortoli

Received: 28 April 2022
Accepted: 7 June 2022
Published: 8 June 2022

Publisher's Note: MDPI stays neutral with regard to jurisdictional claims in published maps and institutional affiliations.

1. Introduction

Fine particle ($PM_{2.5}$) lowers visibility [1,2], affects human health [3], modifies cloud properties [4], and exerts radiative forcing on the Earth's surface and at the top of the atmosphere [5]. Accurate estimates of high spatio-temporal resolution of surface $PM_{2.5}$ concentrations is critical for assessing its effects. However, national-wide in situ measurements of surface $PM_{2.5}$ in China were unavailable until 2013. To study the driving forces of long-term variations of surface $PM_{2.5}$, many studies use long-term measurements of fog–haze days [6], visibilities [7], and conducive weather conditions [8] as approximations. Specifically, Niu et al. showed that in the past three decades, the doubled frequencies of fog events in wintertime over eastern-central China was strongly related to the weakening of the East Asian winter monsoon (EAWM), which showed decreased surface wind speed and number of cold air outbreaks and increased relative humidity and frequency of light wind events [6]. Li et al. found that the number of wintertime fog–haze days correlates with the inter-annual variations of the winter monsoon index, with a correlation coefficient of -0.41 [9]. Shi et al. also found that wind speed change in the lower troposphere explains 81.6% of the visibility variance between 1960 and 2014 in Eastern China [7]. Lower wind speed decreased dust emissions, and this decrease moderated the wintertime land–sea surface air temperature difference and further decreased wind speed [10]. However, these meteorological approximations only partly reflect the aerosol variations, and their in situ observations are sparse.

Many studies use satellite observations of aerosol optical depth (AOD, [11]) to retrieve surface $PM_{2.5}$. AOD are observed with large spatio-temporal coverage by remote sensing instruments on board satellites. For example, the total Ozone Mapping Spectroradiometer [12], the Cloud-Aerosol Lidar with Orthogonal Polarization (CALIOP, [13]), and the moderate resolution imaging spectrora-diometer [14]. A number of studies [15–17] have developed different methods, such as statistical relations [18] and machine learning [15], to derive surface $PM_{2.5}$ with high spatiotemporal resolution in China to study the long-term exposure of population to surface $PM_{2.5}$. However, studies showed that AOD–surface $PM_{2.5}$ relations show large spatial and temporal variations (e.g., r = 0.17–0.57, varying within regions in China [19]). Meteorological elements, such as cloud cover, wind speed, boundary layer height, and relative humidity, are used to correct the AOD–surface $PM_{2.5}$ relation for better prediction of surface $PM_{2.5}$ ([11,20–22]). However, to the knowledge of the authors, no study has systematically quantified the different responses of AOD and surface $PM_{2.5}$ to changes in meteorological elements to better understand the AOD–surface $PM_{2.5}$ relations.

In this study, we systematically quantify the contributions of meteorology changes to the inter-annual variations of AOD and surface $PM_{2.5}$ in different seasons and regions in China between 2006 and 2017 by fixing emissions at the 2006 level using a nested global 3D chemical transport model, GEOS-Chem. We study the relationship of AOD and surface $PM_{2.5}$ in different regions and seasons and their relationship with mesoscale weather patterns. We further identify the major meteorological elements that control the inter-annual variations of AOD and surface $PM_{2.5}$ in key regions during different seasons using multiple linear regressions.

2. Materials and Methods

2.1. Model Description

We used the nested 3D chemical transport model, GEOS-Chem version 11.01, over Asia to simulate surface $PM_{2.5}$ and AOD. The nested model has a horizontal resolution of 0.5° latitude × 0.667° longitude, with boundary conditions archived from global simulations at 2° latitude × 2.5° longitude. We simulated AOD in Asia between 2006 and 2017 [23] with a model spin-up of one month. The model was driven with Modern-Era Retrospective analysis for Research and Application, Version 2 (MERRA-2) meteorological fields. We ran GEOS-Chem with full gaseous chemistry and online aerosol calculations, including sulphate-nitrate-ammonium particles [24], black carbon (BC, [25,26]), primary [27] and secondary organic carbon (OC, [28]), natural dusts [29–31], and sea salts [32]. The model coupled aerosol and gas-phase chemistry through nitrate and ammonium partitioning, sulphur chemistry, secondary organic aerosol formation, and uptake of acidic gases by sea salt and dust [24]. We used monthly Asian anthropogenic emissions of SO_2, NOx, BC, OC, NMVOCs, and NH_3 from [33]. We developed anthropogenic emission inventories of these gases and aerosols from industrial, transport, residential, and agricultural sectors in China between 2006 and 2017 using a bottom-up approach [34]. We used daily open fire emissions from the Global Fire Emissions Database, Version 4. Dust emissions followed Fairlie et al. [29] with an improved dust size distribution scheme from [31]. Dry deposition of aerosols followed a resistance-in-series method [35] with updates of dry deposition velocity [36]. Wet removal of aerosols in convective updrafts and large-scale precipitation followed Liu et al. [37], with updates of below-cloud and in-cloud scavenging in ice clouds [38,39] and in-cloud scavenging in mixed-phase clouds [40].

Assuming external mixing, AOD at wavelength λ in each layer was estimated as the sum of AODs of each component *i*

$$\tau_\lambda = \sum_{i=1}^{n} \frac{3}{4}\frac{m_i Q_{\lambda,i}}{\rho_i r_{eff,i}} = \sum_{i=1}^{n} m_i \beta_{\lambda,i}$$

where τ_λ is AOD at wavelength λ, n is the number of aerosol components, m_i is aerosol mass concentration of component i, $Q_{\lambda,i}$ is extinction efficiency factor at wavelength λ calculated with Mie theory, ρ_i is aerosol mass density, $r_{eff,i}$ is particle effective radius. We accounted for the hygroscopicity growth of aerosol particles, as all parameters in the above equation are functions of relative humidity for hydrophilic aerosol components. We used the updated aerosol size distribution and refractive index [41] to calculate $Q_{\lambda,i}$ and $r_{eff,i}$ in a Mie code. Uncertainties of the model-simulated AOD stemmed from aerosol vertical profiles and assumptions on aerosol physical and chemical properties, including mixing state, density, refractive index, and hygroscopic growth [42]. Among the aerosol physical and chemical properties, the mixing state is the most important factor, causing an uncertainty of between 30 and 35% on the simulated AOD [42]. Uncertainties caused by other properties were less than 10% [42]. We used in situ station radiometer AOD measurements, AOD measurements from a moderate resolution imaging spectrora-diometer, and surface in situ measurements of $PM_{2.5}$ to validate model simulations (see details in Supplementary Materials S1 and S2).

We conducted two experiments to quantify the contributions of meteorological changes in AOD and surface $PM_{2.5}$ in China. In the BASE simulation, we used varying meteorology and emissions from 2006 to 2017. The observed and simulated inter-annual variations and trends of annual mean AOD and surface $PM_{2.5}$ are discussed in Supplementary Materials S3. To investigate the contribution of meteorology, we kept emissions at the level of 2006 and varying meteorology in simulation FIXEMISS. The inter-annual variations of AOD and surface $PM_{2.5}$ in this experiment reflect the effects of varying meteorology. Comparing trends of AOD and surface $PM_{2.5}$ simulated by the two experiments shows the relative contribution of meteorology changes in these variations.

2.2. Multiple Linear Regression

We analysed results in three hot spot key regions, including the North China Plain (NCP: 35–41°N, 110–120°E), the Yangtze River Delta (YRD: 27–35°N, 116–122°E), and the Pearl River Delta (PRD: 22–25°N, 110–117°E) as shown in Figure S1. We built multiple linear regression models for all the target regions during the four seasons to investigate the contribution of meteorological elements to variations of AOD and surface $PM_{2.5}$. The candidate variables included temperature (T), zonal wind speed (U), meridional wind speed (V), vertical air movement (O), relative humidity (RH), potential vorticity (PV) at surface, 850 hPa and 500 hPa, pressure at surface (PS), sea level pressure (SLP), tropopause pressure (TROPPT), planetary boundary layer height (PBLH), and precipitation (PREC). The list of abbreviations is shown in Table S2. To avoid redundancy, we use adjusted (adj) the R^2 criterion to determine the best subset of regressors.

$$\text{adj } R^2 = 1 - \frac{(1-R^2)(n-1)}{n-k-1} \tag{1}$$

where k is the number of model parameters, n is the number of pairs of data in the data set, R^2 is the coefficient of determination because the inclusion of the penalty term $n-k-1$, adj R^2 decreases when redundant variables are included. We used a stepwise procedure to select variables. Subsets with the smallest number of variables and largest adj R^2 are regarded as the best predictors. GEOS-Chem-simulated and the best multiple linear regression model-fitted AOD and surface $PM_{2.5}$ are shown in Figures 1 and 2.

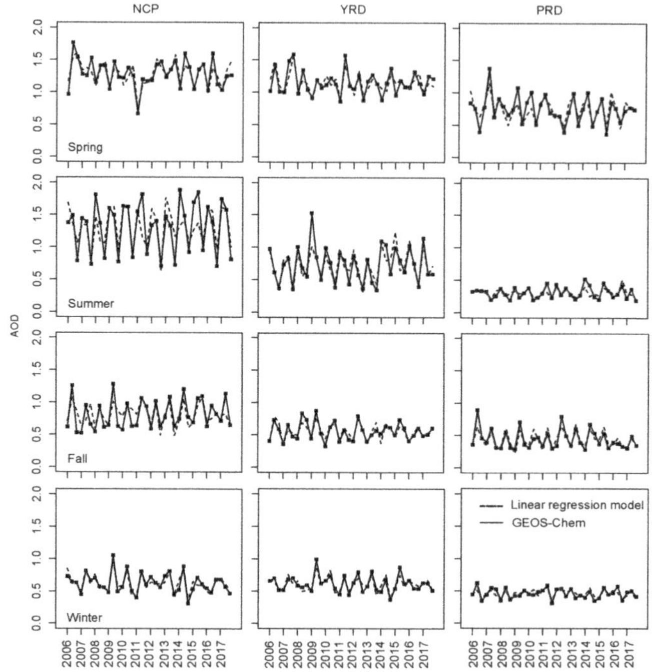

Figure 1. GEOS-Chem-simulated (solid line) and multiple linear regression-predicted (dashed line) monthly mean AOD in NCP, YRD, and PRD.

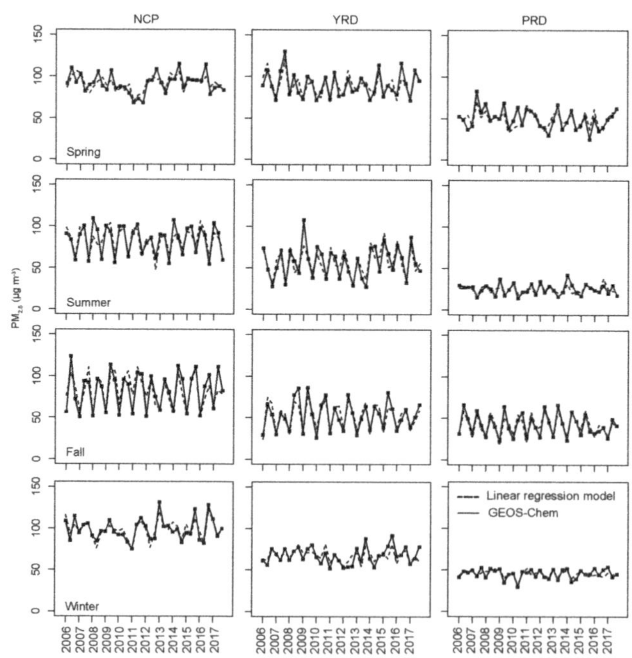

Figure 2. GEOS-Chem-simulated (solid line) and multiple linear regression-predicted (dashed line) monthly mean $PM_{2.5}$ in NCP, YRD, and PRD.

3. Results

3.1. Meteorology Changes Have Larger Influences on Inter-Annual Variations of AOD than Those of Surface $PM_{2.5}$

GEOS-Chem shows that the influence of meteorology changes on AOD trends between 2006 and 2017 is relatively larger than their influences on surface $PM_{2.5}$ (Table 1). Contributions from meteorology changes to downward trends of annual mean AOD are 33% in PRD, larger than that of surface $PM_{2.5}$ (23%). In YRD, meteorology changes reduce AOD by 10%, but are counter-productive with regard to surface $PM_{2.5}$ reduction. Meteorology changes show negligible effects on trends of AOD and surface $PM_{2.5}$ in NCP. In spring, meteorology changes between 2006 and 2017 are favorable to AOD and surface $PM_{2.5}$ reduction. GEOS-Chem attributes 36%, 22%, and 50% of AOD reduction in NCP, YRD, and PRD, respectively, to meteorology changes during the 12 years. However, for surface $PM_{2.5}$ reduction, the contributions of meteorology changes are much smaller: 1% in NCP, 12% in YRD, and 40% in PRD. In summer, meteorology changes show adverse effects on aerosol reduction during the 12 years. AOD enhancements due to meteorology change offset 42%, 25%, and 33% of the reduction caused by anthropogenic emission control in NCP, YRD, and PRD, respectively. Influences of meteorology changes on surface $PM_{2.5}$ are much smaller in NCP (5%) and PRD (10%). In fall, the model suggests a significant adverse effect of meteorology on aerosols in NCP between 2006 and 2017 (+0.011 yr^{-1} and +0.26 $\mu g\ m^{-3}\ yr^{-1}$). These enhancements offset the reduction caused by anthropogenic emission control by 85% for AOD and by 11% for surface $PM_{2.5}$, respectively. Meteorology changes contribute 25% and 40% to the total reduction in AOD in YRD and PRD (Table 1). Their contributions to surface $PM_{2.5}$ reductions are much lower, with 8% in NCP and 35% in YRD.

Table 1. GEOS-Chem-simulated inter-annual trends of annual and seasonal mean AOD (yr^{-1}, % yr^{-1} in parentheses) and surface $PM_{2.5}$ ($\mu g\ m^{-3}\ yr^{-1}$ and % yr^{-1} in parentheses) in NCP, YRD, and PRD between 2006 and 2017.

			NCP	YRD	PRD
Annual	BASE	AOD	−0.020 (−2.1) *	−0.020 (−2.9) *	−0.012 (−2.7) *
		$PM_{2.5}$	−2.61 (−3.4) *	−2.12 (−2.9) *	−1.45 (−2.7) *
	FIXEMISS	AOD	0.001 (0.1)	−0.002 (−0.2)	−0.004 (−0.8) *
		$PM_{2.5}$	0.14 (0.2)	0.11 (0.2)	−0.33 (−0.8) *
Spring	BASE	AOD	−0.028 (−2.3) +	−0.027 (−2.6) *	−0.015 (−2.1) #
		$PM_{2.5}$	−2.57 (−3.1) *	−3.13 (−4.0) *	−2.01 (−4.5) *
	FIXEMISS	AOD	−0.010 (−0.8)	−0.006 (−0.5)	−0.009 (−1.2)
		$PM_{2.5}$	−0.03 (0.0)	−0.38 (−0.4)	−0.81 (−1.6)
Summer	BASE	AOD	−0.021 (−1.8) #	−0.018 (−2.8) #	−0.004 (−1.6)
		$PM_{2.5}$	−2.33 (−3.1) *	−1.53 (−3.1) +	−0.81 (−3.8) *
	FIXEMISS	AOD	0.011 (0.9)	0.006 (0.8)	0.002 (0.5)
		$PM_{2.5}$	0.12 (0.1)	0.67 (1.2)	0.09 (0.0)
Fall	BASE	AOD	−0.005 (−0.7)	−0.016 (−3.2) *	−0.017 (−4.4) +
		$PM_{2.5}$	−2.11 (−2.9) *	−1.67 (−3.7) *	−1.70 (−4.6) *
	FIXEMISS	AOD	0.011 (1.4) #	−0.004 (−0.6)	−0.007 (−1.7)
		$PM_{2.5}$	0.26 (0.3)	−0.13 (−0.2)	−0.60 (−1.3) #
Winter	BASE	AOD	−0.025 (−4.5) *	−0.021 (−3.9) *	−0.013 (−3.1) *
		$PM_{2.5}$	−3.45 (−4.2) *	−2.14 (−3.7) *	−1.27 (−3.3) *
	FIXEMISS	AOD	−0.007 (−1.2) *	−0.004 (−0.6)	−0.001 (−0.2)
		$PM_{2.5}$	0.23 (0.2)	0.27 (0.4)	0.09 (0.2)

Significant at 90% level ($0.05 < p < 0.1$); + significant at 95% level ($0.01 < p < 0.05$); * significant at 99% level ($p < 0.01$).

The larger influence of meteorology changes on AOD than surface $PM_{2.5}$ is attributable to several reasons. First, emissions have larger influence on surface $PM_{2.5}$ than aloft. Second, $PM_{2.5}$ concentration decreases with increasing altitude. Mathematically, the same amount of meteorology changes show relatively larger effects on smaller $PM_{2.5}$ concentrations. Third, observations show that change rates of meteorological elements are amplified with elevation [43,44]. Warming [44] and change of wind speed [43] are more rapid at higher elevations.

3.2. Meteorology Changes in Spring Are Beneficial to Aerosol Reduction

We used multiple linear regression to identify major meteorological elements that possibly have large influences on inter-annual variations of AOD and surface $PM_{2.5}$. We found that meteorological elements that might influence inter-annual variations of AOD and surface $PM_{2.5}$ in spring are similar and mainly related to RH and wind speed. In addition, increasing wind speed in spring is, possibly, the main reason for the beneficial effects of meteorology changes in AOD and surface $PM_{2.5}$ reduction. We found that the inter-annual variations of AOD in NCP is strongly related to the variations of T at 850 hPa, surface RH, and O at 850 hPa, explaining 53% of AOD variations between 2006 and 2017 (Table 2). For surface $PM_{2.5}$, the latter two elements explain 53% of the inter-annual variations (Table 3). In YRD, wind speeds at different altitudes are important factors controlling the inter-annual variations of both AOD and surface $PM_{2.5}$. Surface $PM_{2.5}$ has a stronger correlation with wind speed. Westerly wind at the surface in YRD increases at a rate of 0.07 m s^{-1} yr^{-1} (5.4% yr^{-1}, $p < 0.05$), which possibly contributed to the reduction in AOD and surface $PM_{2.5}$ in this region. In PRD, the contributions from meteorology (−0.009 yr^{-1}, 1.7% of 12-year mean AOD) and emission reduction (−0.010 yr^{-1}) are comparable. Multiple linear regression suggests that AOD in PRD is strongly correlated to surface meridional wind velocity and zonal wind velocity difference between the surface and 850 hPa. The former increased at a rate of 0.04 m s^{-1} yr^{-1} (1.1% of 12-year mean, $p = 0.13$) between 2006 and 2017, possibly causing decreasing aerosol concentration. The latter increased at a rate of 0.05 m s^{-1} yr^{-1}, indicating that dynamic instability was enhanced over the 12 years, favorable for pollution mitigation.

Table 2. Meteorological parameters that explain the AOD variations in NCP, YRD, and PRD in different seasons.

	NCP		YRD		PRD	
	Variables	**adj R^2**	**Variables**	**adj R^2**	**Variables**	**adj R^2**
Spring	T_{850hPa}	0.34	$V_{surface}$	0.18	V_{850hPa}	0.11
	$RH_{surface}$	0.48	$U_{surface}$	0.36	$dU_{surf\text{-}850hPa}$	0.42
	O_{850hPa}	0.53	TROPPT	0.42	PBLH	0.56
	$dT_{surf\text{-}850hPa}$	0.59	$PV_{surface}$	0.48	PV_{500hPa}	0.63
	PBLH	0.67	$dV_{850hPa\text{–}500hPa}$	0.53		
			O_{500hPa}	0.62		
Summer	$dV_{surf\text{-}850hPa}$	0.31	U_{500hPa}	0.72	T_{500hPa}	0.21
	$V_{surface}$	0.64			PREC	0.31
	O_{850hPa}	0.70			V_{500hPa}	0.38
	RH_{850hPa}	0.72			O_{850hPa}	0.46
Fall	O_{850hPa}	0.13	SLP	0.47	PS	0.12
	SLP	0.28	PV_{850hPa}	0.57	SLP	0.59
	$PV_{surface}$	0.36	$dV_{850hPa\text{–}500hPa}$	0.72	U_{500hPa}	0.70
	$RH_{surface}$	0.45				
Winter	V_{850hPa}	0.74	RH_{850hPa}	0.55	RH_{850hPa}	0.28
	O_{500hPa}	0.80	PV_{500hPa}	0.64	PV_{500hPa}	0.45
	TROPPT	0.83	$V_{surface}$	0.73	U_{500hPa}	0.52
					PBLH	0.56

Table 3. Meteorological elements that explain inter-annual variations of surface $PM_{2.5}$ in NCP, YRD, and PRD between 2006 and 2017 in different seasons.

	NCP		YRD		PRD	
	Variables	**adj R^2**	**Variables**	**adj R^2**	**Variables**	**adj R^2**
Spring	$RH_{surface}$	0.32	$V_{surface}$	0.44	V_{850hPa}	0.40
	O_{850hPa}	0.53	$U_{surface}$	0.59	$dU_{surf\text{-}850hPa}$	0.59
			O_{500hPa}	0.61		
			V_{500hPa}	0.68		
Summer	$dV_{surf\text{-}850hPa}$	0.26	U_{500hPa}	0.70	T_{500hPa}	0.13
	$V_{surface}$	0.60			PREC	0.24
	PREC	0.67			V_{500hPa}	0.34
					O_{850hPa}	0.40
Fall	SLP	0.45	SLP	0.66	PS	0.46
	$PV_{surface}$	0.48	PV_{850hPa}	0.72	SLP	0.76
	$RH_{surface}$	0.61	$V_{surface}$	0.78	U_{500hPa}	0.81
	O_{850hPa}	0.65				
Winter	$RH_{surface}$	0.48	$U_{surface}$	0.24	$V_{surface}$	0.19
	PBLH	0.66	$dU_{850hPa\text{–}500hPa}$	0.33	T_{850hPa}	0.36
	$dT_{850hPa\text{–}500hPa}$	0.76	TROPPT	0.36	PV_{850hPa}	0.49
			V_{500hPa_NCP}	0.45		

We estimated the correlation of AOD (surface $PM_{2.5}$) among different regions to investigate spatial variations of aerosols. We found that AOD in NCP and YRD in spring are highly correlated (r = 0.77, Table 4), but surface $PM_{2.5}$ in the two regions are not (r = 0.09). This pattern of correlations in the two regions is related to the activity of the West Pacific Sub-Tropical High system (WPSTH, Figure 3). AOD in NCP and YRD show similar correlation with the area (r = 0.29 and 0.31) and strength (r = 0.24 and 0.24) of the WPSTH. Surface $PM_{2.5}$ in YRD is also related to the two indices (r = 0.33 and 0.25), but surface $PM_{2.5}$ in NCP is not (r = −0.02 and 0.05). Very few studies have investigated the effects of meteorology on the distribution of aerosols in China in spring. A recent study [45] showed that the activity of the WPSTH and Northeast Asia anticyclone system are important to the distribution of $PM_{2.5}$ in NCP. They showed that the climatology of the winds at 850 hPa in spring over NCP and YRD is northwesterly. In an anomalous southeasterly wind year, wind speed is reduced and RH increases in Eastern China, resulting in high aerosol concentrations in the region.

Table 4. Correlation coefficients of seasonal mean AOD and surface $PM_{2.5}$ between 2006 and 2017 among NCP, YRD, and PRD.

		NCP~YRD	**NCP~PRD**	**PRD~YRD**
Spring	AOD	0.50	−0.13	0.31
	$PM_{2.5}$	0.09	−0.29	0.10
Summer	AOD	0.78	0.30	0.27
	$PM_{2.5}$	0.83	0.22	0.28
Fall	AOD	0.55	0.69	0.66
	$PM_{2.5}$	0.77	0.79	0.76
Winter	AOD	0.77	0.15	0.39
	$PM_{2.5}$	0.43	0.14	0.29

3.3. Meteorology Changes in Summer Are Unfavourable to Aerosol Reduction

GEOS-Chem shows that meteorology changes in summer between 2006 and 2017 offset aerosol pollution control efforts in NCP, YRD, and PRD (Table 1). We used two indices to describe the East Asian summer monsoon (EASM) in this study (Figure 4). Index 1 is the mean meridional wind speed at 850 hPa in Eastern Asia (20–40°N, 110–125°E), reflecting activity of the monsoon system in the whole region [46]. Index 2 is the zonal wind speed difference at 200 hPa between Northern (40–50°N, 110–150°E) and Southern (25–35°N, 110–150°E) China, reflecting the different variations between the two regions [47]. We found that AOD and surface $PM_{2.5}$ in the three regions are moderately to strongly correlated with both EASM Index 1 and Index 2 (Table 5). Both indices show that EASM weakened between 2006 and 2017 (Index 1: -0.023 m s^{-1} yr^{-1}; Index 2: -0.004 m s^{-1} yr^{-1}, Figure 4). As a result, AOD in the three target regions increased at rates from 0.006–0.011 yr^{-1}, and surface $PM_{2.5}$ increased at rates from 0.09–0.67 μg m^{-3} yr^{-1} (Table 1). If the enhancement of surface $PM_{2.5}$ was totally attributed to wind speed changes, the sensitivity is from 0.09–0.67 μg m^{-3} %$^{-1}$, in general agreement with a recent estimate (0–0.5 μg m^{-3} %$^{-1}$, [48]).

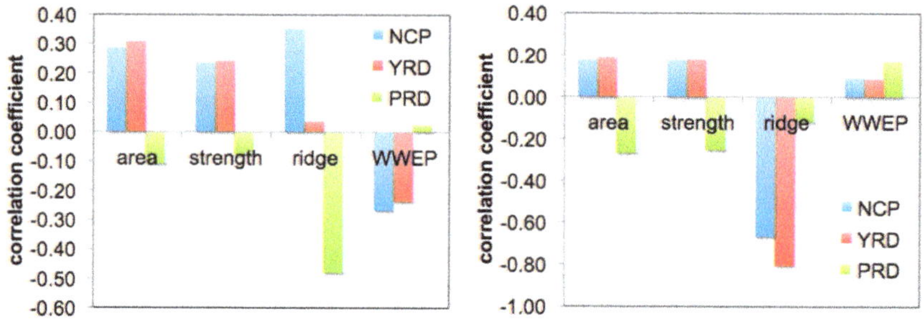

Figure 3. Correlation coefficients of AOD (**a**) and surface $PM_{2.5}$ (**b**) with area, strength, ridge position, and western extension index of Western Pacific sub-tropical high system in spring.

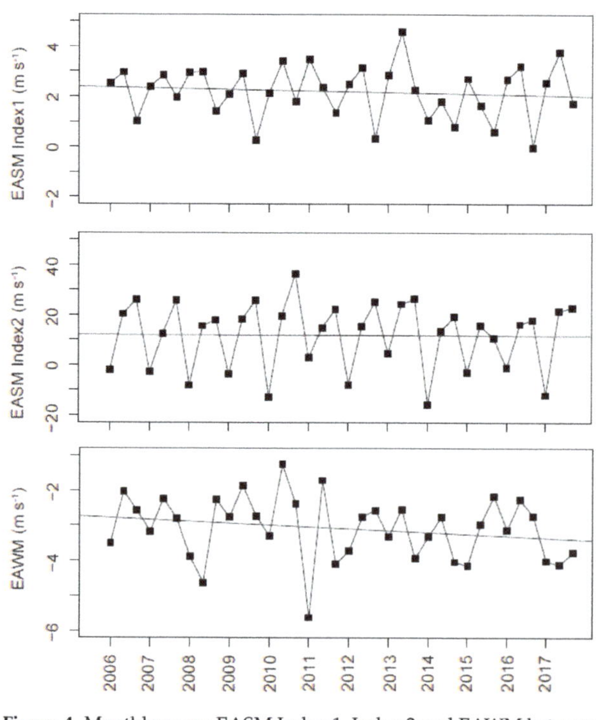

Figure 4. Monthly mean EASM Index 1, Index 2 and EAWM between 2006 and 2017.

Table 5. Correlation coefficients of East Asia monsoon system with AOD and surface $PM_{2.5}$ in NCP, YRD, and PRD in summer and winter between 2006 and 2017.

		NCP	YRD	PRD
EASM Index1	AOD	0.61	0.35	−0.29
	$PM_{2.5}$	0.57	0.27	−0.30
EASM Index2	AOD	−0.67	−0.82	−0.33
	$PM_{2.5}$	−0.67	−0.76	−0.45
EAWM	AOD	0.50	0.51	0.15
	$PM_{2.5}$	0.42	0.31	0.18

In each region, AOD and surface $PM_{2.5}$ are strongly correlated (r = 0.93–0.98) because of the strong vertical mixing in summer. AOD in NCP and YRD are strongly correlated in summer (r = 0.78). However, their correlations with AOD in PRD are relatively low (r = 0.30 and 0.27). Similar relationships of surface $PM_{2.5}$ among the three regions are also shown. In different regions, the correlations of aerosols with EASM indices are different. In NCP, both AOD and surface $PM_{2.5}$ are strongly positively related to EASM Index 1 (r = 0.57–0.61) and negatively related to Index 2 (r = −0.67). EASM Index 1 and Index 2 together explain from 65–69% of the inter-annual variations of AOD and surface $PM_{2.5}$, indicating that both the activity of EASM in the whole Eastern China and the difference between the Northern and Southern China are critical to the inter-annual variations of aerosols in NCP. In contrast, AOD and surface $PM_{2.5}$ in YRD are strongly correlated with EASM Index 2 (r = −0.76−0.82) but weakly correlated with Index 1 (r = 0.27–0.35), indicating that the zonal wind velocity difference at 200 hPa between Northern and Southern China is possibly more important to the inter-annual variations of aerosols in YRD.

We found that EASM Index 2 is strongly correlated with the position of the ridge of the WPSTH system (r = 0.89), indicating that the zonal wind difference between Northern and Southern China is possibly affected by the activity of the WPSTH system. A study [49] showed that, in addition to wind velocity, the ridge position of the WPSTH also strongly affects the distribution of precipitation in China in summer. The climatological pattern of precipitation is more present north of the Yangtze River region and less in Southern China. When the ridge shifts southward, the distribution of precipitation is the opposite. When the ridge shifts northward, two rain-belts show in Southern and Northern China. The humid maritime air mass brought to inland China by EASM shows dual effects on aerosols. The abundant water vapour enhances the hygroscopic growth of sulphate-nitrate-ammonia and, thus, increases aerosol concentrations, while large precipitation removes aerosols from the atmosphere and reduces aerosol concentrations. In NCP, AOD is more affected by RH, while surface $PM_{2.5}$ is more affected by precipitation [50]. Both EASM Index 1 and Index 2 are weakly correlated with AOD and surface $PM_{2.5}$ (r = −0.29–0.45) in PRD, indicating that EASM has limited influence on aerosol distribution in this region. Surface $PM_{2.5}$ shows similar relationships. Regression analysis shows that major meteorological elements in PRD only explains ~40% of the inter-annual variations of aerosols in this region. Meteorology influences on aerosols in PRD need further investigation.

3.4. Meteorology Changes in Fall Show Different Effects on Trends of Aerosols in Different Key Regions

GEOS-Chem suggests adverse effects of meteorology changes on aerosol reduction in NCP, but beneficial effects in YRD and PRD between 2006 and 2017. The significant enhancements of aerosols in NCP (+0.011 yr^{-1} and +0.26 $\mu g\ m^{-3}\ yr^{-1}$) offset the reduction caused by anthropogenic emission control by 85% for AOD and by 11% for surface $PM_{2.5}$, respectively. We found that PV in NCP decreased at a rate of −0.02 PVU yr^{-1} ($p < 0.05$), and RH at 850 hPa increased at a rate of 0.002 yr^{-1}. Both changes favored the enhancement of aerosol concentration in this region. In contrast, meteorology changes contributed from 8–25% and 35–40% to the total reduction in aerosols in YRD and PRD (Table 1). In YRD, PV at 850 hPa became stronger over the years at a rate of +0.01 PVU yr^{-1} (p = 0.004), likely reducing aerosols in this region. In PRD, meridional wind speed at 500 hPa increased (+0.07 m s^{-1} yr^{-1}), related to the decline in aerosols in this region.

GEOS-Chem shows that AOD in NCP, YRD, and PRD have strong correlations in fall (r = 0.55–0.69). The surface $PM_{2.5}$ in the three regions show even larger correlation coefficients (r = 0.76–0.79). In addition, AOD and surface $PM_{2.5}$ in each region are also highly correlated (r = 0.81–0.94), similar to summer. Major meteorological elements that affect AOD and surface $PM_{2.5}$ in each region are similar. A recent study [51] showed that in fall, Eastern China is dominated by large-scale stable circulation patterns without frequent disturbances of small-scale weather systems. Vertically, downward motion dominates. Multiple linear regressions suggest that sea level pressure is the major controlling factor that affects the inter-annual variations of AOD and surface $PM_{2.5}$, particularly for the latter, in each region (Tables 2 and 3). We found that sea level pressure among the three regions is also moderately to highly correlated (r = 0.50–0.75), which partly explains the correlation of aerosols among the three regions.

3.5. Meteorology Changes in Winter Show Opposite Effects on Trends of AOD and Surface $PM_{2.5}$

GEOS-Chem suggests that from 8–30% of the downward trends in AOD in NCP (−0.024 yr^{-1}), YRD (−0.021 yr^{-1}), and PRD (−0.013 yr^{-1}) are attributable to meteorology changes. In contrast, meteorology changes show adverse effects on surface $PM_{2.5}$ reduction over the 12 years, offsetting from 7–9% of the reductions caused by anthropogenic emission control. The opposite effects of meteorology changes on AOD and surface $PM_{2.5}$ variations are mainly due to the isolation of $PM_{2.5}$ at the surface and aloft by the stable boundary layer in winter. The boundary layer is more stable and vertical mixing is weaker in winter than in other seasons. Thus, the correlations of surface $PM_{2.5}$ and AOD in winter in the three regions are much weaker than those during other seasons, with correlation coefficients of 0.44 in NCP, −0.01 in YRD, and 0.21 in PRD. This pattern is possibly partly explained with the dual effects of EAWM on haze–fog variations in Eastern China [52]. Cold wave activity in winter advects aerosol away and cleans up the region. However, the activity of the Siberian High system may reduce the near surface wind speed and enhance the stratification stability, thus, favoring pollution accumulation.

Major meteorological elements controlling the inter-annual variations of AOD and surface $PM_{2.5}$ are completely different. In NCP, a local northerly wind speed at 850 hPa explains 74% of the inter-annual variations of AOD between 2006 and 2017 (Table 2). This wind speed increases at a rate of +0.15 m s^{-1} yr^{-1} ($p < 0.1$), partly explaining the decrease in AOD in the region (−0.007 yr^{-1}, $p < 0.05$). The increasing wind speed in NCP is related to the enhanced Siberian High due to the rapid warming of the Barents-Kara Sea region [53]. Different from AOD, inter-annual variations of surface $PM_{2.5}$ in NCP are mainly affected by surface RH, boundary layer height, and the temperature difference between 850 hPa and 500 hPa. This temperature difference increases over the 12 years, although it is statistically insignificant (+0.1 K yr^{-1}, p = 0.33), favouring aerosol accumulation. These changes are in-line with surface $PM_{2.5}$ enhancement in NCP during these years (+0.23 μg m^{-3}, p = 0.70).

AOD in YRD is strongly correlated with AOD in NCP (r = 0.77), but surface $PM_{2.5}$ in the two regions are only weakly correlated (r = 0.43). We used meridional wind speed at 850 hPa in Eastern Asia (25–50°N, 115–145°E) as an indicator of the strength of EAWM ([46], Figure 4). The EAWM index is moderately correlated with AOD in NCP (r = 0.52) and YRD (r = 0.52), but relatively weakly related to surface $PM_{2.5}$ in NCP (r = 0.42) and YRD (r = 0.31). This correlation pattern suggests that surface $PM_{2.5}$ in YRD is less affected by the 850 hPa wind speed in Eastern Asia.

Multiple linear regressions suggest that the inter-annual variations of AOD in YRD are mainly affected by RH at 850 hPa, potential vorticity at 500 hPa, and surface meridional wind velocity. We found that RH at 850 hPa decreased (−0.003 yr^{-1}, p = 0.29), but surface wind increased (0.046 m s^{-1} yr^{-1}, p = 0.29), favouring aerosol accumulation and, thus, possibly enhancing AOD reduction. Meteorological elements determining the variations of surface $PM_{2.5}$ include surface zonal wind velocity, dynamic instability, tropopause height, and meridional wind speed at 500 hPa in NCP. The former three elements explain 36% of the inter-annual variations of surface $PM_{2.5}$. Including the last element explains 9% more variations, suggesting that transport from NCP to YRD is likely important to surface $PM_{2.5}$ in YRD in winter. We found that meridional wind speed at 500 hPa in NCP increases over the years (+0.15 m s^{-1} yr^{-1}, $p = 0.03$), indicating that transport from NCP to YRD is possibly increasing. As a result, surface $PM_{2.5}$ in YRD is increasing.

Similar as in NCP and YRD, meteorological elements that influence the inter-annual variations of AOD and surface $PM_{2.5}$ are also completely different in PRD. Increasing zonal wind speed at 500 hPa (+0.17 m s^{-1} yr^{-1}, 0.8% of 12-year mean), and decreasing potential vorticity at 500 hPa (−0.004 PVU yr^{-1}) and PBLH (−2.9 m yr^{-1}), are related to the decrease in AOD in this region. Meteorological elements affecting surface $PM_{2.5}$ include surface meridional velocity, temperature at 850 hPa, and potential vorticity.

4. Discussion

We showed that the weakening of EASM and the enhancing of AOD and surface $PM_{2.5}$ between 2006 and 2017 are statistically insignificant, but the trends are still worth notice because they are in-line with the inter-decadal trend as reported by previous studies [54]. Zhu et al. [54] showed that the decadal-scale-weakening of EASM (index: +0.31 between 1948 and 1979 versus −0.32 between 1980 and 2010) within the last thirty years led to increases in aerosol concentration in Northern China by 20%. In addition, the monsoon system also affects the spatial distribution of aerosols. During an active monsoon year, AOD had a positive anomaly in NCP and a negative anomaly in PRD. During a weak monsoon year, the anomalies were the opposite [55,56].

We found that the weakening of EAWM enhances surface $PM_{2.5}$ in the three key regions, in general agreement with [57], which showed that with fixed emissions, meteorological conditions led to an increase in haze in Beijing during winter between 2002 and 2016. In contrast, Wang et al. [58] found that EAWM was significantly anti-correlated with surface $PM_{2.5}$ in Beijing between 2005 and 2016, with a correlation coefficient of ~0.75. The difference between the two studies can be attributed to several reasons. First, we analyzed a much larger region, NCP, in this study than in [58], which focused on a city Beijing. Second, we separated the contribution from anthropogenic emissions and meteorology using a chemical transport model, but Wang et al. [58] used surface in situ observations, which do not distinguish the contributions of meteorology changes from anthropogenic emissions control. Thus, their strong correlation is possibly due to the concurrent downward trends of EAWM index and surface $PM_{2.5}$. Stronger EAWM circulation brings more cold and dry air to NCP and YRD [50] and cleans up the regions. A weaker monsoon barely reaches YRD and even farther north, and favors the accumulation of pollutants [6]. The long-term weakening trend and the inter-annual variations of EAWM were further related to the subtropical Western Pacific Sea surface temperature anomaly [59,60], Arctic sea ice, Eurasian snow [61,62], and El Niño–Southern Oscillation [63]. It is predicted that EAWM would keep the weakening trend in the future (2050–2099), with increased frequency and persistence of conducive weather conditions [8]. This suggests that future meteorology conditions are possibly unfavorable to pollution dissipation.

Very few studies have investigated the influence of meteorology on aerosols in China in spring and fall. $PM_{2.5}$ in Eastern China is related to the inter-annual variations of Asia Polar Vortex intensity [64], North Atlantic Oscillation, and the North Atlantic Sea surface temperature [45] in spring. The influence of synoptic systems on AOD distribution in fall China was investigated [51] and it was found that heavy pollution events with high AOD (>0.6) in Eastern China is associated with a uniform surface pressure field or a steady westerly in the middle troposphere, while clean episodes (AOD < 0.4) occur when strong northwest cold air advection prevails or air masses are transported from sea to land. Further related, by [65], are the haze days in fall to the abnormally warming sea surface temperature over the North Atlantic subtropical and the Western North Pacific.

Most of the previous studies [15,20,51] use cloud cover, wind speed, PBLH, and RH to correct $PM_{2.5}$ retrieval. Specifically, ref. [43] found large spatio-temporal diurnal variations of correlation of AOD and $PM_{2.5}$ in China using measurement data and found that the distribution was strongly affected by cloud fraction, PBLH, and RH. Gong et al. also found that vertical correction by PBLH was important to $PM_{2.5}$ retrieval in Northwestern China [22]. In other regions, vertical correction via CALIOP ratio is recommended [22]. We also suggest that these elements are important to AOD and surface $PM_{2.5}$, but we recommend the inclusion sea level pressure and surface pressure in fall.

Our findings have implications for future surface $PM_{2.5}$ retrieval from satellite-observed AOD. We investigated the controlling meteorological elements of AOD and surface $PM_{2.5}$ variations over 12 years, including the long-term trends and inter-annual variations. We found that the controlling meteorological elements vary with regions and seasons. Thus, surface $PM_{2.5}$ retrieval from satellite AOD should probably consider using different meteorological elements in different seasons. In addition, GEOS-Chem simulation with fixed anthropogenic emissions at the 2006 level showed that meteorology changes throughout the 12 years reduces AOD but enhances surface $PM_{2.5}$ in China during winter, and a multiple linear regression model suggests that the controlling meteorological elements of AOD and surface $PM_{2.5}$ are completely different. Thus, previous estimates, which used meteorological elements to correct surface $PM_{2.5}$ retrieval in winter, should be used with caution, as the long-term trend of surface $PM_{2.5}$ is possibly overestimated. We suggest the use other correction schemes to correct surface $PM_{2.5}$ retrieval in the future, such as the CALIOP ratio and correlation coefficient of AOD and surface $PM_{2.5}$.

5. Conclusions

We studied the effects of meteorology changes on trends in AOD and surface $PM_{2.5}$ in the key regions NCP, YRD, and PRD in China between 2006 and 2017 using a 3D chemical transport model, GEOS-Chem, by fixing emissions at the 2006 level. We further identified major meteorological elements controlling the inter-annual variations of AOD and surface $PM_{2.5}$ using multiple linear regressions.

We found that meteorology changes made larger contributions to trends in AOD than surface $PM_{2.5}$ during spring, summer, and fall between 2006 and 2017. Meteorological changes contributed from 22–50% of AOD reduction in spring, larger than their contributions to surface $PM_{2.5}$ (1–40%). The decrease in aerosols is possibly related to an increase in westerly wind speed (0.07 ms^{-1} yr^{-1}, 5.4% yr^{-1}, $p < 0.05$) in YRD and an increase in meridional wind velocity (0.04 ms^{-1} yr^{-1}, 1.1% of 12-year mean) and dynamic instability (0.05 ms^{-1} yr^{-1}) in PRD. In summer, meteorological changes offset from 25–42% of AOD reduction caused by anthropogenic emission changes. For surface $PM_{2.5}$, the contributions were from 5–10%. The adverse effects are possibly related to the weakening of EASM. In fall, meteorology changes offset 85% of AOD reduction and 11% of surface $PM_{2.5}$ reduction induced by emission changes in NCP. In contrast, from 25–40% of AOD reduction and 8–35% of surface $PM_{2.5}$ reduction is attributed to meteorology changes. Sea level pressure and surface pressure are critical to aerosol distribution in fall. In winter, meteorology changes were beneficial to AOD decreasing, but were unfavourable to surface $PM_{2.5}$ reductions in NCP, YRD, and PRD in between 2006 and 2017. The stable boundary layer in winter suppressed vertical mixing, resulting in a weak correlation of AOD and surface $PM_{2.5}$ in each region. Thus, meteorological elements controlling the inter-annual variations of $PM_{2.5}$ and AOD in each region were completely different. The northerly wind speed at 850 hPa explained 72% of the inter-annual variations of AOD in NCP. The increase in this wind (-0.045 ms^{-1} yr^{-1}, $p < 0.1$) lowered AOD in this region (-0.007 yr^{-1}). In other regions, the trends were statistical insignificant. Thus, previous estimates, which used meteorological elements to correct surface $PM_{2.5}$ retrieval in winter, should be used with caution. Our study provides possible meteorological elements to correct surface $PM_{2.5}$ retrieval from satellite AOD measurements on a seasonal scale.

Supplementary Materials: The following supporting information can be downloaded at: https://www.mdpi.com/article/10.3390/rs14122762/s1, Figure S1: SONET sites and key regions in China: North China Plain (NCP, 35–41°N, 110–120°E), the Yangtze River Delta (YRD, 27–35°N, 116–122°E), and the Pearl River Delta (PRD, 22–25°N, 110–117°E); Figure S2: Monthly mean AOD from SONET (blue line) MODIS (red line) and GEOS-Chem (black line) averaged for 2013–2015. Beijing and Songshan are in the NCP region. Shanghai, Zhoushan, Nanjing and Hefei are in the YRD region. Guangzhou is in the PRD region; Figure S3: GEOS-Chem simulated monthly mean AOD components averaged for 2013–2015; Figure S4: Observed (red line) and GEOS-Chem simulated (BASE, black line) monthly mean surface PM2.5 concentrations (μg m-3) in NCP, YRD and PRD in 2013–2017; Figure S5: Observed (red line) and GEOS-Chem simulated (BASE, black line) annual and seasonal mean surface $PM_{2.5}$ concentrations (μg m^{-3}) in NCP, YRD and PRD in 2013–2017. The vertical lines are standard deviations of daily means in each season in each year; Figure S6: Ratio of annual and seasonal mean AOD relative to their values in 2006 from MODIS (red line) and GEOS-Chem simulations in 2006–2017. Three experiments are shown: varying meteorology and varying emissions (BASE, black line), varying meteorological fields with fixed emissions in 2006 (FIXEMISS, purple line), varying emissions with meteorological fields fixed in 2009 (FIXMET, blue line). See text for details. Figure S7: Similar as Figure S6, but for surface $PM_{2.5}$. Table S1: Statistics of MODIS observed and GEOS-Chem simulated AOD compared to SONET AOD observations at 16 sites; Table S2: List of abbreviations. References [14,30,31,39,66–76] are cited in the Supplementary Materials.

Author Contributions: Conceptualization, L.Q. and S.W.; methodology, D.Y.; software, L.Q.; formal analysis, L.Q.; data curation, H.Z. and D.D.; writing—original draft preparation, L.Q.; writing—review and editing, L.Q. and S.W.; funding acquisition, L.Q. and S.W.; supervision, S.W. All authors have read and agreed to the published version of the manuscript.

Funding: This work was funded by the National Natural Science Foundation of China (No. 21806088), Beijing Natural Science Foundation (No. 8222066), and Fundamental Research Funds for the Central Universities (No. FRF-TP-20-056A1).

Data Availability Statement: Data is contained within the article or Supplementary Materials.

Acknowledgments: We thank the Samsung Advanced Institute of Technology and National Environmental and Energy Science and Technology International Cooperation Base. Shuxiao Wang acknowledges the support from the Tencent Foundation through the XPLORER PRIZE. The simulations were completed on the "Explorer 100" cluster system of Tsinghua National Laboratory for Information Science and Technology.

Conflicts of Interest: The authors declare no conflict of interest.

References

1. Noll, K.E.; Mueller, P.K.; Imada, M. Visibility and Aerosol Concentration in Urban Air. *Atmos. Environ.* **1968**, *2*, 465–475. [CrossRef]
2. Zhang, Q.; Quan, J.; Tie, X.; Li, X.; Liu, Q.; Gao, Y.; Zhao, D. Effects of meteorology and secondary particle formation on visibility during heavy haze events in Beijing, China. *Sci. Total Environ.* **2015**, *502*, 578–584. [CrossRef] [PubMed]
3. Lelieveld, J.; Barlas, C.; Giannadaki, D.; Pozzer, A. Model calculated global, regional and megacity premature mortality due to air pollution. *Atmos. Chem. Phys.* **2013**, *13*, 7023–7037. [CrossRef]
4. Ramanathan, V.; Crutzen, P.J.; Kiehl, J.R.; Rosenfeld, D. Aerosols, climate, and the hydrological cycle. *Science* **2001**, *294*, 2119–2124. [CrossRef]
5. Andreae, M.O.; Jones, C.D.; Cox, P.M. Strong present-day aerosol cooling implies a hot future. *Nature* **2005**, *435*, 1187–1190. [CrossRef]
6. Niu, F.; Li, Z.; Li, C.; Lee, K.-H.; Wang, M. Increase of wintertime fog in China: Potential impacts of weakening of the Eastern Asian monsoon circulation and increasing aerosol loading. *J. Geophys. Res.* **2010**, *115*, D00K20. [CrossRef]
7. Shi, P.; Zhang, G.; Kong, F.; Chen, D.; Azorin-Molina, C.; Guijarro, J.A. Variability of winter haze over the Beijing-Tianjin-Hebei region tied to wind speed in the lower troposphere and particulate sources. *Atmos. Res.* **2019**, *215*, 1–11. [CrossRef]
8. Cai, W.; Li, K.; Liao, H.; Wang, H.; Wu, L. Weather conditions conducive to Beijing severe haze more frequent under climate change. *Nat. Clim. Chang.* **2017**, *7*, 257–262. [CrossRef]
9. Li, Z.; Lau, W.K.M.; Ramanathan, V.; Wu, G.; Ding, Y.; Manoj, M.G.; Liu, J.; Qian, Y.; Li, J.; Zhou, T.; et al. Aerosol and monsoon climate interactions over Asia. *Rev. Geophys.* **2016**, *54*, 866–929. [CrossRef]
10. Yang, Y.; Russell, L.M.; Lou, S.; Liao, H.; Guo, J.; Liu, Y.; Singh, B.; Ghan, S.J. Dust-wind interactions can intensify aerosol pollution over eastern China. *Nat. Commun.* **2017**, *8*, 15333. [CrossRef]

11. Geng, G.; Zhang, Q.; Martin, R.V.; van Donkelaar, A.; Huo, H.; Che, H.; Lin, J.; He, K. Estimating long-term PM2.5 concentrations in China using satellite-based aerosol optical depth and a chemical transport model. *Remote Sens. Environ.* **2015**, *166*, 262–270. [CrossRef]
12. Torres, O.; Bhartia, P.K.; Herman, J.R.; Sinyuk, A.; Ginoux, P.; Holben, B. A long-term record of aerosol optical depth from TOMS observations and comparison to AERONET measurements. *J. Atmos. Sci.* **2002**, *59*, 398–413. [CrossRef]
13. Stephens, G.L.; Vane, D.G.; Boain, R.J.; Mace, G.G.; Sassen, K.; Wang, Z.E.; Illingworth, A.J.; O'Connor, E.J.; Rossow, W.B.; Durden, S.L.; et al. The cloudsat mission and the a-train—A new dimension of space-based observations of clouds and precipitation. *Bull. Am. Meteorol. Soc.* **2002**, *83*, 1771–1790. [CrossRef]
14. Bilal, M.; Nazeer, M.; Nichol, J.; Qiu, Z.; Wang, L.; Bleiweiss, M.; Shen, X.; Campbell, J.; Lolli, S. Evaluation of Terra-MODIS C6 and C6.1 Aerosol Products against Beijing, XiangHe, and Xinglong AERONET Sites in China during 2004–2014. *Remote Sens.* **2019**, *11*, 486. [CrossRef]
15. Xue, T.; Zheng, Y.; Tong, D.; Zheng, B.; Li, X.; Zhu, T.; Zhang, Q. Spatiotemporal continuous estimates of PM2.5 concentrations in China, 2000–2016: A machine learning method with inputs from satellites, chemical transport model, and ground observations. *Environ. Int.* **2019**, *123*, 345–357. [CrossRef]
16. Ma, Z.; Hu, X.; Huang, L.; Bi, J.; Liu, Y. Estimating Ground-Level PM2.5 in China Using Satellite Remote Sensing. *Environ. Sci. Technol.* **2014**, *48*, 7436–7444. [CrossRef]
17. Peng, J.; Chen, S.; Lü, H.; Liu, Y.; Wu, J. Spatiotemporal patterns of remotely sensed PM2.5 concentration in China from 1999 to 2011. *Remote Sens. Environ.* **2016**, *174*, 109–121. [CrossRef]
18. Lin, C.; Li, Y.; Yuan, Z.; Lau, A.K.H.; Li, C.; Fung, J.C.H. Using satellite remote sensing data to estimate the high-resolution distribution of ground-level PM2.5. *Remote Sens. Environ.* **2015**, *156*, 117–128. [CrossRef]
19. Guo, X.Y.; Wang, L.; Tian, L.D.; Li, X.P. Elevation-dependent reductions in wind speed over and around the Tibetan Plateau. *Int. J. Climatol.* **2017**, *37*, 1117–1126. [CrossRef]
20. Wang, Z.; Chen, L.; Tao, J.; Zhang, Y.; Su, L. Satellite-based estimation of regional particulate matter (PM) in Beijing using vertical-and-RH correcting method. *Remote Sens. Environ.* **2010**, *114*, 50–63. [CrossRef]
21. Liu, Y.; Paciorek, C.J.; Koutrakis, P. Estimating Regional Spatial and Temporal Variability of PM2.5 Concentrations Using Satellite Data, Meteorology, and Land Use Information. *Environ. Health Perspect.* **2009**, *117*, 886–892. [CrossRef] [PubMed]
22. Gong, W.; Huang, Y.; Zhang, T.; Zhu, Z.; Ji, Y.; Xiang, H. Impact and Suggestion of Column-to-Surface Vertical Correction Scheme on the Relationship between Satellite AOD and Ground-Level PM2.5 in China. *Remote Sens.* **2017**, *9*, 1038. [CrossRef]
23. Ma, Z.; Liu, R.; Liu, Y.; Bi, J. Effects of air pollution control policies on PM2.5 pollution improvement in China from 2005 to 2017: A satellite-based perspective. *Atmos. Chem. Phys.* **2019**, *19*, 6861–6877. [CrossRef]
24. Park, R.J.; Jacob, D.J.; Field, B.D.; Yantosca, R.M.; Chin, M. Natural and transboundary pollution influences on sulfate-nitrate-ammonium aerosols in the United States: Implications for policy. *J. Geophys. Res. Atmos.* **2004**, *109*, D15204. [CrossRef]
25. Qi, L.; Li, Q.; Li, Y.; He, C. Factors controlling black carbon distribution in the Arctic. *Atmos. Chem. Phys.* **2017**, *17*, 1037–1059. [CrossRef]
26. van der A, R.J.; Mijling, B.; Ding, J.; Koukouli, M.E.; Liu, F.; Li, Q.; Mao, H.; Theys, N. Cleaning up the air: Effectiveness of air quality policy for SO_2 and NO_x emissions in China. *Atmos. Chem. Phys.* **2017**, *17*, 1775–1789. [CrossRef]
27. Park, R.J.; Jacob, D.J.; Chin, M.; Martin, R.V. Sources of carbonaceous aerosols over the United States and implications for natural visibility. *J. Geophys. Res. Atmos.* **2003**, *108*, 4355. [CrossRef]
28. Pye, H.O.T.; Chan, A.W.H.; Barkley, M.P.; Seinfeld, J.H. Global modeling of organic aerosol: The importance of reactive nitrogen (NO_x and NO_3). *Atmos. Chem. Phys.* **2010**, *10*, 11261–11276. [CrossRef]
29. Fairlie, T.D.; Jacob, D.J.; Park, R.J. The impact of transpacific transport of mineral dust in the United States. *Atmos. Environ.* **2007**, *41*, 1251–1266. [CrossRef]
30. Fairlie, T.D.; Jacob, D.J.; Dibb, J.E.; Alexander, B.; Avery, M.A.; van Donkelaar, A.; Zhang, L. Impact of mineral dust on nitrate, sulfate, and ozone in transpacific Asian pollution plumes. *Atmos. Chem. Phys.* **2010**, *10*, 3999–4012. [CrossRef]
31. Zhang, L.; Kok, J.F.; Henze, D.K.; Li, Q.; Zhao, C. Improving simulations of fine dust surface concentrations over the western United States by optimizing the particle size distribution. *Geophys. Res. Lett.* **2013**, *40*, 3270–3275. [CrossRef]
32. Jaeglé, L.; Quinn, P.K.; Bates, T.S.; Alexander, B.; Lin, J.T. Global distribution of sea salt aerosols: New constraints from in situ and remote sensing observations. *Atmos. Chem. Phys.* **2011**, *11*, 3137–3157. [CrossRef]
33. Li, M.; Zhang, Q.; Kurokawa, J.-I.; Woo, J.-H.; He, K.; Lu, Z.; Ohara, T.; Song, Y.; Streets, D.G.; Carmichael, G.R.; et al. MIX: A mosaic Asian anthropogenic emission inventory under the international collaboration framework of the MICS-Asia and HTAP. *Atmos. Chem. Phys.* **2017**, *17*, 935–963. [CrossRef]
34. Zheng, H.; Zhao, B.; Wang, S.; Wang, T.; Ding, D.; Chang, X.; Liu, K.; Xing, J.; Dong, Z.; Aunan, K.; et al. Transition in source contributions of PM2.5 exposure and associated premature mortality in China during 2005–2015. *Environ. Int.* **2019**, *132*, 105111. [CrossRef] [PubMed]
35. Wesely, M.L. Parameterization of surface resistances to gaseous dry deposition in regional-scale numerical models. *Atmos. Environ.* **1989**, *23*, 1293–1304. [CrossRef]
36. Zhang, L.; Gong, S.; Padro, J.; Barrie, L. A size-segregated particle dry deposition scheme for an atmospheric aerosol module. *Atmos. Environ.* **2001**, *35*, 549–560. [CrossRef]

37. Liu, H.; Jacob, D.J.; Bey, I.; Yantosca, R.M. Constraints from 210Pb and 7Be on wet deposition and transport in a global three-dimensional chemical tracer model driven by assimilated meteorological fields. *J. Geophys. Res. Atmos.* **2001**, *106*, 12109–12128. [CrossRef]
38. Wang, Q.; Jacob, D.J.; Fisher, J.A.; Mao, J.; Leibensperger, E.M.; Carouge, C.C.; Le Sager, P.; Kondo, Y.; Jimenez, J.L.; Cubison, M.J.; et al. Sources of carbonaceous aerosols and deposited black carbon in the Arctic in winter-spring: Implications for radiative forcing. *Atmos. Chem. Phys.* **2011**, *11*, 12453–12473. [CrossRef]
39. Wang, Q.; Jacob, D.J.; Spackman, J.R.; Perring, A.E.; Schwarz, J.P.; Moteki, N.; Marais, E.A.; Ge, C.; Wang, J.; Barrett, S.R.H. Global budget and radiative forcing of black carbon aerosol: Constraints from pole-to-pole (HIPPO) observations across the Pacific. *J. Geophys. Res. Atmos.* **2014**, *119*, 195–206. [CrossRef]
40. Qi, L.; Li, Q.; He, C.; Wang, X.; Huang, J. Effects of the Wegener-Bergeron-Findeisen process on global black carbon distribution. *Atmos. Chem. Phys.* **2017**, *17*, 7459–7479. [CrossRef]
41. Drury, E.; Jacob, D.J.; Spurr, R.J.D.; Wang, J.; Shinozuka, Y.; Anderson, B.E.; Clarke, A.D.; Dibb, J.; McNaughton, C.; Weber, R. Synthesis of satellite (MODIS), aircraft (ICARTT), and surface (IMPROVE, EPA-AQS, AERONET) aerosol observations over eastern North America to improve MODIS aerosol retrievals and constrain surface aerosol concentrations and sources. *J. Geophys. Res. Atmos.* **2010**, *115*, D14204. [CrossRef]
42. Curci, G.; Hogrefe, C.; Bianconi, R.; Im, U.; Balzarini, A.; Baró, R.; Brunner, D.; Forkel, R.; Giordano, L.; Hirtl, M.; et al. Uncertainties of simulated aerosol optical properties induced by assumptions on aerosol physical and chemical properties: An AQMEII-2 perspective. *Atmos. Environ.* **2015**, *115*, 541–552. [CrossRef]
43. Guo, J.; Xia, F.; Zhang, Y.; Liu, H.; Li, J.; Lou, M.; He, J.; Yan, Y.; Wang, F.; Min, M.; et al. Impact of diurnal variability and meteorological factors on the PM2.5—AOD relationship: Implications for PM2.5 remote sensing. *Environ. Pollut.* **2017**, *221*, 94–104. [CrossRef] [PubMed]
44. Pepin, N.; Bradley, R.S.; Diaz, H.F.; Baraer, M.; Caceres, E.B.; Forsythe, N.; Fowler, H.; Greenwood, G.; Hashmi, M.Z.; Liu, X.D.; et al. Elevation-dependent warming in mountain regions of the world. *Nat. Clim. Chang.* **2015**, *5*, 424–430. [CrossRef]
45. Chen, S.; Guo, J.; Song, L.; Li, J.; Liu, L.; Cohen, J.B. Inter-annual variation of the spring haze pollution over the North China Plain: Roles of atmospheric circulation and sea surface temperature. *Int. J. Climatol.* **2019**, *39*, 783–798. [CrossRef]
46. Wang, H.; Jiang, D. A new East Asian winter monsoon intensity index and atmospheric circulation comparison between strong and weak composite. *Quat. Sci.* **2004**, *24*, 19–27. (In Chinese)
47. Lau, K.M.; Kim, K.M.; Yang, S. Dynamical and boundary forcing characteristics of regional components of the Asian summer monsoon. *J. Clim.* **2000**, *13*, 2461–2482. [CrossRef]
48. Shi, Z.; Huang, L.; Li, J.; Ying, Q.; Zhang, H.; Hu, J. Sensitivity Analysis of the Surface Ozone and Fine Particulate Matter to Meteorological Parameters in China. *Atmos. Chem. Phys.* **2020**, *20*, 13455–13466. [CrossRef]
49. Zhao, J.; Feng, G.; Yang, J.; Zhi, R.; Wang, Q. Analysis of the distribution of the large-scale drought/flood of summer in China under different types of the western Pacific sub-tropical high. *Acta Meteorol. Sin.* **2012**, *70*, 1021–1031. (In Chinese)
50. Li, Q.; Zhang, R.; Wang, Y. Interannual variation of the wintertime fog-haze days across central and eastern China and its relation with East Asian winter monsoon. *Int. J. Climatol.* **2016**, *36*, 346–354. [CrossRef]
51. Zheng, X.Y.; Fu, Y.F.; Yang, Y.J.; Liu, G.S. Impact of atmospheric circulations on aerosol distributions in autumn over eastern China: Observational evidence. *Atmos. Chem. Phys.* **2015**, *15*, 12115–12138. [CrossRef]
52. Liu, Q.; Sheng, L.; Cao, Z.; Diao, Y.; Wang, W.; Zhou, Y. Dual effects of the winter monsoon on haze-fog variations in eastern China. *J. Geophys. Res. Atmos.* **2017**, *122*, 5857–5869. [CrossRef]
53. Yi, K.; Liu, J.; Wang, X.; Ma, J.; Hu, J.; Wan, Y.; Xu, J.; Yang, H.; Liu, H.; Xiang, S. A combined Arctic-tropical climate pattern controlling the inter-annual climate variability of wintertime PM2.5 over the North China Plain. *Environ. Pollut.* **2019**, *245*, 607–615. [CrossRef] [PubMed]
54. Zhu, J.; Liao, H.; Li, J. Increases in aerosol concentrations over eastern China due to the decadal-scale weakening of the East Asian summer monsoon. *Geophys. Res. Lett.* **2012**, *39*, L09809. [CrossRef]
55. Liu, X.; Yan, L.; Yang, P.; Yin, Z.-Y.; North, G.R. Influence of Indian Summer Monsoon on Aerosol Loading in East Asia. *J. Appl. Meteorol. Climatol.* **2011**, *50*, 523–533. [CrossRef]
56. Yan, L.; Liu, X.; Yang, P.; Yin, Z.-Y.; North, G.R. Study of the Impact of Summer Monsoon Circulation on Spatial Distribution of Aerosols in East Asia Based on Numerical Simulations. *J. Appl. Meteorol. Climatol.* **2011**, *50*, 2270–2282. [CrossRef]
57. Gao, M.; Liu, Z.; Zheng, B.; Ji, D.; Sherman, P.; Song, S.; Xin, J.; Liu, C.; Wang, Y.; Zhang, Q. China's emission control strategies have suppressed unfavorable influences of climate on wintertime PM2.5 concentrations in Beijing since 2002. *Atmos. Chem. Phys.* **2020**, *20*, 1497–1505. [CrossRef]
58. Wang, C.; An, X.; Zhang, P.; Sun, Z.; Cui, M.; Ma, L. Comparing the impact of strong and weak East Asian winter monsoon on PM2.5 concentration in Beijing. *Atmos. Res.* **2019**, *215*, 165–177. [CrossRef]
59. Yin, Z.; Wang, H. The relationship between the subtopical Western Pacific SST and haze over North-Central North China Plain. *Int. J. Climatol.* **2016**, *36*, 3479–3491. [CrossRef]
60. Pei, L.; Yan, Z.; Sun, Z.; Miao, S.; Yao, Y. Increasing persistent haze in Beijing: Potential impacts of weakening East Asian winter monsoons associated with northwestern Pacific sea surface temperature trends. *Atmos. Chem. Phys.* **2018**, *18*, 3173–3183. [CrossRef]

61. Zou, Y.; Wang, Y.; Zhang, Y.; Koo, J.-H. Arctic sea ice, Eurasia snow, and extreme winter haze in China. *Sci. Adv.* **2017**, *3*, e1602751. [CrossRef] [PubMed]
62. Yin, Z.; Wang, H. The strengthening relationship between Eurasian snow cover and December haze days in central North China after the mid-1990s. *Atmos. Chem. Phys.* **2018**, *18*, 4753–4763. [CrossRef]
63. Zhao, S.; Zhang, H.; Xie, B. The effects of El Niño–Southern Oscillation on the winter haze pollution of China. *Atmos. Chem. Phys.* **2018**, *18*, 1863–1877. [CrossRef]
64. He, Q.; Geng, F.; Li, C.; Mu, H.; Zhou, G.; Liu, X.; Gao, W.; Wang, Y.; Cheng, T. Long-term variation of satellite-based PM2.5 and influence factors over East China. *Sci. Rep.* **2018**, *8*, 11764. [CrossRef]
65. Wang, J.; Zhu, Z.; Qi, L.; Zhao, Q.; He, J.; Wang, J.X.L. Two pathways of how remote SST anomalies drive the interannual variability of autumnal haze days in the Beijing-Tianjin-Hebei region, China. *Atmos. Chem. Phys.* **2019**, *19*, 1521–1535. [CrossRef]
66. Li, Z.Q.; Xu, H.; Li, K.T.; Li, D.H.; Xie, Y.S.; Li, L.; Zhang, Y.; Gu, X.F.; Zhao, W.; Tian, Q.J.; et al. Comprehensive Study of Optical, Physical, Chemical, and Radiative Properties of Total Columnar Atmospheric Aerosols over China: An Overview of Sun–Sky Radiometer Observation Network (SONET) Measurements. *Bull. Am. Meteorol. Soc.* **2018**, *99*, 739–755. [CrossRef]
67. Eck, T.F.; Holben, B.N.; Reid, J.S.; Dubovik, O.; Smirnov, A.; O'Neill, N.T.; Slutsker, I.; Kinne, S. Wavelength dependence of the optical depth of biomass burning, urban, and desert dust aerosols. *J. Geophys. Res.* **1999**, *104*, 31333–31349. [CrossRef]
68. Cheng, T.; Chen, H.; Gu, X.; Yu, T.; Guo, J.; Guo, H. The inter-comparison of MODIS, MISR and GOCART aerosol products against AERONET data over China. *J. Quant. Spectrosc. Radiat. Transf.* **2012**, *113*, 2135–2145. [CrossRef]
69. He, Q.; Li, C.; Tang, X.; Li, H.; Geng, F.; Wu, Y. Validation of MODIS derived aerosol optical depth over the Yangtze River Delta in China. *Remote Sens. Environ.* **2010**, *114*, 1649–1661. [CrossRef]
70. Li, Z.; Zhao, X.; Kahn, R.; Mishchenko, M.; Remer, L.; Lee, K.; Wang, M.; Laszlo, I.; Nakajima, T.; Maring, H. Uncertainties in satellite remote sensing of aerosols and impact on monitoring its long-term trend: A review and perspective. *Ann. Geophys.* **2009**, *27*, 2755–2770. [CrossRef]
71. Harrison, D.; Maggs, R.; Booker, J. *UK Equivalence Programme for Monitoring of Particulate Matter*; BV/AQ/AD202209/DH/2396; Bureau Veritas: London, UK, 2006.
72. Li, S.; Yu, C.; Chen, L.; Tao, J.; Letu, H.; Ge, W.; Si, Y.; Liu, Y. Inter-comparison of model-simulated and satellite-retrieved componential aerosol optical depths in China. *Atmos. Environ.* **2016**, *141*, 320–332. [CrossRef]
73. Li, S.; Zhang, L.; Cai, K.; Ge, W.; Zhang, X. Comparisons of the vertical distributions of aerosols in the CALIPSO and GEOS-Chem datasets in China. *Atmos. Environ. X* **2019**, *3*, 100036. [CrossRef]
74. Hammer, M.S.; Martin, R.V.; van Donkelaar, A.; Buchard, V.; Torres, O.; Ridley, D.A.; Spurr, R.J.D. Interpreting the ultraviolet aerosol index observed with the OMI satellite instrument to understand absorption by organic aerosols: Implications for atmospheric oxidation and direct radiative effects. *Atmos. Chem. Phys.* **2016**, *16*, 2507–2523. [CrossRef]
75. Li, S.; Chen, L.; Fan, M.; Tao, J.; Wang, Z.; Yu, C.; Si, Y.; Letu, H.; Liu, Y. Estimation of GEOS-Chem and GOCART Simulated Aerosol Profiles Using CALIPSO Observations over the Contiguous United States. *Aerosol Air Qual. Res.* **2017**, *16*, 3256–3265. [CrossRef]
76. Sogacheva, L.; Rodriguez, E.; Kolmonen, P.; Virtanen, T.H.; Saponaro, G.; de Leeuw, G.; Georgoulias, A.K.; Alexandri, G.; Kourtidis, K.; van der A, R.J. Spatial and seasonal variations of aerosols over China from two decades of multi-satellite observations-Part 2: AOD time series for 1995–2017 combined from ATSR ADV and MODIS C6.1 and AOD tendency estimations. *Atmos. Chem. Phys.* **2018**, *18*, 16631–16652. [CrossRef]

remote sensing

MDPI

Article

Ambient Formaldehyde over the United States from Ground-Based (AQS) and Satellite (OMI) Observations

Peidong Wang [1,2], Tracey Holloway [2,3,*], Matilyn Bindl [2], Monica Harkey [2] and Isabelle De Smedt [4]

1 Department of Earth, Atmospheric, and Planetary Sciences, Massachusetts Institute of Technology, Cambridge, MA 02139, USA; pdwang@mit.edu
2 Nelson Institute Center for Sustainability and the Global Environment (SAGE), University of Wisconsin-Madison, Madison, WI 53706, USA; mbindl@wisc.edu (M.B.); mkharkey@wisc.edu (M.H.)
3 Department of Atmospheric and Oceanic Sciences, University of Wisconsin-Madison, Madison, WI 53706, USA
4 Royal Belgian Institute for Space Aeronomy (BIRA-IASB), 1180 Brussels, Belgium; isabelle.desmedt@aeronomie.be
* Correspondence: taholloway@wisc.edu

Abstract: This study evaluates formaldehyde (HCHO) over the U.S. from 2006 to 2015 by comparing ground monitor data from the Air Quality System (AQS) and a satellite retrieval from the Ozone Monitoring Instrument (OMI). Our comparison focuses on the utility of satellite data to inform patterns, trends, and processes of ground-based HCHO across the U.S. We find that cities with higher levels of biogenic volatile organic compound (BVOC) emissions, including primary HCHO, exhibit larger HCHO diurnal amplitudes in surface observations. These differences in hour-to-hour variability in surface HCHO suggests that satellite agreement with ground-based data may depend on the distribution of emission sources. On a seasonal basis, OMI exhibits the highest correlation with AQS in summer and the lowest correlation in winter. The ratios of HCHO in summer versus other seasons show pronounced seasonal variability in OMI, likely due to seasonal changes in the vertical HCHO distribution. The seasonal variability in HCHO from satellite is more pronounced than at the surface, with seasonal variability 20–100% larger in satellite than surface observations. The seasonal variability also has a latitude dependency, with more variability in higher latitude regions. OMI agrees with AQS on the interannual variability in certain periods, whereas AQS and OMI do not show a consistent decadal trend. This is possibly due to a rather large interannual variability in HCHO, which makes the small decadal drift less significant. Temperature also explains part of the interannual variabilities. Small temperature variations in the western U.S. are reflected with more quiescent HCHO interannual variability in that region. The decrease in summertime HCHO in the southeast U.S. could also be partially explained by a small and negative trend in local temperatures.

Keywords: formaldehyde; trend; OMI; satellite; monitor; annual; seasonal; temperature

Citation: Wang, P.; Holloway, T.; Bindl, M.; Harkey, M.; De Smedt, I. Ambient Formaldehyde over the United States from Ground-Based (AQS) and Satellite (OMI) Observations. *Remote Sens.* **2022**, *14*, 2191. https://doi.org/10.3390/rs14092191

Academic Editors: Maria João Costa and Daniele Bortoli

Received: 24 March 2022
Accepted: 26 April 2022
Published: 4 May 2022

Publisher's Note: MDPI stays neutral with regard to jurisdictional claims in published maps and institutional affiliations.

1. Introduction

Formaldehyde (HCHO) is a carcinogen and mutagen that has been categorized by the U.S. Environmental Protection Agency (EPA) as one of 187 Hazardous Air Pollutants (HAPs). HCHO can be either primary (emitted) or secondary (chemically produced). Global background HCHO concentration primarily comes from the oxidation of methane or methanol [1,2]. In the continental boundary layer (PBL), HCHO is most often formed by the oxidation of non-methane volatile organic compounds (VOCs) [3,4]. The dominant biogenic precursor of HCHO is isoprene, which comes from vegetation and can quickly oxidize to form HCHO [5,6]. The reaction of HCHO with OH, as well as HCHO photolysis [7], among other loss pathways, results in a lifetime of HCHO on the order of hours [8]. On shorter time scales, biomass burning and wildfires are also sources for instantaneous elevated HCHO [9]. The major emission sources for anthropogenic VOCs include fuelwood production, utilization of gasoline, and biomass burning [10,11].

To support air quality management and public health assessments, the U.S. maintains several networks and programs to monitor HCHO at the surface. These include the Interagency Monitoring of Protected Visual Environments (IMPROVE), National Air Toxics Trends Stations (NATTS), and Photochemical Assessment Monitoring Stations (PAMS) networks. Data from these networks are archived through the EPA Air Quality System (AQS) Ambient Monitoring Archive (AMA) for hazardous air pollutants (available at https://www3.epa.gov/ttnamti1/toxdat.html#data; accessed on 30 August 2017). Several methods are used for in situ measurements of HCHO, including spectroscopic, colorimetric, chromatographic, and fluorometric techniques, which are discussed in detail by [12]. Among these techniques, the EPA commonly uses the chromatographic technique with 2,4-Dinitrophenylhydrazine (DNPH) [13] to measure HCHO, though there may be some bias with this method, such as interference from ozone, nitrogen dioxides, and water vapor [14–17].

Between 2006 and 2015, there were 338 ground-based monitors operated by states, local agencies, and tribes throughout the U.S. in compliance with EPA standards on archiving and measuring HCHO. Of these 338 total stations, the measurement frequency of HCHO varies, with stations reporting either 1 h (10 stations), 3 h (41 stations), or 24 h concentrations (322 stations). Most stations measured HCHO by collecting air samples via cartridges coated with DNPH and analyzing those samples using the High-Performance Liquid Chromatography (HPLC) method. Even among these monitors, there are significant gaps in the data records of every station.

Satellite observations offer the potential to complement this limited surface monitoring network. Satellite retrievals of HCHO bear relevance to air quality planning as an indicator of HCHO exposure, as well as VOC emissions, VOC reactivity, and associated ozone formation. This study compares satellite and ground-based HCHO with the goal of assessing spatial and temporal HCHO patterns and the appropriate role of satellite vertical column density (VCD) as a potential proxy for near-surface HCHO.

Currently, four polar-orbiting satellite instruments detect HCHO, including: the Ozone Monitoring Instrument (OMI) onboard the Aura satellite [18], which has a local overpass time in the early afternoon at 13:30 and a spatial resolution of 24×13 km^2; the Tropospheric Monitoring Instrument (TROPOMI) onboard the Sentinel-5 Precursor, with the same overpass time as OMI, but finer resolution [19] of 3.5×5.5 km^2 as of August 2019; the Ozone Mapping and Profiler Suite (OMPS) on the Suomi NPP satellite [20], which also has an overpass time of 13:30, but with a spatial resolution of 50×50 km^2; and the Global Ozone Monitoring Experiment-2 (GOME-2) on the Metop satellite series [21], with a local overpass time of 09:30 and a spatial resolution of 80×40 km^2. In addition to these polar-orbiting satellites, a number of instruments are planned for or have recently reached geostationary orbit: Tropospheric Emissions: Monitoring of Pollution (TEMPO), from the National Aeronautics and Space Administration (NASA; spatial resolution: 2 km $\times$ 4.5 km) in 2022; Sentinel-4 from the European Space Agency (ESA; spatial resolution: 8 km $\times$ 8 km) in 2023; and Geostationary Environment Monitoring Spectrometer (GEMS), from the Korea Aerospace Research Institute (KARI; spatial resolution: 7 km $\times$ 8 km), which launched successfully in early 2020. These satellites will provide continuous observation of HCHO over North America, Europe, and Asia, respectively [22–24].

Satellite observations of HCHO have been used to advance the understanding of atmospheric chemistry, e.g., refs. [25–33], as well as for air quality management to protect public health [34–37]. Because of HCHO's strong detectability from space, its local footprint due to a short atmospheric lifetime, and its high yield from VOCs, HCHO has been used as an indicator of total VOCs in the atmosphere [38–40]. Combined with satellite derived NO_2, HCHO has been used to support the assessment of the ozone production regime [36,41,42] and has even been used in decision-making contexts [43].

As satellite data usage expands, there is interest in the relevance of satellite products to better characterize emissions and near-surface concentrations. Much of work has focused on satellite observations of NO_2. For example, ref. [44] used satellite observations of NO_2 to constrain emissions of NO_x ($NO + NO_2$) at the surface, ref. [45] used a model-derived scaling factor to scale satellite observations of NO_2 to near-surface amounts, and ref. [46] found similar responses to weather variables for both surface and column NO_2. Ref. [47] found a good correlation between surface and column NO_2, discovering that both datasets captured weekly cycles over Leicester, England; ref. [48] found strong seasonal and weekly cycles in both datasets over Israeli cities in 2006; and ref. [49] found that both datasets showed a small weekly cycle in NO_2 in Beijing. Similar work with satellite HCHO includes studies by [50,51], who evaluated relationships among 17 years of satellite-based HCHO, biogenic isoprene emissions, and land cover datasets, by [52–54], who characterized anthropogenic emissions, and by [34], who used satellite HCHO to evaluate differing chemical transport model configurations over the continental U.S., and following [45], explored the utility of scaling satellite HCHO to near-surface amounts, and by [55], who derived surface HCHO amounts from TROPOMI and surface monitor observations using a neural network technique.

Fewer studies have compared satellite and surface HCHO observations. The majority of previous studies using both in situ and remote measurements of HCHO have focused on global patterns and VOC emissions. Between 2004 and 2014, ref. [26] found that OMI HCHO decreased in the eastern U.S., central South America, and across Europe, but increased in India and central-eastern China. That same study found that HCHO is highest in the early afternoon in the mid-latitudes by differencing the morning overpass of GOME-2 with the afternoon overpass of OMI [26]. Several studies have used satellite-based HCHO observations to infer the spatial distribution of isoprene over the U.S., e.g., [4,38,56]. Between 2005 and 2014, OMI HCHO increased over the U.S. overall, but decreased in the southeast [57]. HCHO trends have been found to be largely dependent on temperature and fire events [26,56,58–61], as well as anthropogenic emission sources [53,57,62,63].

This study informs potential applications of satellite-based HCHO within the health and air quality communities, which focus on near-surface concentrations. We evaluate the diurnal, seasonal, and interannual trends for HCHO over the U.S. by comparing HCHO from satellite retrievals with those from ground-based measurements. We assess HCHO data availability from EPA monitoring stations (Section 2.1), diurnal cycles (Section 3), seasonal variability (Section 4), and interannual trends (Section 5). In Section 6, we connect our results with previous findings to evaluate mechanisms that could potentially explain some of the observed behaviors. Finally, we conclude our results in Section 7.

2. Data and Methods

We evaluated HCHO for the U.S. between 2006 and 2015 from ground measurements and satellite retrievals. Ground monitor data for HCHO comes from the EPA Air Quality System (AQS) Ambient Monitoring Archive (AMA) for hazardous air pollutants, which includes data collected from the IMPROVE, NATTS, and PAMS networks. Satellite HCHO observations have been retrieved from the measurements of the NASA Earth Observing System's OMI instrument as a part of the Quality Assurance for Essential Climate Variables project (QA4ECV; available at http://www.qa4ecv.eu/; accessed on 24 July 2019).

2.1. Ground-Based Measurements

In this study, we used AQS HCHO concentrations ($\mu g\ m^{-3}$) which have been converted to local meteorological conditions (using local pressure and temperature) from standard conditions. Detailed conversion is available in the Quality Assurance Summary Report for HAPs (https://www3.epa.gov/ttn/amtic/files/toxdata/techmemo2017.pdf; accessed on 30 August 2017). As described in this report, we eliminated data that are flagged as non-detect or below measurement detection limit. Of the 338 monitors in the AMA between 2006 and 2015, we only used data from sites with the DNPH/HPLC method.

We choose to use publicly available data in a manner consistent with the use of monitoring data by air quality managers and by the EPA for the National Air Toxics Assessment.

There are significant gaps in the data records of every station, which we characterize as the percentage of available data for each available reporting frequency (Table 1). Percent availability is calculated as the number of available measurements divided by the total number of possible measurements at each station's measurement frequency. We find that no station offered more than 50% data coverage across the 10-year period of analysis. Ten stations measured HCHO at an hourly frequency, with one site, located at St. Louis, Missouri, having data coverage of 38% of all days. The other nine stations have below 1% of data coverage. Forty-one sites measured HCHO with a 3 h frequency, but only two stations in Los Angeles County in California (Burbank and Pico Rivera) had data coverage over more than 10% of the days in the study period, and these sites only had measurements in July, August, and September.

Table 1. Number of HCHO ground monitoring stations in the United States, with data distributed by the EPA AQS. Stations are grouped based on data availability, and by measurement frequency (hourly, 3 h, and 24 h). Data availability is calculated as the percentage of measurements available from 2006 to 2015 relative to the potential number of measurements during this period at the monitor's reporting frequency. The three sites used for analysis of the diurnal HCHO cycle are marked in bold, which include the 1 h site with 30–40% data availability (St. Louis, MO, USA), and the two 3 h site with 10–20% data availability (Burbank, CA, USA; Pico Rivera, CA, USA). These three sites are marked as triangle in Figure 1.

Percent Available	1 h Frequency	3 h Frequency	24 h Frequency
0–10%	9	39	230
10–20%	0	**2**	70
20–30%	0	0	18
30–40%	**1**	0	3
40–50%	0	0	1
Total stations	10	41	322

For our diurnal analysis, we relied on the three stations that offered >10% data coverage in a 1 h or 3 h measurement frequency. For our seasonal and interannual analyses, we used data from sites with a 24 h measurement frequency and six or more samplings of each season throughout 2006 to 2015 continuously. For comparison, only 37 sites have one or more measurements per month, so we chose the seasonal average basis to include more monitors. The threshold of six samples each season was selected to balance temporal and geographic coverage of monitors. For thresholds less than six samples per season, more monitors would be available (58 monitors at a threshold of five samples per season; 64 monitors with a threshold of one sample per season). For thresholds greater than six samples per season, fewer monitors would be available (41 monitors at a threshold of seven samples per season; 10 monitors with a threshold of ten samples per season). We removed five outlier stations which had significantly large maximum over median values in each seasonal average (larger than two standard deviations among all stations), since these sites might not be representative for interannual trend studies and cannot represent regional conditions. This approach yielded 45 ground monitor stations for the seasonal and interannual analyses. The black symbols in Figure 1 identify the locations of the AQS stations used in this study: circles (45 sites) indicate the stations used in seasonal and interannual analyses; triangles (3 sites) indicate sites included in the diurnal analysis.

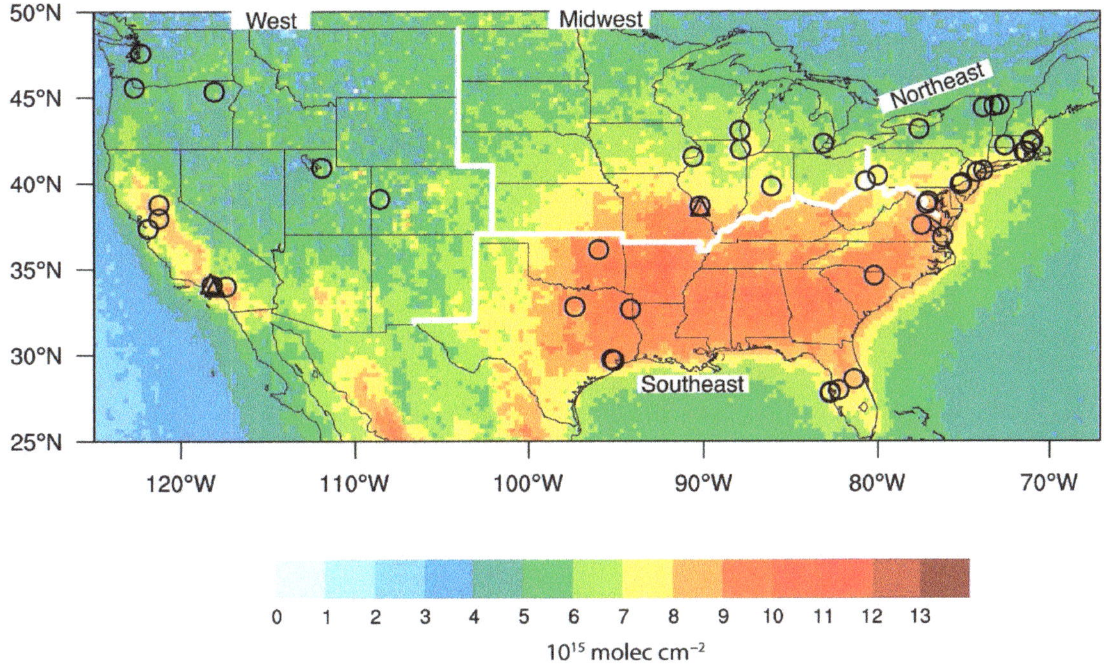

Figure 1. 2006–2015 annual average HCHO vertical column density from OMI. Four U.S. regions are designated for analysis, with region name aside the map. Overall, for studying diurnal patterns, there are three sites: two in the Western U.S. and one in the Midwest. These stations are labeled as triangles. For seasonal and interannual studies, there are 45 AQS stations (10 in the west, 6 in the Midwest, 14 in the southeast and 15 in the northeast) that are labeled as circles. For more details on monitoring sites, please see Section 2.1 and Table 1.

2.2. Satellite Observations

We used the HCHO Level-3 product, with horizontal resolution at 0.25° × 0.25°, from OMI onboard the Aura satellite, for which the U.S. overpass occurs in the early afternoon. We used the OMI retrieval algorithm from the EU FP7-project QA4ECV (hereafter abbreviated OMI unless otherwise specified) [64,65]. The QA4ECV algorithm utilizes a fitting window ranging from 328.5 to 359 nm from OMI. We removed data with solar zenith angles greater than 70 degrees and cloud fractions greater than 40%. We also removed data that are quality flagged or influenced by the OMI row anomaly (http://projects.knmi.nl/omi/research/product/rowanomaly-background; accessed on 24 July 2019). Detailed descriptions of this algorithm are described by [66].

Compared to other instruments with data covering any of our 2006–2015 study years (GOME2A, GOME2B, OMPS), OMI offers the highest spatial retrieval resolution of HCHO at 24 × 13 km^2 at nadir, as discussed by [67]. Aura's early afternoon overpass time corresponds with the average daily peak amount of HCHO at mid-latitudes [26,68]. Following [69], who compared trends in satellite- and ground-based observations of NO_2, we use OMI HCHO VCD. While the total vertical column density indicates the number of molecules between the satellite and ground, tropospheric HCHO accounts for the majority of the total column amount. We use OMI HCHO observations for all seasons from 2006 to 2015 and compare with AQS measurements to evaluate OMI's ability to indicate surface HCHO trends.

OMI has exhibited a positive drift since 2008, possibly due to instrumental degradation [70,71]. The QA4ECV algorithm applied a background correction over the remote

Pacific to reduce HCHO slant column uncertainty. Note that this approach assumes that the remote Pacific HCHO is only due to the oxidation of methane [66].

Ref. [56] notes an instrument detection threshold of ~4×10^{15} molec cm^{-2}; here, we present all values for thoroughness and for the purpose of evaluating whether winter HCHO values from OMI agree with AQS observations [26,72].

Figure 1 shows oversampled OMI 2006–2015 averaged HCHO for the continental U.S. On average, HCHO column amounts are higher where precursor emissions of isoprene are high, e.g., [38]. In particular, the southeastern U.S. shows elevated HCHO column amounts ($\geq 9 \times 10^{15}$ molec cm^{-2}). Amounts in other regions are lower, except in areas in the Western U.S. corresponding to mountainous terrain and national parks, where average amounts exceed 8×10^{15} molec cm^{-2}. High values may be caused by isoprene emissions and/or anthropogenic emissions associated with industries, such as oil and gas extraction, or caused by direct and precursor emissions from fires [73].

Unless otherwise noted, all analyses were conducted with seasonal average data over a three-month period (DJF: December January February; MAM: March April May; JJA: June July August; SON: September October November). Although the QA4ECV data would be appropriate for monthly average analysis, we chose to average by season due to the limited availability of data from the AQS monitors. We checked the impact of constraining AQS HCHO data to include only measurements taken when local cloudiness was <40%. As we screened for cloudiness in the OMI dataset, we found a high correlation (r = 0.999) between the seasonal mean of all 24 h ground measurements and the seasonal mean of AQS measurements under the same OMI viewing conditions described above, with the subset data having a small mean bias (-0.005 µg m^{-3}; Supplementary Material Figure S1). In order to obtain continuous seasonal data, we used all available AQS data with a 24 h measurement frequency.

2.3. Emissions Data

We compared AQS and OMI HCHO with the EPA's National Emissions Inventory (NEI, available at https://www.epa.gov/; accessed on 23 October 2017). Every three years, the NEI reports the annual summations from different emissions sources for various air pollutants. We considered NEI HCHO emissions from biogenic, anthropogenic, and wildfire sources to assess their contributions to observed HCHO abundance. Biogenic HCHO comes from vegetation and soil, anthropogenic HCHO is mostly due to fuel combustion and transportation, and wildfire is a single category that contributes significantly to HCHO amounts in the western U.S. We considered the 2008 NEI in this study for the diurnal section in which all three diurnal sites have only 2008 HCHO in common. This NEI includes biogenic emissions calculated using the Biogenic Emission Inventory System version 3.14 (BEIS) with land use data from the Biogenic Emissions Land use Database version 3 (BELD3) [74]. We used 2008 NEI county-level data for each station in our diurnal analysis and state-level data for studying seasonal and interannual trends.

3. Diurnal Cycle of HCHO

Given the limitations of the observational dataset, we focused our analysis on summer months, when data are consistently available. For our evaluation of HCHO diurnal cycles, we used data from three urban AQS sites (St. Louis, Burbank, and Pico Rivera) with 1- and 3 h measurement frequencies available in July, August, and September (JAS) from 2006 to 2010. The St. Louis site is located in St. Louis County in Missouri, while both the Burbank and Pico Rivera sites are located in Los Angeles County in California. We excluded days without full-day measurements (8 measurements per day for three-hour frequency and 24 measurements per day for hourly frequency). Figure 2 shows JAS mean HCHO for each year at three diurnal sites (note the different y-axes for each plot, due to the large differences in HCHO levels across the three sites).

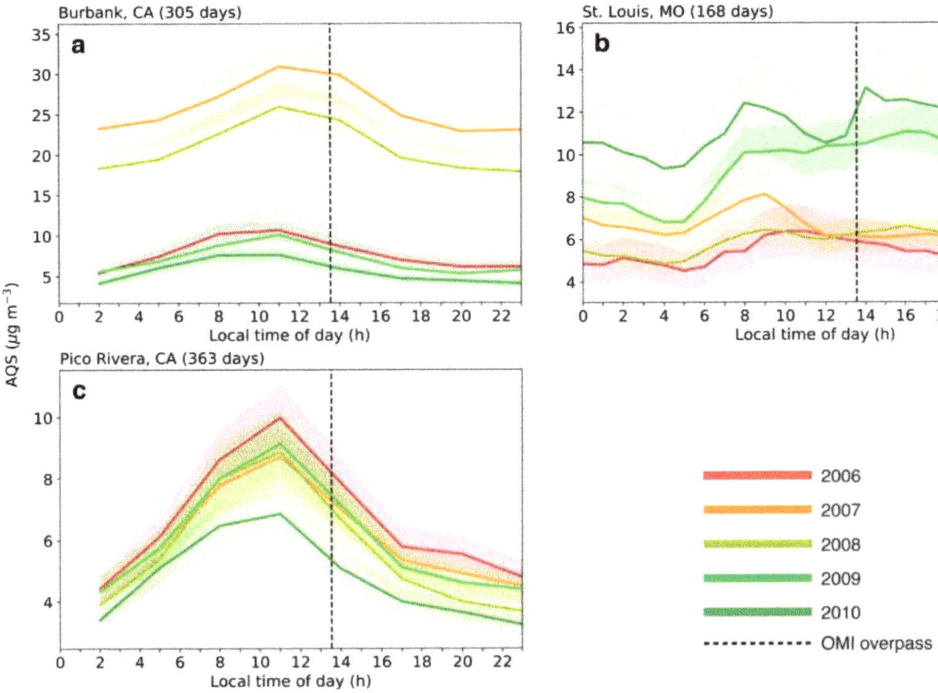

Figure 2. HCHO mixing ratios from ground-based AQS monitors at three sites ((**a**): Burbank, CA; (**b**): St. Louis, MO; (**c**): Pico Rivera, CA) from 2006 to 2010. Solid lines show June, August, September (JAS) average diurnal cycle with complete measurements (24 measurements per day for 1 h site or 8 measurements per day for 3 h site). The shaded area indicates standard deviation in that averaging year. Different years are labeled in different colors, indicated in the bottom right. OMI overpass time (13:30 local time) is labeled with a vertical dashed line at each site.

In general, diurnal patterns in HCHO are affected by direct and precursor emissions, chemical reactions, and vertical and horizontal mixing. Biogenic precursor emissions are expected to peak in the mid- to late afternoon when photosynthetically active radiation and temperature are high [75]. Formaldehyde yields from these precursors tend to peak mid-day with elevated isoprene oxidation [76]. Vertical mixing of trace gases such as HCHO and its precursors will also vary as diurnal heating drives changes in mixed layer depth, e.g., ref. [77].

From 2006 to 2010, Burbank and Pico Rivera show clear diurnal trends with peaks ranging between 5 and 30 $\mu g\ m^{-3}$ around 11:00 LT (Figure 2a,c). Observations at St. Louis indicate a less significant diurnal pattern, with peak values at 14:00 and 20:00 LT in 2010, 9:00 LT in 2007, and all years showing minima between 4:00 and 5:00 LT (Figure 2b). It is noteworthy that for both 2007 and 2008, Burbank shows HCHO values over 15 $\mu g\ m^{-3}$, which are almost three-times greater than concentrations observed in other years (Figure 2a). Since Burbank is 5 km southwest of the Angeles National Forest, these high concentrations could be due to significant wildfire emissions in 2007 and 2008. This wildfire enhancement is consistent with the NEI, which indicates that wildfires yielded 296 tons of HCHO for Los Angeles County in 2008, compared to 48 tons in 2011 and 36 tons in 2014. However, Pico Rivera, another station in Los Angeles County, did not record anomalously high HCHO in 2007 and 2008 (Figure 2c). This could be due to the fact that the Pico Rivera monitor is 20 km south of the Angeles National Forest. In Figure 2, we overlaid the OMI overpass time (13:30 LT) for each AQS site and found that with the exception of the St. Louis site in 2009 and 2010, OMI generally passes the sites after the peak in HCHO.

To better describe the amount of diurnal variability at each site, we quantified the amplitude of the diurnal pattern in AQS data, presented in Table 2. We differentiated between absolute amplitude (A_{abs}) and relative amplitude (A_{rel}) and calculated mean amplitudes from 2006 to 2010 at the three sites. Amplitudes for 2008 were compared with NEI emissions reported for that year. We defined A_{abs} as the difference in value between the daily maximum and minimum, in $\mu g\ m^{-3}$, and A_{rel} as the ratio between A_{abs} and the daily average value to represent the difference between the two, expressed as a percentage. Overall, Burbank and Pico Rivera, with clear diurnal patterns, show larger amplitudes (A_{abs} of 5.94 $\mu g\ m^{-3}$ and 4.70 $\mu g\ m^{-3}$, respectively, and A_{rel} of 53.92% and 80.47%, respectively) than St. Louis (A_{abs} of 2.74 $\mu g\ m^{-3}$ and A_{rel} of 34.70%; Table 2). These differences are likely the result of differences in HCHO sources. We extracted NEI emissions for total VOCs and for HCHO in 2008 where the three sites have data in common (Table 3). In both Los Angeles and St. Louis counties, direct HCHO emissions only account for a small portion compared to total VOC emissions, regardless of emission sector. Overall Los Angeles has four times more total VOC emissions and ten times more direct HCHO emissions than those in St. Louis. Both counties also indicate more anthropogenic sources of HCHO and total VOCs than biogenic sources, with different anthropogenic to biogenic ratios.

Table 2. 2006–2010 and 2008 (NEI reported year) mean June, August, September (JAS) diurnal HCHO amplitudes at the three AQS sites in Figure 2. A_{abs} is the absolute amplitude, and A_{rel} is the relative amplitude.

	A_{abs} ($\mu g\ m^{-3}$)		A_{rel} (%)	
	2006–2010	**2008**	**2006–2010**	**2008**
Burbank, CA	5.94	8.05	53.92	38.65
Pico Rivera, CA	4.70	5.13	80.47	91.62
St. Louis, MO	2.74	1.76	34.70	29.86

Table 3. 2008 NEI total VOC and HCHO emissions at St. Louis and Los Angeles county, along with anthropogenic to biogenic emissions ratio.

	Total VOC Emissions (Kilotons)		**HCHO Emissions (Tons)**	
	St. Louis	**Los Angeles**	**St. Louis**	**Los Angeles**
Biogenic	6.48	70.68	101.43	1213.80
Anthropogenic	36.33	111.75	338.58	2733.73
Wild Fire	0	4.76	0	295.65
Anthro/Bio Ratio	5.61	1.58	3.34	2.25

For total VOC emissions, St. Louis has above five-times more anthropogenic emissions than biogenic emissions, in which the major anthropogenic source is on-road, non-diesel, and light-duty vehicles. Direct HCHO emissions at St. Louis also have a higher ratio between anthropogenic and biogenic sources, in which on-road vehicles from the mobile sector contribute to more than half of the anthropogenic HCHO emissions. Although Los Angeles has much higher emissions than St. Louis, it has a lower ratio of anthropogenic to biogenic VOC and HCHO emissions (~2 versus ~5 in St. Louis). As indicated by [78], the reactivity of anthropogenic VOCs (emissions mostly coming from motor vehicles) remains consistent with temperature, but the reactivity of biogenic VOCs grows exponentially with temperature. Therefore, higher contributions of anthropogenic emissions could explain the lack of a diurnal cycle in St. Louis County compared to the two sites in Los Angeles County.

However, since we only use three sites and they are all in urban areas, these results might have a sample size and location bias. Hourly retrievals of HCHO from geostationary satellites, when available, could be used to evaluate how anthropogenic versus biogenic emissions affect the diurnal cycle of HCHO over wider areas.

4. Regional HCHO Seasonality Analysis

To assess seasonal patterns in HCHO, we divided the continental U.S. into four geographical regions (shown in Figure 1) generally following regions defined by the U.S. Census. There are 6 AQS stations in the Midwest; these have a humid continental climate (hot summer) and crops as the dominant Plant Functional Type (PFT). There are 15 AQS sites in the Northeast; these have humid continental climate with a cool summer, and broadleaf trees as the dominant PFT. The southeast has 14 stations, a humid subtropical climate, and a mixture of broadleaf and fine leaf trees, shrubs, crops, and grass as the PFTs. The west has 10 sites, a mixture of Mediterranean, semi-arid, and desert climate, and shrubs and grassland as the PFTs. Since AQS sites are not evenly distributed in each zone, they might not be representative of the entire region. For consistency, we used OMI coincident pixels at these 45 AQS sites for all of the AQS-OMI comparison analyses.

4.1. Overall AQS-OMI Seasonal Correlation

Figure 3 compares HCHO AQS and OMI observations for each season at 45 ground monitor stations. Abundance of HCHO varies seasonally, with greater amounts of biogenic precursors in warm seasons at all sampling sites (corresponding with the 45 ground monitors). To obtain continuous winter averages, we combined January and February data with December data from the previous year. For these seasonal evaluations, we considered 2007 to be the first year of analysis, which includes December 2006. Each symbol represents seasonal HCHO averaged from 2007 to 2015 at one AQS station, with symbols color-coded by region.

As shown in Figure 3, correlations between AQS and OMI peak in summer and drop to a minimum in winter. This result is consistent with [35], who reported a larger contribution of near-surface HCHO in summer months, and a lower vertical gradient in winter. The summer has a larger fraction of column HCHO in the boundary layer, consistent with the positive correlation between AQS and OMI in warm months.

We find a positive correlation between AQS and OMI HCHO in all four geographical regions for every season except for winter, where the correlations becomes negative or insignificant for most regions. This could be due to the fact that winter has (1) greater solar zenith angles in the northern hemisphere, (2) frequent cloud coverage, and (3) lower HCHO emissions that are below the OMI detection limit. AQS and OMI have consistently high agreement in the Northeast ($r = 0.49$, 0.65, 0.87, and 0.56 for DJF, MAM, JJA, and SON, respectively), and a slightly weaker but consistent correlation in the west ($r = 0.25$, 0.61, 0.71, and 0.67 for DJF, MAM, JJA, and SON, respectively). AQS and OMI show high agreement in the Midwest and southeast during JJA, but weak or even negative correlation in other seasons.

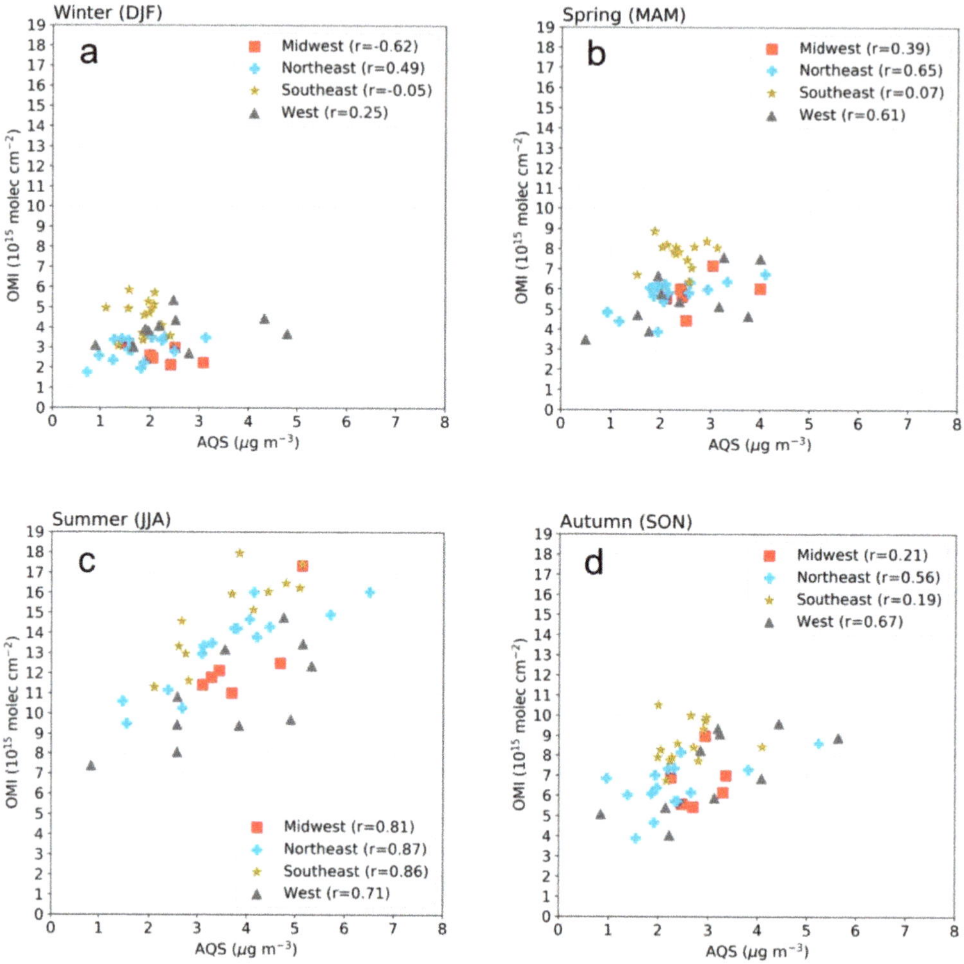

Figure 3. Spatial correlation of 2007–2015 average HCHO from 45 AQS sites (*x*-axis) and coincident OMI pixels (*y*-axis). Each point represents a single site, with OMI and AQS values averaged over the season. Points are color coded to reflect their region, and spatial correlation coefficient r values in different regions are shown in the legend. Each panel represents average values for: (**a**) winter, (**b**) spring, (**c**) summer, and (**d**) autumn.

4.2. Seasonal Variability

We evaluated seasonal variability by comparing the HCHO ratios of summer to the other three seasons. Previous studies indicate that in the U.S., summertime HCHO amounts are higher than in winter due to higher summertime temperatures, leading to an increase in biogenic VOC emissions and HCHO production [4,58,72]. This increase is characterized by the seasonal variability in surface-level HCHO, estimated by the GEOS-Chem model using the ratio of yearly mean to summer amounts. For both AQS (measured in μg m^{-3}) and OMI (measured in 10^{15} molec cm^{-2}), we calculated the unitless ratio of the summer JJA average to winter DJF average, JJA to spring MAM average, and JJA to autumn SON average. These ratios are given in Figure 4 for each region as box plots.

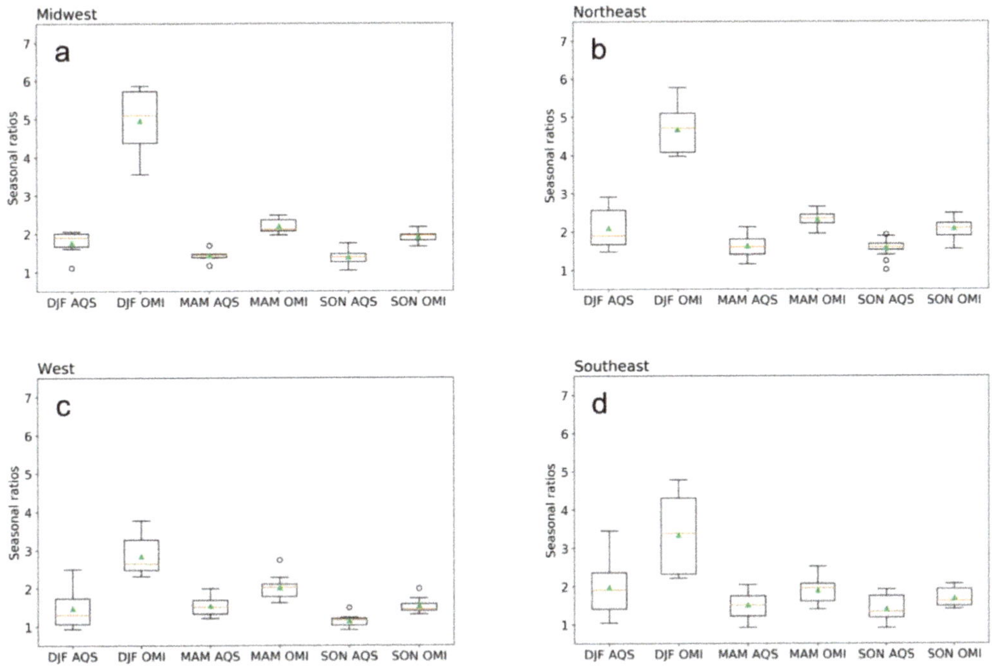

Figure 4. Summer to all other season ratios from AQS and OMI for the Midwest (**a**), northeast (**b**), west (**c**), and southeast (**d**). Red lines indicate regional median, green triangles indicate regional mean, and cycles represent extreme values.

Overall, as seen in Figure 4, both AQS and OMI HCHO have summer to other season ratios >1 in all four regions, with peak ratios in JJA/DJF and similar ratios between JJA/MAM and JJA/SON, indicating a clear seasonal cycle with maximum HCHO in summer, minimum HCHO in winter, and similar amounts in spring and autumn. For all four regions, OMI shows more pronounced ratios than AQS, which reflects the greater variability in column amounts compared to near-surface amounts seen in Figure 3. Among those regions, OMI ratios mostly bias high in the higher latitude regions in the Midwest and Northeast by a factor or 1.3 to 2.8 depending on different seasons compared to AQS, despite these regions having strong correlation between AQS and OMI in summer from Figure 3. OMI ratios have less overestimation in the two lower latitude regions in the southeast and west, though still bias high with a factor between 1.2 to 1.3, with larger bias in winter (up to 1.9 greater than AQS ratios).

5. Interannual Trends

In Section 4.2, we showed that HCHO exhibits a strong seasonal cycle. To assess HCHO's interannual variability between 2006 and 2015, we deseasonalized the seasonal mean values for each region, using observations from all 45 AQS stations and collocated OMI data. We also calculated the line of best fit for the deseasonalized data in each region. Endpoints are SON average 2006 and SON 2015 to avoid seasonality affecting the trend. Figure 5 shows the mean deseasonalized HCHO for AQS and OMI in each region (solid lines). These plots are overlaid with collocated seasonal average 2 m temperature from the North American Regional Reanalysis (NARR) [79], to indicate the role of temperature in the variability of each data set. The slopes from the linear regression and the associated standard errors are indicated in each panel.

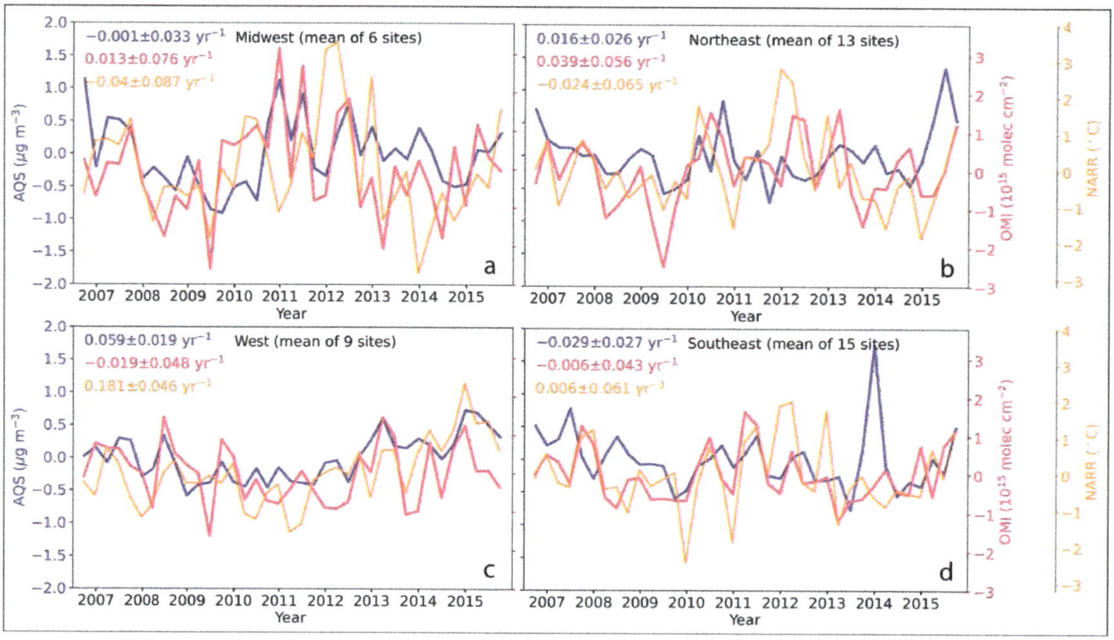

Figure 5. 2006–2015 AQS and OMI HCHO deseasonalized interannual trends averaged by season in the (**a**) Midwest, (**b**) northeast, (**c**) west, and (**d**) southeast. Blue lines indicate AQS trends, red lines represent OMI trends, and yellow represents 2 m temperature from the North American Regional Reanalysis (NARR) data, sampled at the AQS monitor locations. Solid lines are the mean seasonal HCHO abundance or temperatures, and shaded areas are the standard deviation. Region name and number of stations in the region are labeled at the top of each figure, along with the slope of the line of best fit and the standard errors associated with the slopes.

After deseasonlizing the data, AQS and OMI show some consistency in certain periods. For example, both AQS and OMI captured the double peak in HCHO in 2010–2012 over the Midwest (Figure 5a). Both AQS and OMI HCHO also show elevated amounts in the west in summer of 2008 (Figure 5c), possibly corresponding to local fire emissions. However, the vertical and horizontal transport of VOC emissions from fires are more likely captured by satellite observations, and then affect regional trends downstream of fires more in the OMI record than in AQS.

During the entire ten-year period, the slopes from the linear regression on AQS and OMI do not aways agree. Since the standard errors associated with the slopes are rather large, this indicates HCHO did not have a significant and monotonic drift during 2006–2015. There are two exceptions: AQS in the west does indicate a significant increasing trend at a rate of 0.059 ± 0.019 μg m^{-3} yr^{-1}, and AQS in the southeast is decreasing at a rate of 0.029 ± 0.027 μg m^{-3} yr^{-1} from 2006 to 2015. The large standard error in the southeast AQS data could partially come from the winter in 2014, when the regional mean is skewed by a few exceptionally high values.

In all regions, the increase in HCHO is partially explained by temperature trends, as shown by the NARR data. The west, which shows a significant increase in AQS HCHO, also shows the greatest warming at a rate of 0.181 ± 0.046 °C yr^{-1}. However, this large increase is not captured by OMI, despite OMI having high agreement in seasonal variability with AQS. The interannual variations in temperature are also more quiescent in the western U.S. than in other regions, which could partially explain smaller year-to-year variations in HCHO in the West.

We used the same best fit linear regression to consider interannual changes for each season (Figure 6) and yearly mean for every grid box from OMI. We see a significant increase in California at around 0.3×10^{15} molec cm^{-2} per year; this is mostly contributed by an increase in MAM. The broader southeast and northeast regions experienced a strong increase in HCHO during DJF but were offset by a strong decrease in JJA. Such a decrease in HCHO could be partially explained by summertime cooling in the southeast U.S. over the past few decades [80,81].

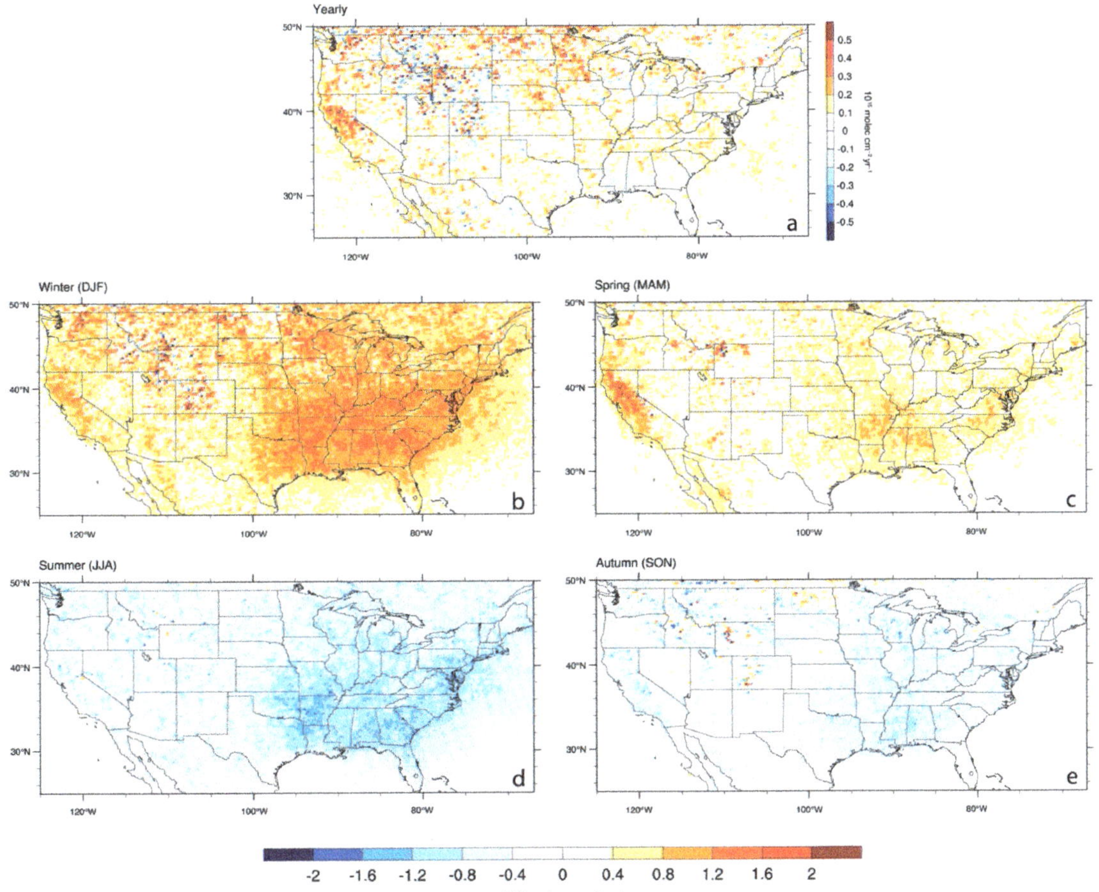

Figure 6. Map of the U.S. with 2006–2015 OMI HCHO trend in (**a**) all seasons, (**b**) winter, (**c**) spring, (**d**) summer, and (**e**) autumn.

Though OMI (co-located with AQS sites) indicates a decreasing trend that is opposite to AQS and NARR in the west (Figure 5c), there is a significant increase in OMI HCHO over large area of California, Oregon and Washington (Figure 6). Thus, using the OMI co-located pixels at AQS sites might not be indicative of the entire region.

6. Discussion

The evaluation of diurnal, seasonal, and interannual HCHO suggests that satellite-derived HCHO serves as a useful indicator for surface HCHO change on seasonal to interannual timescales. While surface-to-column agreement varies in space and time, the combined analysis of these two datasets informs the chemical and meteorological processes that impact HCHO.

This study builds on the earlier work of [72], with a greater focus on seasonality, temperature, and comparison between satellite and ground-based data. Whereas ref. [72] evaluated May-September change of temperature-corrected HCHO, we examine all seasons without removing temperature effects with the goal of characterizing the degree to which satellite data can inform near-surface trends and patterns.

The weighting function for OMI HCHO peaks in the upper troposphere [26], so it is well established that satellite data does not capture the same surface air as measured by the AQS monitors. With such differing physical characteristics between observation methods, it would be challenging to reconstruct ground level HCHO merely from satellite data. For example, ref. [82] reviewed methodologies of calculating surface level PM2.5 from satellite observed aerosol optical depth, which requires either a chemistry transport model or a statistical model based on empirical relations from the existing data. Future work to derive surface level HCHO from satellite observations could be made possible with such large datasets, such as the newly available EPA Air Quality Time Series (EQATES) project (available online, https://www.epa.gov/cmaq/equates; accessed on 21 April 2022).

Additionally, validation studies have shown that satellite HCHO tends to be biased low when column amounts are high (e.g., summer) and biased high when column amounts are low (e.g., winter), which would tend to dampen the OMI HCHO seasonal cycle [83–85]. Our observational results corroborate [35], who found that surface HCHO is a more significant contributor to column HCHO in the summer, based on model simulations. Despite the low agreement between AQS and OMI data in the winter, both datasets capture a consistent seasonal cycle and consistent interannual trends over certain periods.

Our findings suggest that temperature sensitivity of column HCHO is greater than near-surface HCHO. The seasonal amplitude of HCHO is higher in OMI data than in AQS in all regions, which may be due to the larger role of secondary HCHO in the column versus the surface. Furthermore, the warming across the 2006–2015 period leads to a stronger increase in the HCHO column in most regions, and a less-pronounced increase (or decrease) in the AQS monitor data, except in the west. The authors in [46] found that column NO_2 is more sensitive to temperature than surface NO_2, both in observations (monitors versus satellite) and in a numerical model; the same appears to be true for HCHO. As discussed by [72,86], trends in HCHO are attributable to a range of land use and emissions changes, independent of temperature. The impact of local emissions is evident in our results as well. In our analysis of diurnal HCHO at three sites with sufficient ground-data (Section 3), we found a mid-day peak in HCHO only at sites dominated by biogenic VOC emissions (Burbank and Pico Rivera in California). The increased temperature in the middle of the day accelerates the emission and oxidation of isoprene, consistent with [76]. Although current-generation satellites provide daily (or less frequent) HCHO data, the sensitivity of HCHO to temperature, including as a function of biogenic VOC emissions versus anthropogenic VOC emissions, would be a valuable application of future hourly HCHO observations from geostationary satellites.

7. Conclusions

As a pollutant with direct health impacts and an indicator of ozone formation, HCHO has emerged as an atmospheric species of interest for air quality management. Ground monitoring data of HCHO is limited, and satellite data may complement the sparse network with spatially continuous information on HCHO abundance. Multiple satellites can detect HCHO, but these data are only beginning to be applied to operational and health-relevant applications. For user communities interested in the interpretation of satellite data, a key question is the agreement in spatial and temporal patterns between ground-based and space-based measurements. Our comparison focuses on the utility of satellite data to inform patterns, trends, and processes of ground-based HCHO across the U.S.

Over our study period, HCHO data were available from 338 sites managed by the EPA. However, only 45 of these had continuous seasonal measurements in the time range, and only three stations provided hourly and three-hour measurement frequencies that had more than 10% data available and passed quality control. Of those three sites, the two sites with larger diurnal amplitudes (in Los Angeles County) had a lower anthropogenic/biogenic ratio for both direct HCHO emissions and for total VOC emissions, as compared to the site with the smaller diurnal amplitude (in St. Louis county). These relative diurnal changes indicate that the origin of VOC emissions may be an important driver of diurnal HCHO patterns. In addition to emissions, chemistry and meteorology also play important roles in affecting the diurnal cycle of HCHO. With the upcoming availability of hourly HCHO data from TEMPO, Sentinel-4, and GEMS, it will be interesting to assess how the diurnal amplitude of HCHO changes between areas dominated by biogenic versus anthropogenic emissions. We expect that areas with larger anthropogenic emissions will exhibit a weaker diurnal signal.

On a seasonal basis, OMI exhibits the highest correlation with AQS in summer and the lowest correlation in winter. These results are consistent with past work indicating that boundary layer HCHO is a greater contributor to the summertime HCHO column and less so in the winter [35]. Combining summer to other season ratios showed OMI bias is high compared to AQS in all regions, but with different factors depending on different regions. Seasonal differences across the regions are likely due to the differences in dominant plant types in each area, as well as VOC emissions in young versus mature trees [87]. The overestimation of the ratios from OMI suggests a more pronounced sensitivity to temperature in the HCHO column than in surface HCHO concentrations.

There are emerging opportunities to study HCHO and its trends in the near future from both ground- and space-based platforms. In 2021, U.S. states, in coordination with the EPA, established 27 new Photochemical Assessment Monitoring Stations (PAMS), which provide hourly ground-based HCHO measurements (official and supporting documents available at https://www.regulations.gov/document/EPA-HQ-OAR-2019-0137-0013; accessed on 22 April 2022). Additionally, ozone nonattainment areas have been required to develop Enhanced Monitoring Plans (EMP), which would expand observations of meteorology and VOCs, potentially including observations of columnar VOCs from PANDORA spectrometers [88]. In addition to these monitoring programs, next generation satellite-based observations are expected to provide high-resolution and hourly column HCHO measurements. Further analysis of space- versus ground-based measurements, using these next-generation platforms, will maximize the relevance of Earth-observing satellites to air quality and public health user communities.

Supplementary Materials: The following supporting information can be downloaded at: https://www.mdpi.com/article/10.3390/rs14092191/s1, Figure S1: All available AQS data from 2006 to 2015 calculated for every season (x-axis) compared with seasonal mean AQS under same OMI viewing conditions (y-axis), with correlation coefficient r = 0.999, mean bias = -0.005 $\mu g\ m^{-3}$. Red dotted line indicates 1-1 ratio.

Author Contributions: Conceptualization, T.H.; Data curation, I.D.S.; Formal analysis, P.W. and M.B.; Funding acquisition, T.H.; Investigation, P.W.; Methodology, T.H.; Resources, M.H. and I.D.S.; Software, P.W.; Supervision, T.H.; Visualization, P.W. and M.B.; Writing—original draft, P.W. and M.H.; Writing—review & editing, T.H., M.B., M.H. and I.D.S. All authors have read and agreed to the published version of the manuscript.

Funding: This research and the APC were funded by the NASA Health and Air Quality Applied Sciences Team (HAQAST), grant numbers NNX16AQ92G and 80NSSC21K0427, as well as the Wisconsin Hilldale Undergraduate/Faculty Research Fellowship.

Data Availability Statement: Data and code for this study is available at https://doi.org/10.5281/zenodo.6499349 (accessed on 15 April 2022).

Acknowledgments: The authors acknowledge the processing of AQS data by Madeleine Strum from EPA. The authors appreciate the editorial contributions by Daegan Miller. Ground monitor data for HCHO is available at EPA AMA (https://www3.epa.gov/ttnamti1/toxdat.html#data; accessed on 30 August 2017), Satellite HCHO observations come from QA4ECV (available at http://www.qa4ecv.eu; accessed on 24 July 2019). EPA's NEI data is available at https://www.epa.gov (accessed on 23 October 2017). NOAA North American Regional Reanalysis data is available at https://www.esrl.noaa.gov (accessed on 28 October 2016).

Conflicts of Interest: The authors declare no conflict of interest.

References

1. Lowe, D.C.; Schmidt, U. Formaldehyde (HCHO) measurements in the nonurban atmosphere. *J. Geophys. Res.* **1983**, *88*, 10844. [CrossRef]
2. Wolfe, G.M.; Kaiser, J.; Hanisco, T.F.; Keutsch, F.N.; de Gouw, J.A.; Gilman, J.B.; Graus, M.; Hatch, C.D.; Holloway, J.; Horowitz, L.W.; et al. Formaldehyde production from isoprene oxidation across NO_x regimes. *Atmos. Chem. Phys.* **2016**, *16*, 2597–2610. [CrossRef] [PubMed]
3. Sillman, S. Chapter 12 The relation between ozone, NOx and hydrocarbons in urban and polluted rural environments. *Dev. Environ. Sci.* **1998**, *1*, 339–385. [CrossRef]
4. Palmer, P.I.; Jacob, D.J.; Fiore, A.M.; Martin, R.V.; Chance, K.; Kurosu, T.P. Mapping isoprene emissions over North America using formaldehyde column observations from space. *J. Geophys. Res. Atmos.* **2003**, *108*, 4180. [CrossRef]
5. Millet, D.B.; Jacob, D.J.; Turquety, S.; Hudman, R.C.; Wu, S.; Fried, A.; Walega, J.; Heikes, B.G.; Blake, D.R.; Singh, H.B.; et al. Formaldehyde distribution over North America: Implications for satellite retrievals of formaldehyde columns and isoprene emission. *J. Geophys. Res. Atmos.* **2006**, *111*, D24S02. [CrossRef]
6. Sprengnether, M.; Demerjian, K.L.; Donahue, N.M.; Anderson, J.G. Product analysis of the OH oxidation of isoprene and 1,3-butadiene in the presence of NO. *J. Geophys. Res. Atmos.* **2002**, *107*, ACH-8. [CrossRef]
7. Levy, H. Photochemistry of the lower troposphere. *Planet. Space Sci.* **1972**, *20*, 919–935. [CrossRef]
8. Atkinson, R. Atmospheric chemistry of VOCs and NOx. *Atmos. Environ.* **2000**, *34*, 2063–2101. [CrossRef]
9. Abbot, D.S.; Palmer, P.I.; Martin, R.V.; Chance, K.V.; Jacob, D.J.; Guenther, A. Seasonal and interannual variability of North American isoprene emissions as determined by formaldehyde column measurements from space. *Geophys. Res. Lett.* **2003**, *30*, 1999–2002. [CrossRef]
10. Holzinger, R.; Warneke, C.; Hansel, A.; Jordan, A.; Lindinger, W.; Scharffe, D.H.; Schade, G.; Crutzen, P.J. Biomass burning as a source of formaldehyde, acetaldehyde, methanol, acetone, acetonitrile, and hydrogen cyanide. *Geophys. Res. Lett.* **1999**, *26*, 1161–1164. [CrossRef]
11. Piccot, S.D.; Watson, J.J.; Jones, J.W. A global inventory of volatile organic compound emissions from anthropogenic sources. *J. Geophys. Res. Atmos.* **1992**, *97*, 9897–9912. [CrossRef]
12. Cárdenas, L.M.; Brassington, D.J.; Allan, B.J.; Coe, H.; Alicke, B.; Platt, U.; Wilson, K.M.; Plane, J.M.C.; Penkett, S.A. Intercomparison of Formaldehyde Measurements in Clean and Polluted Atmospheres. *J. Atmos. Chem.* **2000**, *37*, 53–80. [CrossRef]
13. U.S. Environmental Protection Agency. Compendium Methods for the Determination of Toxic Organic Compounds in Ambient Air: Compendium Method TO-11A. 1999. Available online: https://www3.epa.gov/ttnamti1/files/ambient/airtox/to-11ar.pdf (accessed on 30 August 2017).
14. Arnts, R.R.; Tejada, S.B. 2,4-Dinitrophenylhydrazine-coated silica gel cartridge method for determination of formaldehyde in air: Identification of an ozone interference. *Environ. Sci. Technol.* **1989**, *23*, 1428–1430. [CrossRef]
15. Karst, U.; Binding, N.; Cammann, K.; Witting, U. Interferences of nitrogen dioxide in the determination of aldehydes and ketones by sampling on 2,4-dinitrophenylhydrazine-coated solid sorbent. *Fresenius' J. Anal. Chem.* **1993**, *345*, 48–52. [CrossRef]
16. Rodier, D.R.; Nondek, L.; Birks, J.W. Evaluation of ozone and water vapor interferences in the derivatization of atmospheric aldehydes with dansylhydrazine. *Environ. Sci. Technol.* **1993**, *27*, 2814–2820. [CrossRef]
17. Achatz, S.; Lörinci, G.; Hertkorn, N.; Gebefügi, I.; Kettrup, A. Disturbance of the determination of aldehydes and ketones: Structural elucidation of degradation products derived from the reaction of 2,4-dinitrophenylhydrazine (DNPH) with ozone. *Fresenius' J. Anal. Chem.* **1999**, *364*, 141–146. [CrossRef]
18. Levelt, P.F.; Van Den Oord, G.H.J.; Dobber, M.R.; Malkki, A.; Visser, H.; De Vries, J.; Stammes, P.; Lundell, J.O.V.; Saari, H. The ozone monitoring instrument. *IEEE Trans. Geosci. Remote. Sens.* **2006**, *44*, 1093–1101. [CrossRef]
19. Veefkind, J.P.; Aben, I.; McMullan, K.; Förster, H.; de Vries, J.; Otter, G.; Claas, J.; Eskes, H.J.; de Haan, J.F.; Kleipool, Q.; et al. TROPOMI on the ESA Sentinel-5 Precursor: A GMES mission for global observations of the atmospheric composition for climate, air quality and ozone layer applications. *Remote Sens. Environ.* **2012**, *120*, 70–83. [CrossRef]
20. Abad, G.G.; Vasilkov, A.; Seftor, C.; Liu, X.; Chance, K. Smithsonian Astrophysical Observatory Ozone Mapping and Profiler Suite (SAO OMPS) formaldehyde retrieval. *Atmos. Meas. Tech.* **2016**, *9*, 2797–2812. [CrossRef]
21. Munro, R.; Lang, R.; Klaes, D.; Poli, G.; Retscher, C.; Lindstrot, R.; Huckle, R.; Lacan, A.; Grzegorski, M.; Holdak, A.; et al. The GOME-2 instrument on the Metop series of satellites: Instrument design, calibration, and level 1 data processing—An overview. *Atmos. Meas. Tech.* **2016**, *9*, 1279–1301. [CrossRef]

22. Kim, J.; Jeong, U.; Ahn, M.-H.; Kim, J.H.; Park, R.J.; Lee, H.; Song, C.H.; Choi, Y.-S.; Lee, K.-H.; Yoo, J.-M.; et al. New Era of Air Quality Monitoring from Space: Geostationary Environment Monitoring Spectrometer (GEMS). *Bull. Am. Meteorol. Soc.* **2020**, *101*, E1–E22. [CrossRef]
23. Kwon, H.-A.; Park, R.J.; Abad, G.G.; Chance, K.; Kurosu, T.P.; Kim, J.; De Smedt, I.; Van Roozendael, M.; Peters, E.; Burrows, J. Description of a formaldehyde retrieval algorithm for the Geostationary Environment Monitoring Spectrometer (GEMS). *Atmos. Meas. Tech.* **2019**, *12*, 3551–3571. [CrossRef]
24. Zoogman, P.; Liu, X.; Suleiman, R.; Pennington, W.; Flittner, D.; Al-Saadi, J.; Hilton, B.; Nicks, D.; Newchurch, M.; Carr, J.; et al. Tropospheric emissions: Monitoring of pollution (TEMPO). *J. Quant. Spectrosc. Radiat. Transf.* **2016**, *186*, 17–39. [CrossRef]
25. Bauwens, M.; Stavrakou, T.; Müller, J.-F.; De Smedt, I.; Van Roozendael, M.; van der Werf, G.R.; Wiedinmyer, C.; Kaiser, J.W.; Sindelarova, K.; Guenther, A. Nine years of global hydrocarbon emissions based on source inversion of OMI formaldehyde observations. *Atmos. Chem. Phys.* **2016**, *16*, 10133–10158. [CrossRef]
26. De Smedt, I.; Stavrakou, T.; Hendrick, F.; Danckaert, T.; Vlemmix, T.; Pinardi, G.; Theys, N.; Lerot, C.; Gielen, C.; Vigouroux, C.; et al. Diurnal, seasonal and long-term variations of global formaldehyde columns inferred from combined OMI and GOME-2 observations. *Atmos. Chem. Phys.* **2015**, *15*, 12519–12545. [CrossRef]
27. Kaiser, J.; Wolfe, G.M.; Bohn, B.; Broch, S.; Fuchs, H.; Ganzeveld, L.N.; Gomm, S.; Häseler, R.; Hofzumahaus, A.; Holland, F.; et al. Evidence for an unidentified non-photochemical ground-level source of formaldehyde in the Po Valley with potential implications for ozone production. *Atmos. Chem. Phys.* **2015**, *15*, 1289–1298. [CrossRef]
28. Kaiser, J.; Wolfe, G.M.; Min, K.E.; Brown, S.S.; Miller, C.C.; Jacob, D.J.; Degouw, J.A.; Graus, M.; Hanisco, T.F.; Holloway, J.; et al. Reassessing the ratio of glyoxal to formaldehyde as an indicator of hydrocarbon precursor speciation. *Atmos. Chem. Phys.* **2015**, *15*, 7571–7583. [CrossRef]
29. Kefauver, S.C.; Filella, I.; Peñuelas, J. Remote sensing of atmospheric biogenic volatile organic compounds (BVOCs) via satellite-based formaldehyde vertical column assessments. *Int. J. Remote Sens.* **2014**, *35*, 7519–7542. [CrossRef]
30. Kwon, H.-A.; Park, R.J.; Jeong, J.I.; Lee, S.; Abad, G.G.; Kurosu, T.P.; Palmer, P.I.; Chance, K. Sensitivity of formaldehyde (HCHO) column measurements from a geostationary satellite to temporal variation of the air mass factor in East Asia. *Atmos. Chem. Phys.* **2017**, *17*, 4673–4686. [CrossRef]
31. Wells, K.C.; Millet, D.B.; Payne, V.H.; Deventer, M.J.; Bates, K.H.; de Gouw, J.A.; Graus, M.; Warneke, C.; Wisthaler, A.; Fuentes, J.D. Satellite isoprene retrievals constrain emissions and atmospheric oxidation. *Nature* **2020**, *585*, 225–233. [CrossRef]
32. Wolfe, G.M.; Nicely, J.M.; Clair, J.M.S.; Hanisco, T.F.; Liao, J.; Oman, L.D.; Brune, W.B.; Miller, D.; Thames, A.; Abad, G.G.; et al. Mapping hydroxyl variability throughout the global remote troposphere via synthesis of airborne and satellite formaldehyde observations. *Proc. Natl. Acad. Sci. USA* **2019**, *116*, 11171–11180. [CrossRef] [PubMed]
33. Zheng, Y.; Unger, N.; Barkley, M.P.; Yue, X. Relationships between photosynthesis and formaldehyde as a probe of isoprene emission. *Atmos. Chem. Phys.* **2015**, *15*, 8559–8576. [CrossRef]
34. Harkey, M.; Holloway, T.; Kim, E.J.; Baker, K.R.; Henderson, B. Satellite Formaldehyde to Support Model Evaluation. *J. Geophys. Res. Atmos.* **2021**, *126*, e2020JD032881. [CrossRef]
35. Jin, X.; Fiore, A.M.; Murray, L.T.; Valin, L.C.; Lamsal, L.N.; Duncan, B.; Boersma, K.F.; De Smedt, I.; Abad, G.G.; Chance, K.; et al. Evaluating a Space-Based Indicator of Surface Ozone-NO_x-VOC Sensitivity Over Midlatitude Source Regions and Application to Decadal Trends. *J. Geophys. Res. Atmos.* **2017**, *122*, 10439–10461. [CrossRef] [PubMed]
36. Jin, X.; Fiore, A.; Boersma, K.F.; De Smedt, I.; Valin, L. Inferring Changes in Summertime Surface Ozone–NO_x–VOC Chemistry over U.S. Urban Areas from Two Decades of Satellite and Ground-Based Observations. *Environ. Sci. Technol.* **2020**, *54*, 6518–6529. [CrossRef] [PubMed]
37. Li, D.; Wang, S.; Xue, R.; Zhu, J.; Zhang, S.; Sun, Z.; Zhou, B. OMI-observed HCHO in Shanghai, China, during 2010–2019 and ozone sensitivity inferred by an improved $HCHO/NO_2$ ratio. *Atmos. Chem. Phys.* **2021**, *21*, 15447–15460. [CrossRef]
38. Palmer, P.; Abbot, D.S.; Fu, T.-M.; Jacob, D.J.; Chance, K.; Kurosu, T.P.; Guenther, A.; Wiedinmyer, C.; Stanton, J.C.; Pilling, M.J.; et al. Quantifying the seasonal and interannual variability of North American isoprene emissions using satellite observations of the formaldehyde column. *J. Geophys. Res. Atmos.* **2006**, *111*, D12315. [CrossRef]
39. Duncan, B.N.; Prados, A.; Lamsal, L.N.; Liu, Y.; Streets, D.G.; Gupta, P.; Hilsenrath, E.; Kahn, R.A.; Nielsen, J.E.; Beyersdorf, A.J.; et al. Satellite data of atmospheric pollution for U.S. air quality applications: Examples of applications, summary of data end-user resources, answers to FAQs, and common mistakes to avoid. *Atmos. Environ.* **2014**, *94*, 647–662. [CrossRef]
40. Hong, Q.; Liu, C.; Hu, Q.; Zhang, Y.; Xing, C.; Su, W.; Ji, X.; Xiao, S. Evaluating the feasibility of formaldehyde derived from hyperspectral remote sensing as a proxy for volatile organic compounds. *Atmos. Res.* **2021**, *264*, 105777. [CrossRef]
41. Martin, R.; Parrish, D.D.; Ryerson, T.B.; Nicks, D.K.; Chance, K.; Kurosu, T.P.; Jacob, D.J.; Sturges, E.D.; Fried, A.; Wert, B.P. Evaluation of GOME satellite measurements of tropospheric NO_2 and HCHO using regional data from aircraft campaigns in the southeastern United States. *J. Geophys. Res. Atmos.* **2004**, *109*, D24307. [CrossRef]
42. Duncan, B.N.; Yoshida, Y.; Olson, J.R.; Sillman, S.; Martin, R.; Lamsal, L.; Hu, Y.; Pickering, K.E.; Retscher, C.; Allen, D.; et al. Application of OMI observations to a space-based indicator of NOx and VOC controls on surface ozone formation. *Atmos. Environ.* **2010**, *44*, 2213–2223. [CrossRef]
43. Witman, S.; Holloway, T.; Reddy, P.J. *Integrating Satellite Data into Air Quality Management: Experience from Colorado*; Air and Waste Management Association: Pittsburgh, PA, USA, 2014; pp. 34–38.

44. Martin, R.V.; Jacob, D.J.; Chance, K.; Kurosu, T.P.; Palmer, P.; Evans, M.J. Global inventory of nitrogen oxide emissions constrained by space-based observations of NO_2 columns. *J. Geophys. Res. Atmos.* **2003**, *108*, 4537. [CrossRef]
45. Lamsal, L.N.; Martin, R.; Van Donkelaar, A.; Steinbacher, M.; Celarier, E.A.; Bucsela, E.; Dunlea, E.J.; Pinto, J.P. Ground-level nitrogen dioxide concentrations inferred from the satellite-borne Ozone Monitoring Instrument. *J. Geophys. Res. Atmos.* **2008**, *113*, D16308. [CrossRef]
46. Harkey, M.; Holloway, T.; Oberman, J.; Scotty, E. An evaluation of CMAQ NO_2 using observed chemistry-meteorology correlations. *J. Geophys. Res. Atmos.* **2015**, *120*, 11775–11797. [CrossRef]
47. Kramer, L.; Leigh, R.J.; Remedios, J.J.; Monks, P.S. Comparison of OMI and ground-based in situ and MAX-DOAS measurements of tropospheric nitrogen dioxide in an urban area. *J. Geophys. Res. Atmos.* **2008**, *113*, D16S39. [CrossRef]
48. Boersma, K.F.; Jacob, D.J.; Trainic, M.; Rudich, Y.; DeSmedt, I.; Dirksen, R.; Eskes, H.J. Validation of urban NO_2 concentrations and their diurnal and seasonal variations observed from the SCIAMACHY and OMI sensors using in situ surface measurements in Israeli cities. *Atmos. Chem. Phys.* **2009**, *9*, 3867–3879. [CrossRef]
49. Ma, J.Z.; Beirle, S.; Jin, J.L.; Shaiganfar, R.; Yan, P.; Wagner, T. Tropospheric NO_2 vertical column densities over Beijing: Results of the first three years of ground-based MAX-DOAS measurements (2008–2011) and satellite validation. *Atmos. Chem. Phys.* **2013**, *13*, 1547–1567. [CrossRef]
50. Opacka, B.; Müller, J.-F.; Stavrakou, T.; Bauwens, M.; Sindelarova, K.; Markova, J.; Guenther, A.B. Global and regional impacts of land cover changes on isoprene emissions derived from spaceborne data and the MEGAN model. *Atmos. Chem. Phys.* **2021**, *21*, 8413–8436. [CrossRef]
51. Wang, H.; Wu, Q.; Guenther, A.B.; Yang, X.; Wang, L.; Xiao, T.; Li, J.; Feng, J.; Xu, Q.; Cheng, H. A long-term estimation of biogenic volatile organic compound (BVOC) emission in China from 2001–2016: The roles of land cover change and climate variability. *Atmos. Chem. Phys.* **2021**, *21*, 4825–4848. [CrossRef]
52. Gopikrishnan, G.; Kuttippurath, J. A decade of satellite observations reveal significant increase in atmospheric formaldehyde from shipping in Indian Ocean. *Atmos. Environ.* **2020**, *246*, 118095. [CrossRef]
53. Shen, L.; Jacob, D.J.; Zhu, L.; Zhang, Q.; Zheng, B.; Sulprizio, M.P.; Li, K.; De Smedt, I.; Abad, G.G.; Cao, H.; et al. The 2005–2016 Trends of Formaldehyde Columns Over China Observed by Satellites: Increasing Anthropogenic Emissions of Volatile Organic Compounds and Decreasing Agricultural Fire Emissions. *Geophys. Res. Lett.* **2019**, *46*, 4468–4475. [CrossRef]
54. Bauwens, M.; Verreyken, B.; Stavrakou, T.; Müller, J.-F.; De Smedt, I. Spaceborne evidence for significant anthropogenic VOC trends in Asian cities over 2005–2019. *Environ. Res. Lett.* **2022**, *17*, 015008. [CrossRef]
55. Guan, J.; Jin, B.; Ding, Y.; Wang, W.; Li, G.; Ciren, P. Global Surface HCHO Distribution Derived from Satellite Observations with Neural Networks Technique. *Remote Sens.* **2021**, *13*, 4055. [CrossRef]
56. Millet, D.B.; Jacob, D.J.; Boersma, K.; Fu, T.-M.; Kurosu, T.P.; Chance, K.; Heald, C.L.; Guenther, A. Spatial distribution of isoprene emissions from North America derived from formaldehyde column measurements by the OMI satellite sensor. *J. Geophys. Res. Atmos.* **2008**, *113*, D16308. [CrossRef]
57. Zhu, L.; Mickley, L.J.; Jacob, D.J.; Marais, E.A.; Sheng, J.; Hu, L.; Abad, G.G.; Chance, K. Long-term (2005–2014) trends in formaldehyde (HCHO) columns across North America as seen by the OMI satellite instrument: Evidence of changing emissions of volatile organic compounds. *Geophys. Res. Lett.* **2017**, *44*, 7079–7086. [CrossRef]
58. Duncan, B.N.; Yoshida, Y.; Damon, M.R.; Douglass, A.R.; Witte, J.C. Temperature dependence of factors controlling isoprene emissions. *Geophys. Res. Lett.* **2009**, *36*, L05813. [CrossRef]
59. Guenther, A.; Geron, C.; Pierce, T.; Lamb, B.; Harley, P.; Fall, R. Natural emissions of non-methane volatile organic compounds, carbon monoxide, and oxides of nitrogen from North America. *Atmos. Environ.* **2000**, *34*, 2205–2230. [CrossRef]
60. Tingey, D.T.; Manning, M.; Grothaus, L.C.; Burns, W.F. The Influence of Light and Temperature on Isoprene Emission Rates from Live Oak. *Physiol. Plant.* **1979**, *47*, 112–118. [CrossRef]
61. Zhu, L.; Jacob, D.J.; Mickley, L.J.; Marais, E.; Cohan, D.; Yoshida, Y.; Duncan, B.N.; Abad, G.G.; Chance, K. Anthropogenic emissions of highly reactive volatile organic compounds in eastern Texas inferred from oversampling of satellite (OMI) measurements of HCHO columns. *Environ. Res. Lett.* **2014**, *9*, 114004. [CrossRef]
62. Lieschke, K.J.; Fisher, J.A.; Paton-Walsh, C.; Jones, N.B.; Greenslade, J.W.; Burden, S.; Griffith, D.W.T. Decreasing Trend in Formaldehyde Detected From 20-Year Record at Wollongong, Southeast Australia. *Geophys. Res. Lett.* **2019**, *46*, 8464–8473. [CrossRef]
63. Morfopoulos, C.; Müller, J.; Stavrakou, T.; Bauwens, M.; De Smedt, I.; Friedlingstein, P.; Prentice, I.C.; Regnier, P. Vegetation responses to climate extremes recorded by remotely sensed atmospheric formaldehyde. *Glob. Chang. Biol.* **2021**, *28*, 1809–1822. [CrossRef] [PubMed]
64. Lorente, A.; Boersma, K.F.; Yu, H.; Dörner, S.; Hilboll, A.; Richter, A.; Liu, M.; Lamsal, L.N.; Barkley, M.; De Smedt, I.; et al. Structural uncertainty in air mass factor calculation for NO_2 and HCHO satellite retrievals. *Atmos. Meas. Tech.* **2017**, *10*, 759–782. [CrossRef]
65. De Smedt, I.; Yu, H.; Richter, A.; Beirle, S.; Eskes, H.; Boersma, K.F.; Van Roozendael, J.; Van Geffen, A.; Lorente, E.; Peters, E. *QA4ECV HCHO Tropospheric Column Data from OMI*; (Version 1.1) [Data set]; Royal Belgian Institute for Space Aeronomy: Uccle, Belgium, 2017. [CrossRef]

66. De Smedt, I.; Theys, N.; Yu, H.; Danckaert, T.; Lerot, C.; Compernolle, S.; Van Roozendael, M.; Richter, A.; Hilboll, A.; Peters, E.; et al. Algorithm theoretical baseline for formaldehyde retrievals from S5P TROPOMI and from the QA4ECV project. *Atmos. Meas. Tech.* **2018**, *11*, 2395–2426. [CrossRef]
67. Zhu, L.; Jacob, D.J.; Kim, P.S.; Fisher, J.A.; Yu, K.; Travis, K.R.; Mickley, L.J.; Yantosca, R.M.; Sulprizio, M.P.; De Smedt, I.; et al. Observing atmospheric formaldehyde (HCHO) from space: Validation and intercomparison of six retrievals from four satellites (OMI, GOME2A, GOME2B, OMPS) with SEAC4RS aircraft observations over the southeast US. *Atmos. Chem. Phys.* **2016**, *16*, 13477–13490. [CrossRef]
68. Franco, B.; Marais, E.A.; Bovy, B.; Bader, W.; Lejeune, B.; Roland, G.; Servais, C.; Mahieu, E. Diurnal cycle and multi-decadal trend of formaldehyde in the remote atmosphere near 46° N. *Atmos. Chem. Phys.* **2016**, *16*, 4171–4189. [CrossRef]
69. Lamsal, L.N.; Duncan, B.N.; Yoshida, Y.; Krotkov, N.A.; Pickering, K.E.; Streets, D.G.; Lu, Z.U.S. NO_2 trends (2005–2013): EPA Air Quality System (AQS) data versus improved observations from the Ozone Monitoring Instrument (OMI). *Atmos. Environ.* **2015**, *110*, 130–143. [CrossRef]
70. Kim, J.H.; Kim, S.M.; Baek, K.H.; Wang, L.; Kurosu, T.; De Smedt, I.; Chance, K.; Newchurch, M.J. Evaluation of satellite-derived HCHO using statistical methods. *Atmos. Chem. Phys. Discuss.* **2011**, *11*, 8003–8025. [CrossRef]
71. Marais, E.A.; Jacob, D.J.; Kurosu, T.P.; Chance, K.; Murphy, J.G.; Reeves, C.; Mills, G.; Casadio, S.; Millet, D.B.; Barkley, M.P.; et al. Isoprene emissions in Africa inferred from OMI observations of formaldehyde columns. *Atmos. Chem. Phys.* **2012**, *12*, 6219–6235. [CrossRef]
72. Zhu, L.; Jacob, D.J.; Keutsch, F.N.; Mickley, L.J.; Scheffe, R.; Strum, M.; Abad, G.G.; Chance, K.; Yang, K.; Rappenglück, B.; et al. Formaldehyde (HCHO) As a Hazardous Air Pollutant: Mapping Surface Air Concentrations from Satellite and Inferring Cancer Risks in the United States. *Environ. Sci. Technol.* **2017**, *51*, 5650–5657. [CrossRef]
73. Stavrakou, T.; Müller, J.-F.; De Smedt, I.; Van Roozendael, M.; van der Werf, G.R.; Giglio, L.; Guenther, A. Evaluating the performance of pyrogenic and biogenic emission inventories against one decade of space-based formaldehyde columns. *Atmos. Chem. Phys.* **2009**, *9*, 1037–1060. [CrossRef]
74. Carlton, A.G.; Baker, K.R. Photochemical Modeling of the Ozark Isoprene Volcano: MEGAN, BEIS, and Their Impacts on Air Quality Predictions. *Environ. Sci. Technol.* **2011**, *45*, 4438–4445. [CrossRef] [PubMed]
75. Fuentes, J.D.; Wang, D.; Neumann, H.H.; Gillespie, T.J.; Hartog, G.D.; Dann, T.F. Ambient biogenic hydrocarbons and isoprene emissions from a mixed deciduous forest. *J. Atmos. Chem.* **1996**, *25*, 67–95. [CrossRef]
76. Choi, W.; Faloona, I.C.; Bouvier-Brown, N.C.; McKay, M.; Goldstein, A.H.; Mao, J.; Brune, W.H.; LaFranchi, B.W.; Cohen, R.C.; Wolfe, G.M.; et al. Observations of elevated formaldehyde over a forest canopy suggest missing sources from rapid oxidation of arboreal hydrocarbons. *Atmos. Chem. Phys.* **2010**, *10*, 8761–8781. [CrossRef]
77. Lin, J.-T.; McElroy, M.B. Impacts of boundary layer mixing on pollutant vertical profiles in the lower troposphere: Implications to satellite remote sensing. *Atmos. Environ.* **2010**, *44*, 1726–1739. [CrossRef]
78. Pusede, S.E.; Gentner, D.R.; Wooldridge, P.J.; Browne, E.C.; Rollins, A.W.; Min, K.-E.; Russell, A.R.; Thomas, J.; Zhang, L.; Brune, W.H.; et al. On the temperature dependence of organic reactivity, nitrogen oxides, ozone production, and the impact of emission controls in San Joaquin Valley, California. *Atmos. Chem. Phys.* **2014**, *14*, 3373–3395. [CrossRef]
79. Mesinger, F.; DiMego, G.; Kalnay, E.; Mitchell, K.; Shafran, P.C.; Ebisuzaki, W.; Jović, D.; Woollen, J.; Rogers, E.; Berbery, E.H.; et al. North American Regional Reanalysis. *Bull. Am. Meteorol. Soc.* **2006**, *87*, 343–360. [CrossRef]
80. Mascioli, N.R.; Previdi, M.; Fiore, A.; Ting, M. Timing and seasonality of the United States 'warming hole'. *Environ. Res. Lett.* **2017**, *12*, 034008. [CrossRef]
81. Partridge, T.F.; Winter, J.M.; Osterberg, E.C.; Hyndman, D.W.; Kendall, A.D.; Magilligan, F.J. Spatially Distinct Seasonal Patterns and Forcings of the U.S. Warming Hole. *Geophys. Res. Lett.* **2018**, *45*, 2055–2063. [CrossRef]
82. Diao, M.; Holloway, T.; Choi, S.; O'Neill, S.M.; Al-Hamdan, M.Z.; Van Donkelaar, A.; Martin, R.V.; Jin, X.; Fiore, A.M.; Henze, D.K.; et al. Methods, availability, and applications of $PM_{2.5}$ exposure estimates derived from ground measurements, satellite, and atmospheric models. *J. Air Waste Manag. Assoc.* **2019**, *69*, 1391–1414. [CrossRef]
83. De Smedt, I.; Pinardi, G.; Vigouroux, C.; Compernolle, S.; Bais, A.; Benavent, N.; Boersma, F.; Chan, K.-L.; Donner, S.; Eichmann, K.-U.; et al. Comparative assessment of TROPOMI and OMI formaldehyde observations and validation against MAX-DOAS network column measurements. *Atmos. Chem. Phys.* **2021**, *21*, 12561–12593. [CrossRef]
84. Vigouroux, C.; Langerock, B.; Aquino, C.A.B.; Blumenstock, T.; Cheng, Z.; De Mazière, M.; De Smedt, I.; Grutter, M.; Hannigan, J.W.; Jones, N.; et al. TROPOMI–Sentinel-5 Precursor formaldehyde validation using an extensive network of ground-based Fourier-transform infrared stations. *Atmos. Meas. Tech.* **2020**, *13*, 3751–3767. [CrossRef]
85. Zhu, L.; Abad, G.G.; Nowlan, C.R.; Miller, C.C.; Chance, K.; Apel, E.C.; DiGangi, J.P.; Fried, A.; Hanisco, T.F.; Hornbrook, R.S.; et al. Validation of satellite formaldehyde (HCHO) retrievals using observations from 12 aircraft campaigns. *Atmos. Chem. Phys.* **2020**, *20*, 12329–12345. [CrossRef]
86. Fan, J.; Ju, T.; Wang, Q.; Gao, H.; Huang, R.; Duan, J. Spatiotemporal variations and potential sources of tropospheric formaldehyde over eastern China based on OMI satellite data. *Atmos. Pollut. Res.* **2020**, *12*, 272–285. [CrossRef]
87. Bracho-Nunez, A.; Welter, S.; Staudt, M.; Kesselmeier, J. Plant-specific volatile organic compound emission rates from young and mature leaves of Mediterranean vegetation. *J. Geophys. Res. Atmos.* **2011**, *116*, 1–13. [CrossRef]
88. U.S. Environmental Protection Agency. Technical Note—Guidance for Developing Enhanced Monitoring Plans. 2018. Available online: https://www.epa.gov/sites/default/files/2019-11/documents/pams_emp_guidance.pdf (accessed on 21 April 2022).

remote sensing MDPI

Article

The Effect of Urban Form on $PM_{2.5}$ Concentration: Evidence from China's 340 Prefecture-Level Cities

Ying Liu [1], Lijie He [2], Wenmin Qin [3], Aiwen Lin [4] and Yanzhao Yang [1,*]

1 Institute of Geographic Sciences and Natural Resources Research, Chinese Academy of Sciences, Beijing 100101, China; liuy.20b@igsnrr.ac.cn
2 College of Public Administration, Huazhong Agricultural University, Wuhan 430079, China; helijie@mail.hzau.edu.cn
3 School of Geography and Information Engineering, China University of Geosciences, Wuhan 430079, China; qinwenmin@cug.edu.cn
4 School of Resource and Environmental Science, Wuhan University, Wuhan 430079, China; awlin@whu.edu.cn
* Correspondence: yangyz@igsnrr.ac.cn

Abstract: Exploring how urban form affects the Particulate Matter 2.5 ($PM_{2.5}$) concentration could help to find environmentally friendly urbanization. According to the definition of geography, this paper constructs a comprehensive urban form evaluation index system applicable to many aspects. Four urban form metrics, as well as road density and five control variables are selected. Based on 2015 data on China's 340 prefecture-level cities, the spatial regression model and geographically weighted regression model were used to explore the relationship between the urban form evaluation index system and $PM_{2.5}$ pollution. The main results show that the spatial distribution of $PM_{2.5}$ in China follows an increasing trend from northwest to southeast. Urban form indicators such as AI, LPI, PLAND, LSI and road density were all significantly related to $PM_{2.5}$ concentrations. More compact urban construction, lower fragmentation of urban land, and lower density of the road network are conducive factors for improving air quality conditions. In addition, affected by seasonal changes, the correlation between urban form and $PM_{2.5}$ concentration in spring and winter is higher than that in summer and winter. This study confirmed that a reasonable urban planning strategies are very important for improving air quality.

Keywords: urban form; $PM_{2.5}$; landscape metrics; geographically weighted regression

Citation: Liu, Y.; He, L.; Qin, W.; Lin, A.; Yang, Y. The Effect of Urban Form on $PM_{2.5}$ Concentration: Evidence from China's 340 Prefecture-Level Cities. *Remote Sens.* **2022**, *14*, 7. https://doi.org/10.3390/rs14010007

Academic Editors: Maria João Costa and Daniele Bortoli

Received: 21 October 2021
Accepted: 17 December 2021
Published: 21 December 2021

1. Introduction

In the past several decades, with the rapid urbanization and industrialization across many regions of the world, atmospheric pollution became increasingly serious and is already a major social problem [1,2]. Especially in China, as the largest developing country, the rate of urbanization increased from 17.9% to 54.8% in years between 1978 and 2015 years and continues to increase [3,4]. The average Particulate Matter 2.5 ($PM_{2.5}$) concentration in cities reached 62 μgm^{-3}, 60.8 μgm^{-3}, and 57 μgm^{-3} in 2013, 2014, and 2015, respectively. China became the most $PM_{2.5}$-polluted area in the world [5]. $PM_{2.5}$ is considered one of the most important pollutants because of its indirect impacts on health, agriculture production, atmospheric visibility, and climate environment [6,7]. In 2001–2006, 165 prefectures' annual $PM_{2.5}$ levels had far and away beyond the national atmosphere quality standard of China (NAQSC, annual average < 35 μgm^{-3}) [8]. Many studies showed that $PM_{2.5}$ is a key atmosphere pollutant threatening public health [9,10]. An increase of $PM_{2.5}$ concentration by 10 $\mu g/m^3$ causes a 0.40% increase in all-course mortality, a 1.43% increase in deaths caused by respiratory failure, and a 0.53% increase in deaths caused by cardiovascular failure [11]. It is estimated that $PM_{2.5}$ pollution caused 1.2 million premature deaths in China in 2010 and nearly 35% of worldwide deaths [12].

To explore the factors that affect $PM_{2.5}$ concentration can help to better analyze the effects of $PM_{2.5}$ pollution. A number of literatures showed that human activities and natural factors act on $PM_{2.5}$ pollution concentrations either through direct or indirect influences. These influences may be social economy [2,13], the industrial structure [5], climate change [14], seasonality [15], or the prevalence of monsoons [16]. For example, Xu found that economic growth had the large impact on $PM_{2.5}$ [5]. Most studies found a clear reverse "U-shaped" curve between economic development levels, urbanization, and atmosphere pollution, and with improving economy levels, most cities are at a stage of increasing pollution levels [2,17]. It was confirmed that $PM_{2.5}$ pollution is influenced by seasons and regions, and the highest levels were found in winter despite differences in temperature and relative humidity among different regions. He and his colleague applied the global regression model, and found that increased fossil energy consumption leads to an increase in $PM_{2.5}$ concentrations, while elevation, precipitation, temperature, and GDP per capita are all likely to reduce the impact of $PM_{2.5}$ pollution [18].

Moreover, many studies showed that urban planning exert a positive effect on the reduction of $PM_{2.5}$ levels over recent years. Examples for these factors are a reasonable urban form, the moderate reduction of road density, building density, population ratios, and improving green spaces [19]. Of course, there are also studies that used comprehensive indicators in the urbanization process to explore the impact on $PM_{2.5}$, for example, the Liveability and Health Index (LHI) [20]. Urban form, which includes a city area as well as its shape and layout, can be defined as urban land use organization and human activity arrangement [21], and it is usually measured by several landscape indicators of a city. $PM_{2.5}$ pollution is affected by vehicle use, green land, pollutant diffusion, and the heat island effect [22]. Research proved that higher urban compactness and less fragmentation (i.e., the largest patch index (LPI)) can reduce $PM_{2.5}$ pollution in China [23,24]. However, other studies argued that motor vehicles are the main cause of atmospheric pollutants emission in cities, and there is a strong correlation between $PM_{2.5}$ and mortality in the traffic emission [25]; thus, a more compact development alone may still increase local $PM_{2.5}$ concentrations and also cause more population to be affected by $PM_{2.5}$ [26]. In the USA, controlling the population, the level of urbanization, and the mixing of different land cover types were found to be important influencing factors between pollutant levels and atmospheric quality [27,28]. In addition, the distance from the main road, the standard deviation of the building floor, and the average floor are the main urban morphological characteristics that affect the spatial variation of a variety of pollutants [29]. In the above analysis, because of the complexity of socioeconomic and natural conditions, the relationship between urban form and $PM_{2.5}$ may be inconsistent and complex. Scientific urban planning can effectively reduce urban $PM_{2.5}$. Therefore, it is necessary to explore the effect of urban form on $PM_{2.5}$.

Investigating $PM_{2.5}$ concentration is important for research. A large number of experiments used to study the relationship between $PM_{2.5}$ and urban form to identify better approaches for reducing atmosphere pollution. Most studies used linear regression models to analyze the urban form indexes that are related to $PM_{2.5}$ pollution and estimated the coefficient of form indexes in the model [30]. In addition, spatial econometric models and the Environmental Kuznets curve (EKC) hypothesis were used to study the socioeconomic and natural factors on urban atmosphere pollution [13,31]. Most studies mainly obtained urban form data through urban land use data and calculated the urban landscape pattern index to represent specific characteristics of the urban form. Most models selected class area (CA), number of patches (NP), patch density (PD), LPI, area-weighted mean shape index (AWMSI), percentage of landscape (PLAND), aggregation index (AI), landscape shape index (LSI), contiguity index (CONTIG), effective mesh size (MESH), interspersion juxtaposition index (IJI), and other landscape pattern indexes [32,33]. $PM_{2.5}$ data are obtained in three main ways: monitoring real-time data through environmental observation sites, monitoring through experimental instruments, and estimating $PM_{2.5}$ concentrations using atmospheric aerosol optical depth (AOD) data obtained from remote sensing images [34]. The latter can compensate for the shortcomings of experimental technology and ground monitoring sites and provides large-scale and real-time continuous observation data [35]. Although many studies explored the correlation between urban form and $PM_{2.5}$ pollution at different scales from different-sized cities to urban agglomerations to countries, more specific urban form indicators need to be analyzed in the case of China. For example, there is a lack of variables explaining the effects of local meteorological conditions on the spatial aggregation and dispersion of $PM_{2.5}$ pollution. The geographical conditions of China are complex, and the urban forms of different geographical locations vary greatly; therefore, it is necessary to develop more specific urban form indicators to evaluate the shape of cities from different aspects and to measure the urban forms of China more adequately.

To develop more effective urban development strategies that can alleviate air pollution, as well as integrate the strengths and weaknesses of previous scientific research, this paper assesses the relationship between urban form and $PM_{2.5}$. For this, multisource data is used to establish an index system of urban form. Estimates of $PM_{2.5}$ concentrations are based on AOD data. Based on 340 prefecture-level cities in China, the linear regression model is applied to study the correlation between urban form and $PM_{2.5}$. Next, a geographically weighted regression (GWR) model is used to analyze the geographical differentiation of the impact urban form exerts on pollutant emissions. $PM_{2.5}$ concentration data and other natural factors are derived from satellite-derived data. Then, a comprehensive evaluation system of urban form indicators was established by using land use data and road network data. Next, the results of the linear regression method and GWR model between urban form and $PM_{2.5}$ concentration are analyzed. The paper ends with a discussion of the research results and presents relevant policy implications.

2. Data and Methods

2.1. Data Sources

Data were collected on the scale of 340 prefecture-level cities to explore the effect of urban form on $PM_{2.5}$ concentration of this study. Table 1 provides the data framework and variables abbreviations for the study. By reference to other research findings as well as our own experimental results [21,36], this research assumes that the urban form will affect the $PM_{2.5}$ concentration through road density, AI, LPI, PLAND, LSI, population density, per capita GDP, and the proportion of secondary industry. Compared with that of other studies, these indicators are combined from the perspective of urban form and economic development. Then, these eight aspects are combined to quantify urban form indicators. In China, different regions have large climatic differences. Both temperature and precipitation significantly impact atmosphere pollution [37], and therefore, these are used as explanatory variables.

Table 1. Abbreviation summary.

Category	Variable Name	Abbreviation
Independent variable	Fine particles matter	$PM_{2.5}$
Explanatory variable	Aggregation index	AI
	Largest patch index	LPI
	Percentage of landscape	PLAND
	Landscape shape index	LSI
	Road density	RD
	GDP per capita	PCGDP
	Population density	PD
	Proportion of the secondary industrial output-value	SIP
	Temperature	TEM
	Precipitation	PRE
Methods	Geographically weighted regression	GWR
	Ordinary least squares	OLS
	Spatial error model	SEM
	Spatial lag model	SLM

2.1.1. $PM_{2.5}$ Concentration Data Estimation

Many studies used AOD to estimate the concentration of $PM_{2.5}$ pollution, and many experiments proved a significant correlation between $PM_{2.5}$ levels and satellite-obtained AOD. Therefore, satellite remote sensing AOD is an effective tool for $PM_{2.5}$ pollution monitoring. In this study, MERRA-2 was mainly used to estimate $PM_{2.5}$ levels. The dataset was compiled by NASA's Goddard Earth Science Data and Information Ser-vice Center (GESDISC, https://daac.gsfc.nasa.gov/ (accessed on 10 August 2021)), used an upgraded version of the Goddard Earth Observing System Model, Version 5 (GEOS-5) data assimilation system. The spatial resolution of MERRA2 data is 0.5° × 0.625°, and the temporal resolution is daily. Combined with the GEOS-5 model, the annual and monthly averages of near-surface $PM_{2.5}$ concentrations were obtained from the AOD observations in the MERRA-2 data set. For data verification, 200 prefecture-level cities were selected, and their urban pollution point data from the China National Environmental Monitoring Centre were compared to the satellite data. As shown in Figure 1, data verification found a high correlation coefficient (R^2 = 0.688) between the $PM_{2.5}$ obtained by satellite and field observations. Thus, satellite data can be used to analyze spatial distribution characteristics. Because of the lack of collection and estimation of nitrate aerosols in the adopted data sources and models, the overall $PM_{2.5}$ value is low (accounting for only 20% of the $PM_{2.5}$ level on severely polluted days), but this was shown to impose little influence on the overall spatial distribution pattern [38].

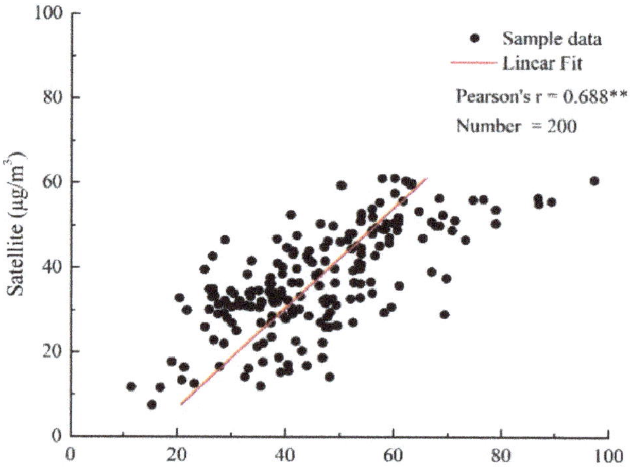

Figure 1. Scatter plot of observation data vs. satellite-retrieved $PM_{2.5}$ data. Note: ** indicate significance levels at 5% levels.

2.1.2. Urban Form Data

Urban form can be defined as the physical characteristics of urban built-up areas, such as their size, shape, and density. Related research on the relationship between urban form and atmospheric quality showed that atmospheric quality is closely related to the fragmentation, size, shape, accessibility, and continuity of the urban form [23]. The landscape patch index is widely used to describe the characteristics of urban land use, as it can scientifically represent the urban form. In the landscape patch of construction land, total built-up area (TA), patch density (PD), mean patch area (MPA), PLAND, LPI, area weighted mean fractal dimension (AWMFD), edge density (ED), LSI, and AI were used to represent the fragmentation, size, shape, accessibility, and continuity of the urban form [39]. Those indices were calculated based on the land use and land cover dataset (1 × 1 km), obtained by remote sensing classification from Landsat 8 data of 2015.

However, multiple collinearity among these indexes is high. To avoid downstream problems, four indicators were selected based on a variance inflation factor of less than 7.5 [40]. AI indicates the degree of land concentration, and values range within 0–100 (the higher the value, the better the urban land use connectivity). LPI represents the proportion of the maximum patch area to the total land area, with a value range of 0–100 (the higher the value, the lower the city continuity and fragmentation). PLAND is used to measure the size of the area occupied by urban construction land in the whole urban landscape, with a value range of 0–100. LSI indicates the complexity of the shape of a city (the larger the value, the more fragmented urban construction land). These above-mentioned four indicators (i.e., AI, LPI, PLAND, and LSI) represent the expansion of urban construction area, the compactness and fragmentation of construction form, and the complexity of the internal landscape.

In addition, road density is used as a representation of the scale of the urban road network, which is a good measure of urban traffic accessibility. A large amount of traffic increases the concentration of atmosphere pollution, especially in areas next to roads; car exhausts discharge into the atmosphere, and the movement of cars transports dust from the ground into the air, which causes atmospheric pollution [41]. The road network of China was downloaded from OpenStreetMap (https://www.openstreetmap.org/ (accessed on 10 August 2021)) and computed through ArcGIS 10.x platform. Table 2 provides the calculation method and simple description of 5 urban form indexes.

Table 2. Main urban form indexes in this study.

Index	Formula	Description
Aggregation index	$\mathrm{AI} = \left[\frac{g_{ii}}{max g_{ii}}\right] \times 100$	Measure of the natural connectivity of urban construction land
Largest patch index	$\mathrm{LPI} = \frac{max_j^i(a_{ij})}{TA}(100)$	Measure of the superiority of urban construction land landscape
Percentage of landscape	$\mathrm{PLAND} = p_t = \frac{\sum_{j=1}^{n} a_{mn}}{P} \times 100$	Measure of the proportion of urban construction land in the entire urban landscape
Landscape shape index	$\mathrm{LSI} = \frac{0.25 \sum_{k=1}^{m} e_{ik}^*}{\sqrt{TA}}$	Measure of the fragmentation and complexity of urban construction land
Road density	$RD = \frac{(length)km}{(area)km^2}$	Measure the ratio of road length to area in city

2.1.3. Control Variables

The level of air pollution emissions was influenced by many variables indirectly related to urban form. Therefore, it is necessary to employ a more accurate statistical assessment of the association between urban form and air pollution using control variables.

As socioeconomic data, this paper mainly selects GDP per capita, population density, and the secondary industry proportion (SIP). GDP per capita refers to the economic development of a city. Economic development is the ultimate goal of urban development. According to the EKC, China's economy is ahead of the EKC peak; thus, economic development causes more energy consumption and construction activities, which may be the main source of urban $PM_{2.5}$ pollution [42]. Population density is defined as the number of people per unit of land area, and a significant correlation between population density and atmospheric pollution was found [22]. This paper mainly adopts the population density of urban built-up areas, which is an important indicator of the current status of urban population distribution. SIP refers to the proportion of the output value of the secondary industry within the total industrial output value, which is an important source of urban atmospheric pollutants. The secondary industry encompasses many energy-intensive industries, mainly fossil-fuel power plants, steel mills, cement plants, and chemical plants [2]. The Overall Energy Balance Sheet for National Bureau of Statistics showed that nearly 70% of China's energy consumption is concentrated in the secondary industry. Therefore, compared with that of other industries, the secondary industry emits more atmospheric pollutants [43].

The data of these three indicators all originate from the "China City Statistical Yearbook" of 2015, where missing and partially imputed data were replaced with relevant data from adjacent years (which was the case in 1.5% of the sample).

As natural factors, temperature (TEM) and precipitation (PRE) were selected to measure the city's climatic characteristics. Meteorological factors play an important role in the concentrations of $PM_{2.5}$ in China (more precipitation and the lower the temperature, the lowed the concentration of pollutants) [37]. Latitude and longitude grid data of China were extracted from the acquired MERRA-2 data set, and interpolation processing and regional statistics at 340 prefecture-level cities were performed in the ArcGIS 10.2 software.

2.2. Statistical Models

China has obvious characteristics of regional differentiation, and atmosphere pollution also presents typical regional characteristics. Air pollution between regions has a strong spatial correlation; thus, the air pollution concentration of a city will affect the air quality of nearby cities.

As typical global linear regression model, the ordinary least squares (OLS) model is a common method to quantify the statistical relationship between independent and dependent variables. OLS can be used to study the correlation between urban form and $PM_{2.5}$. However, the OLS model ignores the influence of spatial heterogeneity, which may lead to evaluation bias [39]. Because of the existing spatial correlation among influencing factors, several spatial regression models were selected to solve the problem by controlling these potential spatial correlations. This paper used the spatial lag model (SLM) and the spatial error model (SEM). SLM explains the influence of variables of the surrounding area by adding lag variables to the model, while SEM considers the spatial dependence of dependent variables (that may otherwise be missed) by adding error terms to the model. The OLS model can be described as:

$$S = \beta_n + \sum_{m=1}^{\rho} \beta_m \alpha_m + \varepsilon \tag{1}$$

where S is the dependent variable, β_n is the intercept, β_m is the regression coefficient corresponding to the explanatory variable m, and ε is the random error value. This model can represent the intensity of the relationship between $PM_{2.5}$ and urban form indicators.

The SLM model can be expressed as:

$$y = \beta_0 + \mu \sum_{i=1}^{p} W_i y + \sum_{i=1}^{p} \beta_i x_i + \varepsilon \tag{2}$$

where μ is the regression coefficient of the spatial lag term, representing the influencing degree of adjacent spatial units on the spatial unit. This value has certain directivity, and the larger the spatial influence degree, the greater the spatial influence degree. W is the spatial weight matrix of $n \times n$, and W_i is the spatial lag dependent variable of the spatial weight matrix W. The parameter β mainly reflects the influence of the independent variable on the dependent variable and the effect of spatial distance on each spatial unit. In this model, inverse distance was used as the weight of the spatial lag term.

The SEM model can be represented as follows:

$$y = \beta_0 + \sum_{i=1}^{p} \beta_i x_i + \omega \sum_{i=1}^{p} W'_i \varepsilon \tag{3}$$

where y is the dependent variable. W is the spatial weight matrix, where the inverse distance was used to calculate the spatial error matrix. β_0 is a normal distributed random error vector. Parameter β_i is the influence coefficient of independent variable x on dependent variable y, and W'_i is the spatial error coefficient of the dependent variable vector, which represents the spatial autocorrelation of the spatial error.

All results of the three regression models can be used to explain the relationship between dependent and independent variables, and the statistical results can be compared by the measurement coefficient (R^2) and Akaike information criterion (AIC) in the model. Both values can be measured relative to a model that is more suitable for this paper. The higher the R^2 value, the smaller the AIC value, indicating that the model is more suitable. All calculation procedures are conducted in Geoda 2017 software.

2.3. Geographically Weighted Regression

In this study, atmospheric pollution presents typical regional characteristics. In other words, the air quality between neighboring cities is geographically closely related. Regression analysis assumes that the regression parameters have no relationship with the geographic location of the sample data, and the spatial characteristics are not considered. The research results do not reflect geographic location characteristics well. In addition, as a spatial autocorrelation index, the results of the bivariate Moran index of $PM_{2.5}$ and 10 indicators are statistically significant (e.g., LPI, AI, and PLAND) and have obvious spatial autocorrelation.

To identify the influence of spatial location, GWR is used to assess the influence of urban morphology of different regions on $PM_{2.5}$ concentration. GWR is an extension of the OLS linear regression model. It uses local regression, embeds spatial position information of the data into the regression parameters, establishes the local weight of the spatial position matrix, and estimates the regression parameters point by point through the local weighted least squares method, to quantify spatial heterogeneity. The model construction is expressed as follows:

$$y_i = \beta_o(u_i, v_i) + \sum_{z-1}^{n} \beta_z(u_i, v_i)x_{iz} + \varepsilon_i \tag{4}$$

where the dependent variable y_i represents the $PM_{2.5}$ concentration of city i, $\beta_o(u_i, v_i)$ represents the constant term of city i, x_{iz} represents the explanatory variable, $\beta_z(u_i, v_i)$ represents the regression parameter of the independent variable at the data sampling point, and ε_i Represents the accumulation error term.

The parameter $\beta_f(u_i, v_i)$ can be estimated by the following formula:

$$\beta_f(u_i, v_i) = \left(X^T W(u_i, v_i) X\right)^{-1} X^T W(u_i, v_i) y \tag{5}$$

where $\beta_f(u_i, v_i)$ is the parameter estimation value of (u_i, v_i), $W(u_i, v_i)$ is an $n \times n$ spatial weight matrix, the nondiagonal original element value of which is 0, and the diagonal element data is the spatial weight of the observation data of city i. The choice of the spatial weight function is the core of GWR model estimation, and directly determines the correctness of the model parameter estimation. To avoid estimation error caused by less sample data around individual sampling points, this model uses the Gaussian kernel function as the spatial weight function:

$$W_{ij} = \begin{cases} \left[1 - (d_{si}/d_{max})^2\right]^2 & d_{si} \le d_{max} \\ 0 & otherwise \end{cases} \tag{6}$$

where d_{si} represents the distance between sampling points s and i, and d_{max} represents the maximum distance between neighboring cities and the city to be assessed.

For the GWR model, bandwidth is important for determining the spatial weight calculation scheme. The smoothness of the model is controlled by bandwidth. Different spatial weighting functions are used to obtain different bandwidths. Fotheringham proposed how to obtain the optimal bandwidth [44]. The standard is the best bandwidth when the AIC of the GWR model is smallest. Therefore, AIC is used to determine the bandwidth.

The formula of the AIC is shown in the following:

$$AIC = -2\text{InL}\theta L\text{,x} + 2q \quad (7)$$

where $L(\hat{\theta}_L, x)$ is the likelihood function of the model, $\hat{\theta}_L$ is the maximum likelihood estimation of θ, x is a random sample, and q is the number of unknown parameters. The GWR tool in ArcGIS 10.2 was used to build the model.

3. Results

3.1. Spatial Distribution Characteristics of $PM_{2.5}$

Figure 2 illustrates the geographic distribution of the average $PM_{2.5}$ concentrations of China's cities in 2015, clearly indicating that the spatial distribution of $PM_{2.5}$ is heterogeneous. Overall, cities in eastern China tend to have higher $PM_{2.5}$ levels than cities in western China, and cities in northern China have higher $PM_{2.5}$ levels than cities in southern China. Specifically, areas with highest $PM_{2.5}$ levels are concentrated in the North China Plain and Sichuan Basin, as well as in parts of the Northwest China. Among these, the high $PM_{2.5}$ level-area of the northwest region is mainly caused by the Taklamakan desert (the world's second largest desert), where perennial wind and sand influx causes extremely rich suspended particulate matter; therefore, the concentration of $PM_{2.5}$ in the desert area is high. The level of economic development in the North China Plains is high. The development of pollution-intensive industries in North China promotes regional economic development, and therefore, man-made atmospheric pollutant emissions are very large. In the southwest region, area with high $PM_{2.5}$ pollution is mainly concentrated in the Sichuan Basin, a region that is particularly affected by humidity and precipitation, which causes rich suspended particles in the atmosphere. Moreover, the special structure of the terrain is not conducive to the spread of pollutants, and the population of the region causes high levels of anthropogenic pollution emissions. Because of its elevation, the Qinghai–Tibet Plateau has a thin atmosphere, these conditions are not conducive for the formation and accumulation of particulate matter in the atmosphere. Therefore, the lowest $PM_{2.5}$ levels were found in the Qinghai–Tibet Plateau. The global Moran's I index for $PM_{2.5}$ levels were 0.765 ($p < 0.01$), indicating a relatively strong positive spatial correlation. Local indicators on $PM_{2.5}$ spatial association (LISA) maps show similar typical distributions, with a high $PM_{2.5}$ cluster in the North China Plains and a low $PM_{2.5}$ cluster in the Northeast China and the Qinghai-Tibet Plateau regions (Figure 3). Seasonally, winter had the highest $PM_{2.5}$ level, followed by spring, autumn, and summer. However, North China has always been a region with severe $PM_{2.5}$ pollution, especially in winter, where the climate is not conducive to the diffusion of atmospheric pollutants [45]. On a seasonal scale, winter had the highest $PM_{2.5}$ level, followed by spring, autumn, and summer. However, North China was always a region with severe $PM_{2.5}$ pollution, especially in winter, where the climate is not conducive to the diffusion of atmospheric pollutants. It is worth noting that Southern China always had low $PM_{2.5}$ pollution because of its advantageous climate (Figure 4).

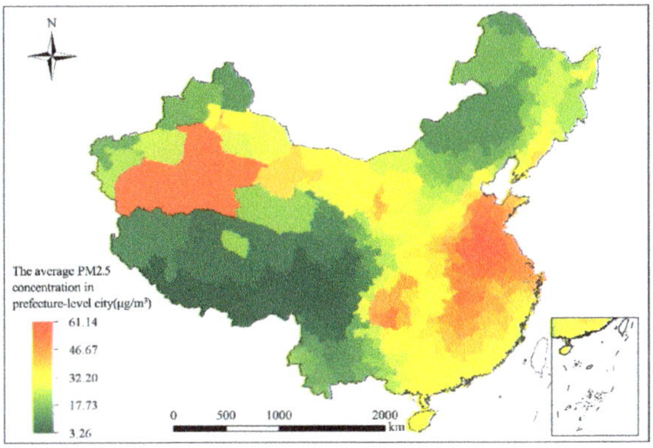

Figure 2. Spatial distribution of $PM_{2.5}$ levels at prefecture-city level in China.

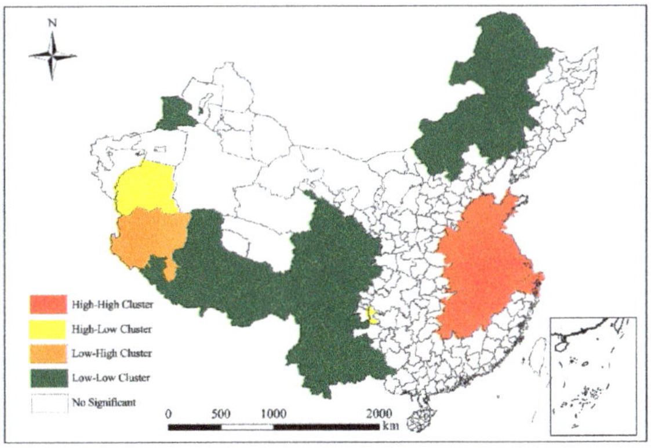

Figure 3. Local indicators on $PM_{2.5}$ spatial association (or LISA) maps of prefecture-level cities in China.

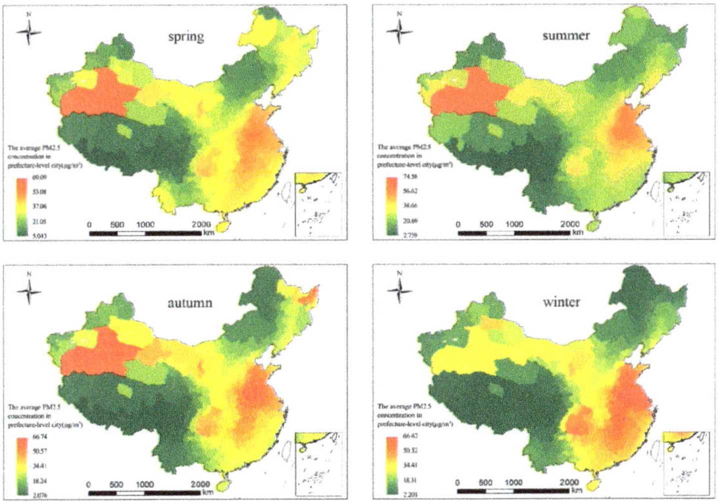

Figure 4. Spatio-temporal distribution of $PM_{2.5}$ levels of prefecture-level cities in spring, summer, autumn, and winter.

3.2. Correlations between Urban Form and $PM_{2.5}$

Table 3 shows the results of global regression model analysis (i.e., OLS, SLM, and SEM). The results of suitability statistics such as R^2, AIC, and log-likelihood imply that the spatial analysis technique is more suitable for this data. R^2 values of OLS, SLM, and SEM models are 0.601, 0.943, and 0.874, respectively. These results show that the spatial effect is important in regression analysis and thus, ignoring the spatial effect will reduce the effectiveness of the model. In addition, R^2, AIC, and log-likelihood test results show that the spatial lag effect is more significant.

Table 3. Global regression results.

Variable	Classic OLS Model		Spatial Lag Model		Spatial Error Model	
	Coefficient	t-Statistic	Coefficient	z-Value	Coefficient	z-Value
LPI	−1.209 ***	−4.652	−0.075 **	−0.620	−0.327 **	−2.241
AI	0.076	1.202	0.068 ***	2.826	0.0424	1.214
PLAND	0.861 ***	5.298	−0.062 **	−0.753	0.226 **	2.454
LSI	0.226 ***	2.664	0.018 **	0.520	0.034 **	0.722
SIP	0.095 *	1.875	−0.006	−0.263	−0.019	−0.661
PRGDP	2.99×10^7	0.013	-1.64×10^6	−0.174	1.55×10^5	1.176
PD	0.0004	0.430	−0.0002	−0.492	0.0003	0.694
TEM	1.737 ***	12.340	2.878 ***	17.104	0.646 ***	6.369
PER	−2.341 ***	−4.759	−2.038 **	−2.397	−1.013 ***	−3.542
RD	−0.191	−0.129	−0.214	−0.320	−1.005	−1.223

Note: ***, **, or * indicate significance levels at the 1%, 5%, and 10% levels, respectively. OLS model: R^2: 0.601, Log likelihood: −1224.72, *p*-value: 0.000; AIC: 2471.45. SLM: R-squared: 0.943, Log likelihood: −954.714, *p*-value: 0.000, AIC: 1931.43. SEM: R-squared: 0.874, Log likelihood: −1057.64; *p*-value: 0.000, AIC: 2139.28.

The results of OLS indicate that most urban form indicators are significantly correlated with $PM_{2.5}$ concentrations, and six urban form indicators showed significantly ($p < 0.01$) positive relationships with city-level annual mean $PM_{2.5}$ levels. Among these six significant factors, LPI, PLAND, and LSI also have a significant impact. LPI has a significant negative correlation with $PM_{2.5}$ levels, indicating that a better continuity of the urban form leads to a lower fragmentation, and a better atmospheric quality. PLAND and LSI were significantly positively correlated with $PM_{2.5}$ levels, indicating that the more complex the urban form, the worse the atmospheric quality. AI indicators on $PM_{2.5}$ concentration is not significant. The four landscape pattern indicators indicate that the fragmentation and complexity of the urban form exert a significant impact on $PM_{2.5}$ levels, and thus, more attention should be focused on urban form area expansion and the internal composition of fragmentation and complexity. RD is negatively correlated with $PM_{2.5}$ levels, and the higher the density of the road network, the lower the atmospheric pollution levels. In addition, there is a significant negative correlation between temperature and precipitation and $PM_{2.5}$ pollution, which confirms that meteorological conditions are conducive to the spread and reduction of atmospheric pollutants. Among other control indicators, SIP has a significantly positive impact on $PM_{2.5}$ pollution, and PD and PRGDP are positively correlated with $PM_{2.5}$ levels. Therefore, the development of the secondary industry has a more significant impact on atmospheric pollution.

The SLM model has the best regression results, indicating that urban form and atmospheric pollution have clear spatial dependence. Among the major urban form indicators, LPI was significantly negatively correlated with $PM_{2.5}$ levels, AI, PLAND, and LSI were significantly positively correlated with $PM_{2.5}$ pollution, and RD was negatively correlated with $PM_{2.5}$ levels. In addition, the correlation between temperature and precipitation on $PM_{2.5}$ was significant (the more precipitation, the lower the temperature), which is beneficial for reducing $PM_{2.5}$ concentrations in the atmosphere. This is consistent with literature. Therefore, clear spatial correlation exists between urban form indicators and $PM_{2.5}$ levels.

the analysis process of the global regression model analysis has limitations. The GWR model can be used for further analysis by adding the relationship of spatial location.

3.3. Spatial Features of Urban Form on $PM_{2.5}$

The coefficient of determination of the GWR model was 0.77, and the results of AIC and variance analysis (F test) showed that the results of the model are statistically significant. All 10 variables show noncollinearity and are used under the AIC minimization standard. The GWR model is superior to the OLS model. Figure 5 shows a distribution map of the regression fitting coefficient R^2 in the regression results for prefecture-level cities. The spatial distribution of R^2 shows that the fitting results of the 10 variables of the urban form range between 0.43 and 0.83, indicating that the 10 indicators selected in this paper exert a stronger comprehensive impact on urban $PM_{2.5}$ levels. Moreover, the R^2 value in the spatial distribution decreases from north to south. Therefore, the urban form system has a stronger explanatory power for the urban form of the northern region. On the one hand, this paper uses temperature and precipitation as control variables of two natural factors. They exert a significant positive effect on reducing atmospheric pollutants. The role of climatic factors is more significant in the south, thus reducing the concentration of urban $PM_{2.5}$ pollution. On the other hand, this may be due to a lack of estimation of nitrate levels in the $PM_{2.5}$ data used in this paper, and therefore, the impact of using a large amount of coal for heating in winter in northern regions may be underestimated, resulting in a low degree of fitting for northern regions. The further south a city is located, the more the explanatory power for $PM_{2.5}$ of the urban form decreases.

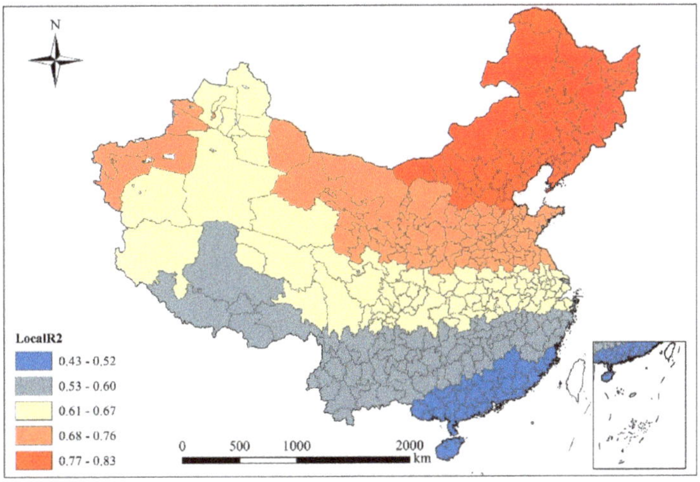

Figure 5. Local R^2 distribution characteristics according to geographically weighted regression (GWR) model.

In the GWR model, each urban form index has a specific regression coefficient for the influence degree of $PM_{2.5}$ levels, and each coefficient has a different spatial distribution law (Figure 6). The spatial distribution can more intuitively depict the influencing effect and changing trend between different urban form indicators and cities. In addition, different indicators exert different positive and negative impacts on $PM_{2.5}$ levels, and their proportions differ. This also indicates that the influence index is not spatially stable and shows spatial heterogeneity (Figure 5). The directions of significant relationships were not the same for the studied cities, even for the same factor.

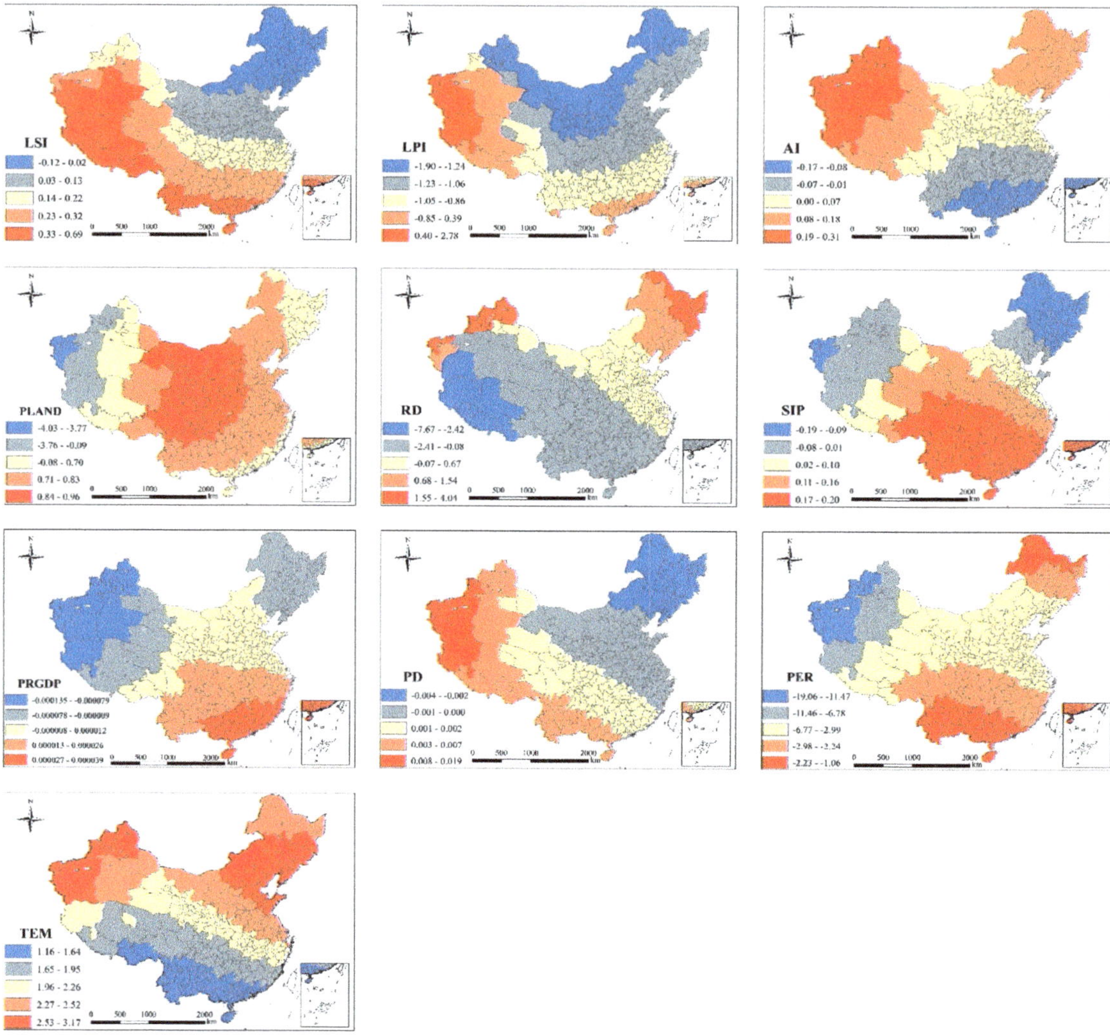

Figure 6. Spatial distribution of local relationship between $PM_{2.5}$ and 10 factors for prefecture-level cities in China.

The regression coefficient of urban form index decreases in the order of RD, PLAND, LPI, LSI, and AI. The correlation between road density and $PM_{2.5}$ levels is highest. Construction dust from the construction phase of roads is the main cause of $PM_{2.5}$ pollution, followed by pollution caused by motor vehicle driving, as well as more harmful gases that are discharged during traffic congestions. The regression coefficient ranges from 7.7 to 4.0 and decreases from northeast to southwest. In China's major urban areas, the density of road networks is negatively correlated with $PM_{2.5}$ concentrations. Improvement of the road network system can effectively reduce traffic congestions and atmospheric pollution. The correlation coefficient between PLAND and $PM_{2.5}$ concentrations is high and negative. This coefficient mainly measures the size of the area occupied by urban construction land in the whole urban landscape. The influencing factor follows a decreasing trend from center to surrounding areas. Cities where construction land is the main land use type are more likely to cause atmospheric pollution. Developed areas in the south are less affected but may be affected by increasing levels of urban construction, where the application of scientific dust reduction measures and the use of green materials are conducive for reducing emissions of atmospheric pollutants such as building dust. Cities in part of the central and western regions are more susceptible to the impact of the area of construction land. Therefore, reasonable increases of urban construction area and improvement of the level of building construction technology (e.g., green building materials and dust reduction) can reduce $PM_{2.5}$ concentrations to a certain extent. LPI exerts a significant influence on urban $PM_{2.5}$ concentrations, where a continuous increase of LPI indicates that the landscape dominance of urban construction land increases, the degree of spatial connection increases, and the intensity of human activities also increases. The values range from -1.9 to 2.8 and increase from north to south with the continuous development of urban construction, which causes more atmospheric pollution. In southern China, driven by the reform and opening up policy, the urbanization level grows faster, and human activities are more intense. This indicates that enhancing the connectivity and dominance of urban construction land has a significant impact on reducing $PM_{2.5}$ concentrations. A higher LSI index indicates stronger fragmentation and more complicated urban areas. Most regression coefficients between LSI and $PM_{2.5}$ concentrations are positive, with values ranging from -0.12 to 0.69, indicating that a higher LSI value represents higher levels of $PM_{2.5}$ pollution. The spatial increase from north to south may be due to the complexity and fragmentation of the shape of the landscape of urban construction land, which leads to an increase of people's daily commuting time and distance, thus also increasing the pollution caused by the heavy use of commuting means. The impact degree of the southern region is large, indicating that the complexity of urban landscape in the southern region is more likely to affect the $PM_{2.5}$ levels. AI is used to measure the agglomeration and compactness of urban construction land. The regression coefficient ranges from -0.17 to 0.31, and it changes from negative to positive from north to south. The higher the compactness of urban construction land, the lower the $PM_{2.5}$ concentrations. Compact urban construction can shorten people's travel distance, improve the efficiency of land use, and reduce energy consumption. Therefore, in the process of urban development, compact and continuous urban construction is conducive to the improvement of urban atmospheric quality.

Among control indicators, the influences of the three socioeconomic factors on urban $PM_{2.5}$ pollution show clear spatial differences. In most urban areas, SIP has a significant positive effect on $PM_{2.5}$ concentrations, indicating that industrial activities aggravate the $PM_{2.5}$ concentrations in Chinese cities, which is consistent with the literature. The effect of SIP on $PM_{2.5}$ concentrations in southern China is strong, indicating that reducing the proportion of output of the secondary industry in southern China can significantly improve atmospheric quality. At the same time, the population density in the southeast coastal areas exerts a greater impact on the $PM_{2.5}$ concentrations. Economic development prompted the migration of many people to the south to work or live. The increased population density has caused more man-made atmospheric pollutant emissions, which strongly impact atmospheric pollution. The overall coefficient of PCGDP is low and its influence is weak. However, in the process of urban development, the improvement of the economic level exerts a significant positive effect on reducing $PM_{2.5}$ pollution levels.

Reasonable increases in the area of urban construction land and improvements of the level of construction, reducing the fragmentation of urban construction land, compacting urban construction, improving traffic accessibility, applying a reasonable road network density are all very beneficial factors for the improvement of urban atmospheric quality. These can, to a certain extent, reduce the concentration of $PM_{2.5}$. However, as urbanization continues to increase, the different geographical location, scale and level of construction, and development of the city should be considered according to local conditions, thus providing planning and construction guidance. Urban form can affect $PM_{2.5}$ concentrations from different aspects. Urban area, geographical location, and economic development level have an impact, and thus also need further discussion and analysis in the future.

4. Discussion

On an annual scale, urban planning factors (e.g., the area of urban construction land, construction fragmentation, and road network density) all exert a specific influence on $PM_{2.5}$ levels. At the same time, the economic development of a city is also closely related to $PM_{2.5}$ levels. China's rapid urbanization led to structural changes of its economy, and large areas of land are being used by energy-intensive and labor-intensive industries. In addition, many people move from rural areas to cities, where the population grows rapidly, which increases the release of large amounts of man-made atmospheric pollutants emission. The study showed that temperature and precipitation (as control variables) were significantly correlated with $PM_{2.5}$ concentrations in 340 cities in China, and climatic factors played a significant role for reducing atmosphere pollution. Therefore, we further discussed this paper discusses the relationship between urban form and $PM_{2.5}$ in different seasons. Seasonality can affect atmospheric quality through changes in precipitation, wind, relative humidity, monsoons, and other diffusion conditions [16,30]. When seasonal variations are considered, different seasons exert different impacts on $PM_{2.5}$ through different urban form indicators. The SLM model achieved the best regression results, and the relationship between urban form and $PM_{2.5}$ in different seasons is discussed through this model (Table 4).

Table 4. Spatial lag model results for different seasons in China during 2015.

Variable	Spring	Summer	Autumn	Winter
LPI	−0.289 ** (−2.044)	−0.489 *** (−3.751)	−0.231(−1.522)	−0.258 * (−1.775)
AI	0.054(1.581)	0.0524 * (1.681)	0.044(1.219)	0.027(−0.773)
PLAND	0.243 *** (2.793)	0.320 *** (3.943)	0.133(1.382)	0.240 *** (−2.652)
LSI	0.055 (1.210)	−0.050(−1.144)	0.049(1.015)	0.026 (−0.555)
RD	−1.573 * (−1.991)	−0.516 (−0.703)	−0.674 (−0.800)	−1.728 ** (−2.139)
SIP	−0.011 (−0.404)	−0.015(−0.597)	−0.026(−0.892)	−0.015(−0.523)
PCGDP	1.04×10^5(0.813)	-5.63×10^6 (−0.480)	6.06×10^6(0.449)	3.20×10^5 ** (−2.410)
PD	4.14×10^5(0.089)	0.0002(0.409)	0.0003 (0.646)	0.0002 (−0.481)
PRE	−0.217 (−1.315)	−0.247 ** (−2.123)	−1.508 *** (−4.530)	−0.392 (−1.062)
TEM	0.543 *** (7.021)	0.484 *** (6.327)	0.621 *** (6.415)	0.282 *** (−4.705)
R^2	0.843	0.892	0.879	0.918
Log likelihood	−1043.98	−1019.82	−1071.09	−1067.61
AIC	2111.96	2063.64	2166.19	2159.22
p-value	0.000	0.000	0.000	0.000

Note: ***, **, or * indicate significance levels at the 1%, 5%, and 10% levels, respectively.

Table 4 shows the results of the regression model, in which the R^2, log-likelihood, AIC, and other statistical results are significantly higher, indicating that the SLM regression model in this study has a relatively good fit. This model can accurately assess the impact of seasonal changes in the urban form on $PM_{2.5}$ concentrations. In addition, the seasonal analysis according to the GWR model shows that the regression coefficients of the impact of each city's form index on the level of $PM_{2.5}$ pollution have different spatial distribution patterns (Figure 7). As shown in Table 4, seasonal variation significantly affects the relationship between urban form and $PM_{2.5}$ concentration.

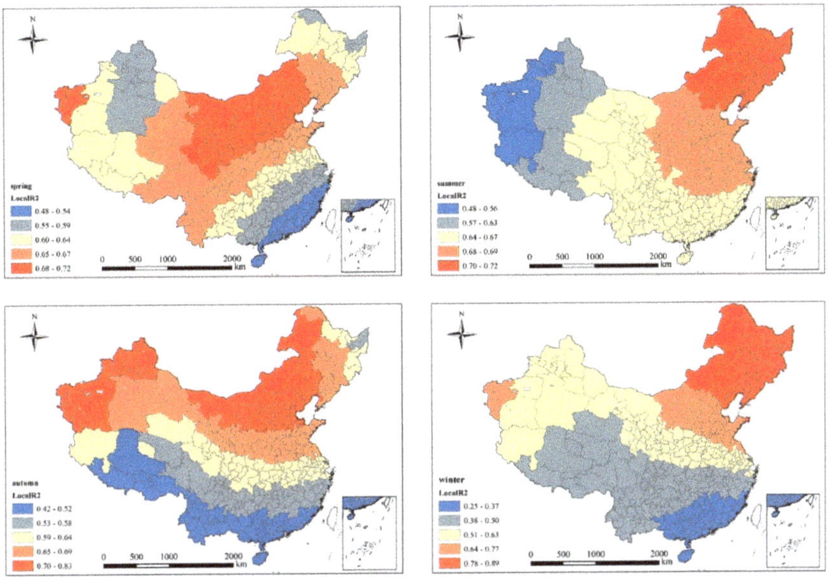

Figure 7. Local R^2 derived from multivariate GWR model in spring, summer, autumn, and winter.

The four main findings of this study are summarized in the following: firstly, a significant correlation exists between urban form and $PM_{2.5}$ concentrations in all four seasons, with the highest R^2 value in winter. Secondly, the temperature and precipitation in the control variables always exerted a significant impact on $PM_{2.5}$ concentrations, while other socioeconomic indicators had no significant impact on $PM_{2.5}$ concentrations. Thirdly, the effect of PLAND and $PM_{2.5}$ concentrations is significantly positively correlated, while that of LPI and $PM_{2.5}$ concentrations is significantly negatively correlated in spring, summer, and winter. Fourthly, a negative correlation was found between the density of urban road network and $PM_{2.5}$ concentrations. Specifically, in spring, LPI, PLAND, and RD significantly impact $PM_{2.5}$ concentrations, with regression coefficients of −0.289, 0.243, and −1.537, respectively. In the summer, urban compactness (AI = 0.052) can decrease the $PM_{2.5}$ concentration to some extent. LPI (−0.489) and PLAND (0.320) were all significantly associated with $PM_{2.5}$ concentrations. In the autumn, the correlation between urban form and $PM_{2.5}$ concentrations was not significant. In the winter, PLAND (0.240) was significantly positively correlated with the $PM_{2.5}$ concentrations. LPI (−0.258) and RD (−1.728) were significantly negatively correlated with $PM_{2.5}$ concentrations.

Therefore, data analysis showed that seasonal change exerts a certain influence on the relationship between urban form indicators and $PM_{2.5}$ concentrations. Especially in spring and winter, increasing the connectivity of urban construction land and improving the efficiency of land use within a city can effectively decrease urban $PM_{2.5}$ concentrations [15]. Moreover, increasing the road connectivity also exerts a significant effect on reducing atmosphere pollution. However, the season of autumn does not exert a significant effect on the relationship between urban form and $PM_{2.5}$ concentrations. In contrast, the indicators between urban form and $PM_{2.5}$ are more significantly correlated in spring and winter, and relatively less in summer and autumn. Firstly, the previous analysis on an annual scale shows that the compact and continuous urban construction land, the reduction of urban land fragmentation, and the reasonable road network density are all conducive to reducing urban atmospheric pollutant emissions. Secondly, seasonal changes are mainly reflected by different climatic variables. The strong Asian monsoon during summer and the subtropical high during autumn result in favorable weather conditions that clean the air from atmospheric pollutants, as they enhance the mobility of the atmosphere above urban centers. The summer monsoon climate moves more precipitation to North China, and at the same time increases the wind speed under the subtropical high pressure. These meteorological conditions are conducive to reducing atmosphere pollutants [16]. Under such climatic conditions, small changes in the structure of building areas or land areas near the ground exert little impact on atmospheric pollutants. Finally, low temperatures in winter and spring cause the atmospheric flow to sink and wind speeds are also lower than in summer [15]. In addition, more coal is burnt in the north in winter, and the resulting atmospheric pollutants cannot easily spread and remain concentrated near the ground. Under these conditions, the irregular urban form near the ground has a more significant impact on $PM_{2.5}$ concentrations. Therefore, seasonal changes exert a significant impact on $PM_{2.5}$ concentrations. When exploring the relationship between urban form and $PM_{2.5}$ concentrations, focusing on the results of spring and winter is more effective.

There are many factors that affect the $PM_{2.5}$ concentration, but our research just focused on the urban form. In addition, industrial areas, greenness area, vegetation coverage, transportation and other factors also have a significant impact on $PM_{2.5}$ concentration. Therefore, there are some limitations of this study. Firstly, more influencing factors should be considered, such as the emission forces, change of pollution effects, LHI [20], meteorological conditions, the development levels of different cities and environmental conditions that surround the region of interest. These factors have an important effect on $PM_{2.5}$ and are also associated with $PM_{2.5}$ through interaction. They can all improve our indicator system. By selecting typical cities for further research, further problems may be identified. Secondly, the classification accuracy of urban land use data can be improved, and the impact of different land use patterns of urban construction land on $PM_{2.5}$ pollution can be explored. For example, extract the industrial area in city for research. Thirdly, experiment with different research methods and data sources. For example, compare the remote sensing estimated $PM_{2.5}$ concentration obtained by different sensors; compare the results of different research models; compare the impact levels of more influencing factors on the $PM_{2.5}$ concentration, and so on.

5. Conclusions

Exploring the relationship between $PM_{2.5}$ pollution and urban form helps to better understand the distribution of $PM_{2.5}$ pollution, provide suggestions for urban planning, and the government with exploring a more sustainable and environmentally friendly development model for cities. Therefore, this study selected 340 prefecture-level cities in China to explore the relationship between urban form and $PM_{2.5}$ pollution via regression analysis and GWR model. The following lists the main conclusions and recommendations:

Firstly, the distribution of $PM_{2.5}$ pollution showed spatial heterogeneity, with an increasing trend from northwest to southeast. Areas with high $PM_{2.5}$ concentrations are mainly located on the North China Plain, which is greatly affected by human activities. The southwest and parts of the northwest are affected by climatic factors, and thus, their $PM_{2.5}$ concentrations are also high. The climatic and human activity conditions on the Qinghai–Tibet Plateau are not conducive to the accumulation of $PM_{2.5}$ pollution, and thus, this area was always less polluted. Affected by seasonal changes, the $PM_{2.5}$ concentration decreases in the respective order of winter, spring, autumn, and summer.

Secondly, most urban form indicators are significantly related to $PM_{2.5}$ concentration. The results of the GWR model show that the spatial distribution decreases from north to south, and the urban form indicators system exerts a stronger influence on cities in northern regions. In general, better continuity is associated with a lower the degree of fragmentation, and a higher compactness and reasonable density of the road network are conducive to reducing the $PM_{2.5}$ concentrations. In addition, meteorological conditions are also conducive to the diffusion and reduction of atmospheric pollutants. Thirdly, on a seasonal scale, seasonal changes impact pollution levels, but not all urban form indicators are significantly related to $PM_{2.5}$ concentrations. Affected by seasonal changes, more urban form indicators were significantly correlated with $PM_{2.5}$ concentrations during spring and winter compared with summer and autumn.

This research used a large data set and confirmed that a good urban form exerts a positive effect on reducing $PM_{2.5}$ concentrations. In the future development, the proportion of the secondary industry in the urban area should be reduced, green industries should be developed, atmospheric pollutant treatment technologies should be improved, and pollution sources should be controlled to reduce emissions. In areas with low pollution values, protection measures should be increased. More importantly, in urban planning, the blind expansion of urban land should be avoided, the compactness of the urban form, the efficiency of land use, and the transportation network in the city should be improved, and various forms of public transportation should be offered.

There are three main limitations of this study. Firstly, many factors influence $PM_{2.5}$ pollution, and the geographical location, climatic conditions, and development levels of different cities are all different. By selecting typical cities for further research, further problems may be identified. Secondly, the classification accuracy of urban land use data can be improved, and the impact of different land use patterns of urban construction land on $PM_{2.5}$ pollution can be explored. Thirdly, different research methods and data can be used to improve the accuracy of pollution source data, as well as research models, and different data methods can be utilized to further explore the estimation of $PM_{2.5}$ levels and relevant influencing factors.

Author Contributions: Conceptualization, Y.L. and L.H.; methodology, Y.L.; software, Y.L.; formal analysis validation, Y.L.; formal analysis, Y.L.; writing—original draft investigation, Y.L.; investigation, L.H.; data curation, Y.L.; visualization, W.Q.; writing—review & editing, A.L. and Y.Y.; supervision, Y.Y.; funding acquisition, Y.Y. All authors have read and agreed to the published version of the manuscript.

Funding: This research was funded by the Strategic Priority Research Program of the Chinese Academy of Sciences (XDA20010201).

Data Availability Statement: MERRA-2 dataset is available at https://daac.gsfc.nasa.gov/ (accessed on 10 August 2021).

Conflicts of Interest: The authors declare no conflict of interest.

References

1. Yang, D.; Ye, C.; Wang, X.; Lu, D.; Xu, J.; Yang, H. Global distribution and evolvement of urbanization and PM2.5 (1998–2015). *Atmos Environ.* **2018**, *182*, 171–178. [CrossRef]
2. Ji, X.; Yao, Y.; Long, X. What causes PM2.5 pollution? Cross-economy empirical analysis from socioeconomic perspective. *Energ. Policy* **2018**, *119*, 458–472. [CrossRef]
3. Zhou, C.; Chen, J.; Wang, S. Examining the effects of socioeconomic development on fine particulate matter (PM2.5) in China's cities using spatial regression and the geographical detector technique. *Sci. Total Environ.* **2018**, *619–620*, 436–445. [CrossRef] [PubMed]
4. Wang, S.; Fang, C.; Guan, X.; Pang, B.; Ma, H. Urbanisation, energy consumption, and carbon dioxide emissions in China: A panel data analysis of China's provinces. *Appl. Energy* **2014**, *136*, 738–749. [CrossRef]
5. Xu, B.; Lin, B. What cause large regional differences in PM2.5 pollutions in China? Evidence from quantile regression model. *J. Clean. Prod.* **2018**, *174*, 447–461. [CrossRef]
6. Luo, K.; Li, G.; Fang, C.; Sun, S. PM2.5 mitigation in China: Socioeconomic determinants of concentrations and differential control policies. *J. Environ. Manag.* **2018**, *213*, 47–55. [CrossRef]
7. Matus, K.; Nam, K.; Selin, N.E.; Lamsal, L.N.; Reilly, J.M.; Paltsev, S. Health damages from air pollution in China. *Glob. Environ. Chang.* **2012**, *22*, 55–66. [CrossRef]
8. He, C.; Han, L.; Zhang, R.Q. More than 500 million Chinese urban residents (14% of the global urban population) are imperiled by fine particulate hazard. *Environ. Pollut.* **2016**, *218*, 558–562. [CrossRef] [PubMed]
9. Ginzburg, A.S.; Semenov, V.A.; Semutnikova, E.G.; Aleshina, M.A.; Zakharova, P.V.; Lezina, E.A. Impact of COVID-19 lockdown on air quality in Moscow. *Dokl. Earth Sci.* **2020**, *495*, 862–866. [CrossRef]
10. Bell, M.L.; Ebisu, K. Environmental Inequality in Exposures to Airborne Particulate Matter Components in the United States. *Environ. Health Persp.* **2012**, *120*, 1699–1704. [CrossRef]
11. Xie, P.; Liu, X.; Liu, Z.; Li, T.; Zhong, L.; Xiang, Y. Human Health Impact of Exposure to Airborne Particulate Matter in Pearl River Delta, China. *Water Air Soil Pollut.* **2011**, *215*, 349–363. [CrossRef]

12. Zhang, Y.; Qu, S.; Zhao, J.; Zhu, G.; Zhang, Y.; Lu, X.; Sabel, C.E.; Wang, H. Quantifying regional consumption-based health impacts attributable to ambient air pollution in China. *Environ. Int.* **2018**, *112*, 100–106. [CrossRef]
13. Hao, Y.; Liu, Y. The influential factors of urban PM2.5 concentrations in China: A spatial econometric analysis. *J. Clean. Prod.* **2016**, *112*, 1443–1453. [CrossRef]
14. O'Lenick, C.R.; Wilhelmi, O.V.; Michael, R.; Hayden, M.H.; Baniassadi, A.; Wiedinmyer, C.; Monaghan, A.J.; Crank, P.J.; Sailor, D.J. Urban heat and air pollution: A framework for integrating population vulnerability and indoor exposure in health risk analyses. *Sci. Total Environ.* **2019**, *660*, 715–723. [CrossRef]
15. Cesari, D.; De Benedetto, G.E.; Bonasoni, P.; Busetto, M.; Dinoi, A.; Merico, E.; Chirizzi, D.; Cristofanelli, P.; Donateo, A.; Grasso, F.M.; et al. Seasonal variability of PM2.5 and PM10 composition and sources in an urban background site in Southern Italy. *Sci. Total Environ.* **2018**, *612*, 202–213. [CrossRef] [PubMed]
16. Liu, Y.; Wu, J.; Yu, D.; Ma, Q. The relationship between urban form and air pollution depends on seasonality and city size. *Environ. Sci. Pollut. R.* **2018**, *25*, 15554–15567. [CrossRef] [PubMed]
17. Cheng, Z.; Li, L.; Liu, J. Identifying the spatial effects and driving factors of urban PM2.5 pollution in China. *Ecol. Indic.* **2017**, *82*, 61–75. [CrossRef]
18. He, J.; Ding, S.; Liu, D. Exploring the spatiotemporal pattern of PM2.5 distribution and its determinants in Chinese cities based on a multilevel analysis approach. *Sci. Total Environ.* **2019**, *659*, 1513–1525. [CrossRef] [PubMed]
19. Guo, L.; Luo, J.; Yuan, M.; Huang, Y.; Shen, H.; Li, T. The influence of urban planning factors on PM2.5 pollution exposure and implications: A case study in China based on remote sensing, LBS, and GIS data. *Sci. Total Environ.* **2019**, *659*, 1585–1596. [CrossRef]
20. Chi, Y.L.; Mak, H.W.L. From Comparative and Statistical Assessments of Liveability and Health Conditions of Districts in Hong Kong towards Future City Development. *Sustainability* **2021**, *13*, 8781. [CrossRef]
21. Fang, C.; Wang, S.; Li, G. Changing urban forms and carbon dioxide emissions in China: A case study of 30 provincial capital cities. *Appl. Energy* **2015**, *158*, 519–531. [CrossRef]
22. Yuan, M.; Huang, Y.; Shen, H.; Li, T. Effects of urban form on haze pollution in China: Spatial regression analysis based on PM2.5 remote sensing data. *Appl. Geogr.* **2018**, *98*, 215–223. [CrossRef]
23. Fan, C.; Tian, L.; Zhou, L.; Hou, D.; Song, Y.; Qiao, X.; Li, J. Examining the impacts of urban form on air pollutant emissions: Evidence from China. *J. Environ. Manag.* **2018**, *212*, 405–414. [CrossRef] [PubMed]
24. Liu, Y.; Arp, H.; Song, X.D.; Song, Y. Research on the relationship between urban form and urban smog in China. Environ. *Plan. B-Urban Anal. City Sci.* **2017**, *44*, 328–342.
25. Mak, H.W.L.; Ng, D.C.Y. Spatial and Socio-Classification of Traffic Pollutant Emissions and Associated Mortality Rates in High-Density Hong Kong via Improved Data Analytic Approaches. *Int. J. Environ. Res. Pub. Health* **2021**, *18*, 6532. [CrossRef] [PubMed]
26. Mansfield, T.J.; Rodriguez, D.A.; Huegy, J.; Gibson, J.M. The Effects of Urban Form on Ambient Air Pollution and Public Health Risk: A Case Study in Raleigh, North Carolina. *Risk Anal.* **2015**, *35*, 901–918. [CrossRef] [PubMed]
27. McCarty, J.; Kaza, N. Urban form and air quality in the United States. *Landsc. Urban Plan.* **2015**, *139*, 168–179. [CrossRef]
28. Sicard, P.; Agathokleous, E.; De Marco, A.; Paoletti, E.; Calatayud, V. Urban population exposure to air pollution in Europe over the last decades. *Environ. Sci. Eur.* **2021**, *33*, 1–12. [CrossRef] [PubMed]
29. Li, C.; Wang, Z.; Li, B.; Peng, Z.; Fu, Q. Investigating the relationship between air pollution variation and urban form. *Build Environ.* **2019**, *147*, 559–568. [CrossRef]
30. Shi, K.; Li, Y.; Chen, Y.; Li, L.; Huang, C. How does the urban form-PM2.5 concentration relationship change seasonally in Chinese cities? A comparative analysis between national and urban agglomeration scales. *J. Clean Prod.* **2019**, *239*, 118088. [CrossRef]
31. Liu, H.; Fang, C.; Zhang, X.; Wang, Z.; Bao, C.; Li, F. The effect of natural and anthropogenic factors on haze pollution in Chinese cities: A spatial econometrics approach. *J. Clean Prod.* **2017**, *165*, 323–333. [CrossRef]
32. Tao, Y.; Zhang, Z.; Ou, W.; Guo, J.; Pueppke, S.G. How does urban form influence PM2.5 concentrations: Insights from 350 different-sized cities in the rapidly urbanizing Yangtze River Delta region of China, 1998–2015. *Cities* **2020**, *98*, 102581. [CrossRef]
33. Li, F.; Zhou, T. Effects of urban form on air quality in China: An analysis based on the spatial autoregressive model. *Cities* **2019**, *89*, 130–140. [CrossRef]
34. Lin, C.; Labzovskii, L.D.; Leung Mak, H.W.; Fung, J.C.H.; Lau, A.K.H.; Kenea, S.T.; Bilal, M.; Vande Hey, J.D.; Lu, X.; Ma, J. Observation of PM2.5 using a combination of satellite remote sensing and low-cost sensor network in Siberian urban areas with limited reference monitoring. *Atmos Environ.* **2020**, *227*, 117410. [CrossRef]
35. Lin, C.; Li, Y.; Lau, A.K.H.; Deng, X.; Tse, T.K.T.; Fung, J.C.H.; Li, C.; Li, Z.; Lu, X.; Zhang, X.; et al. Estimation of long-term population exposure to PM2.5 for dense urban areas using 1-km MODIS data. *Remote Sens. Environ.* **2016**, *179*, 13–22. [CrossRef]
36. Shi, K.; Wang, H.; Yang, Q.; Wang, L.; Sun, X.; Li, Y. Exploring the relationships between urban forms and fine particulate (PM2.5) concentration in China: A multi-perspective study. *J. Clean Prod.* **2019**, *231*, 990–1004. [CrossRef]
37. Liu, Q.; Wang, S.; Zhang, W.; Li, J.; Dong, G. The effect of natural and anthropogenic factors on PM2.5: Empirical evidence from Chinese cities with different income levels. *Sci. Total Environ.* **2019**, *653*, 157–167. [CrossRef] [PubMed]
38. Buchard, V.; Da Silva, A.M.; Randles, C.A.; Colarco, P.; Ferrare, R.; Hair, J.; Hostetler, C.; Tackett, J.; Winker, D. Evaluation of the surface PM2.5 in Version 1 of the NASA MERRA Aerosol Reanalysis over the United States. *Atmos Environ.* **2016**, *125*, 100–111. [CrossRef]

39. Du, Y.; Wan, Q.; Liu, H.; Liu, H.; Kapsar, K.; Peng, J. How does urbanization influence PM2.5 concentrations? Perspective of spillover effect of multi-dimensional urbanization impact. *J. Clean Prod.* **2019**, *220*, 974–983. [CrossRef]
40. Guo, Z.; Hu, D.; Zhang, Z.; Zhang, P.; Zhang, X. Material metabolism and lifecycle GHG emissions of urban road system (URS). *J. Clean Prod.* **2017**, *165*, 243–253. [CrossRef]
41. Zhao, Y.; Wang, S.; Zhou, C. Understanding the relation between urbanization and the eco-environment in China's Yangtze River Delta using an improved EKC model and coupling analysis. *Sci. Total Environ.* **2016**, *571*, 862–875. [CrossRef] [PubMed]
42. Wang, S.; Liu, X.; Yang, X.; Zou, B.; Wang, J. Spatial variations of PM2.5 in Chinese cities for the joint impacts of human activities and natural conditions: A global and local regression perspective. *J. Clean Prod.* **2018**, *203*, 143–152. [CrossRef]
43. Hurvich, C.M.; Simonoff, J.S.; Tsai, C.L. Smoothing parameter selection in nonparametric regression using an improved Akaike information criterion. *J. R. Stat. Soc. B* **1998**, *60*, 271–293. [CrossRef]
44. Wang, S.; Zhou, C.; Wang, Z.; Feng, K.; Hubacek, K. The characteristics and drivers of fine particulate matter ($PM_{2.5}$) distribution in China. *J. Clean Prod.* **2017**, *142*, 1800–1809. [CrossRef]
45. Ibrahim, S.; Landa, M.; Pešek, O.; Pavelka, K.; Halounova, L. Space-Time Machine Learning Models to Analyze COVID-19 Pandemic Lockdown Effects on Aerosol Optical Depth over Europe. *J. Remote Sens.* **2021**, *13*, 3027. [CrossRef]

remote sensing

MDPI

Article

Space-Time Machine Learning Models to Analyze COVID-19 Pandemic Lockdown Effects on Aerosol Optical Depth over Europe

Saleem Ibrahim *, Martin Landa, Ondřej Pešek, Karel Pavelka and Lena Halounova

Department of Geomatics, Faculty of Civil Engineering, Czech Technical University in Prague, 166 29 Prague, Czech Republic; martin.landa@fsv.cvut.cz (M.L.); ondrej.pesek@fsv.cvut.cz (O.P.); karel.pavelka@fsv.cvut.cz (K.P.); lena.halounova@fsv.cvut.cz (L.H.)

* Correspondence: saleem.ibrahim@fsv.cvut.cz

Abstract: The recent COVID-19 pandemic affected various aspects of life. Several studies established the consequences of pandemic lockdown on air quality using satellite remote sensing. However, such studies have limitations, including low spatial resolution or incomplete spatial coverage. Therefore, in this paper, we propose a machine learning-based scheme to solve the pre-mentioned limitations by training an optimized space-time extra trees model for each year of the study period. The results have shown that our trained models reach a prediction accuracy up to 95% when predicting the missing values in the MODIS MCD19A2 Aerosol Optical Depth (AOD) product. The outcome of the mentioned scheme was a geo-harmonized atmospheric dataset for aerosol optical depth at 550 nm with 1 km spatial resolution and full coverage over Europe. As an application, we used the proposed machine learning based prediction approach in AOD levels analysis. We compared the mean AOD levels between the lockdown period from March to June in 2020 and the mean AOD values of the same period for the past 5 years. We found that AOD levels dropped over most European countries in 2020 but increased in several eastern and western countries. The Netherlands had the most significant average decrease in AOD levels (19%), while Spain had the highest average increase (10%). Moreover, we analyzed the relationship between the relative percentage difference of AOD and four meteorological variables. We found a positive correlation between AOD and relative humidity and a negative correlation between AOD and wind speed. The value of the proposed prediction scheme is further emphasized by taking into consideration that the reconstructed dataset can be used for future air quality studies concerning Europe.

Keywords: aerosol optical depth; CAMS; COVID-19; machine learning; MODIS

Citation: Ibrahim, S.; Landa, M.; Pešek, O.; Pavelka, K.; Halounova, L. Space-Time Machine Learning Models to Analyze COVID-19 Pandemic Lockdown Effects on Aerosol Optical Depth over Europe. *Remote Sens.* **2021**, *13*, 3027. https://doi.org/10.3390/rs13153027

Academic Editors: Maria João Costa and Daniele Bortoli

Received: 22 June 2021
Accepted: 29 July 2021
Published: 2 August 2021

1. Introduction

The Severe Acute Respiratory Syndrome-COronaVIrus Diseases 2019 (SARS-COVID-19) pandemic made humanity reconsider how to adapt their daily activities. By late June 2020, the EU average infection rate was around 160 per million inhabitants [1]. In general, most European countries started applying restrictions in March 2020. These restrictions included lockdown, contain, various kinds of curfew, mandatory face masks, etc. By 18 March 2020, more than 250 million people in Europe were in lockdown [2].

Despite the unfortunate losses in human lives and the economy, there could be a bright side to this pandemic when it comes to air quality. Some studies showed that air quality has improved under the applied restrictions. For example, only two weeks of lockdown reduced urban air pollution in Spain, with essential differences among pollutants. The most considerable reduction was in black carbon and Nitrogen Dioxide (NO_2) by 45–51% [3].

According to data released in 2019–2020 by the National Aeronautics and Space Administration (NASA) and the European Space Agency (ESA), NO_2 was reduced up to 30% in some regions that were highly affected by COVID-19 lockdowns such as Wuhan in

China, Italy, Spain, and the USA [4]. Similar results were found in Poland when comparing air quality observations for the year 2020 in five major cities with the same time periods as in the previous two years. In addition, AOD concentrations were reduced in April and May of 2020 by nearly 23% and 18% as compared to 2018–2019 [5].

During the lockdown in China, there was a significant drop in NO_2 (−37%), SO_2 (−64%), and AOD (−8%) for the year 2020, when compared with the 11 year mean average (2009–2019) [6]. Another study of the eastern part of China, where AOD levels are usually high (AOD > 0.7), showed that the emission of pollutants in the first three months of 2020 has decreased when compared to the same period of the previous year [7]. In India, the AOD level was greatly decreased (~45%) during the COVID-19 lockdown periods compared to the mean AOD level in the previous 20 years [8]. Similarly, significant reductions in black carbon concentration (~8.4%) and AOD (10.8%) were observed in southern India during the first lockdown period (25 March–14 April 2020) when compared to the pre-lockdown period (1–24 March 2020) over the selected measuring location [9].

In this study, we focused on AOD, which is defined as a measure of the columnar atmospheric aerosol content. High AOD concentrations have a negative impact on all living things by affecting the respiratory system and reducing naked eye visibility. AOD is measured either from ground-based stations or retrieved by satellites measurements. AOD satellite-based products provide a vast spatial coverage compared to the limited number of ground stations [10].

Due to the correlation between AOD and particulate matter (PM), AOD satellite products are commonly used to retrieve surface PM [11–13]. This justifies the increasing interest in AOD satellite products. Many sensors retrieve AOD at different spatial and temporal resolutions [14], such as the Total Ozone Mapping Spectrometer (TOMS) [15], the Ozone Monitoring Instrument (OMI) [16], the Sea-viewing Wide Field-of-view Sensor (SeaWiFS) [17], the Geostationary Operational Environmental Satellite (GOES) [18], the Advanced Himawari Imager (AHI) [19], the Multi-angle Imaging SpectroRadiometer (MISR) [20], and the widely used Moderate Resolution Imaging Spectroradiometer (MODIS) which we used in our study.

MODIS instrumentations have been carried on both the Terra and Aqua satellites in sun-synchronous polar orbits, since 1999 and 2002, respectively. They can record the earth's surface reflectance and emittance with a 2330 km swath every one to two days [21]. MODIS measures 36 spectral bands between 0.4 and 14.4 μm wavelengths at many different spatial resolutions that provide a great opportunity to study the aerosol thickness and parameters characterizing aerosol size from space with good accuracy and on a worldwide scale.

MODIS provides various AOD products based on different aerosol retrieval algorithms. The most common algorithms are the Dark Target (DT) [22,23], the Deep Blue (DB) [24,25], and the Multi-Angle Implementation of Atmospheric Correction for MODIS (MAIAC) [26] which is the algorithm used to generate the MODIS MCD19A2 product with 1 km spatial resolution.

However, AOD satellite-based products have a great number of gaps due to cloud cover and snow reflectance. An analysis of the spatial and temporal distribution of clouds retrieved by MODIS over 12 years of continuous observations from the Terra satellite and over 9 years from the Aqua satellite showed that clouds cover ~67% of the earth's surface worldwide and ~55% over land [27]. To solve this issue, it has become common to use machine learning and deep learning algorithms in developing models that fill the gaps in satellite-based products either by removing the clouds [28], applying spatiotemporal interpolation [29], or merging different sources of data to predict gaps-free images [30]. Therefore, in this study, we propose a machine learning-based scheme to fill the gaps in MODIS MAIAC AOD retrievals and to generate daily, full coverage, high-resolution AOD maps over Europe. Such maps will minimize time series analysis bias and uncertainty while investigating the influence of COVID-19 lockdown on AOD levels.

2. Material and Data

2.1. Study Area and Period

The study area is shown in Figure 1. It includes the "Continental EU," hence EEA (European Economic Area), and the United Kingdom, Switzerland, Serbia, Bosnia and Herzegovina, Montenegro, Kosovo, North Macedonia, and Albania [31]. In this paper, we refer to the area of study as "Europe" located inside this coordinates box 26° W, 72° N, 42° E, and 36° S. The total study area covers 13,391,504 of 1 km grid cells; 5,450,009 of the total cell number are located over land. The study period covers the months of March–June from the years 2015–2020.

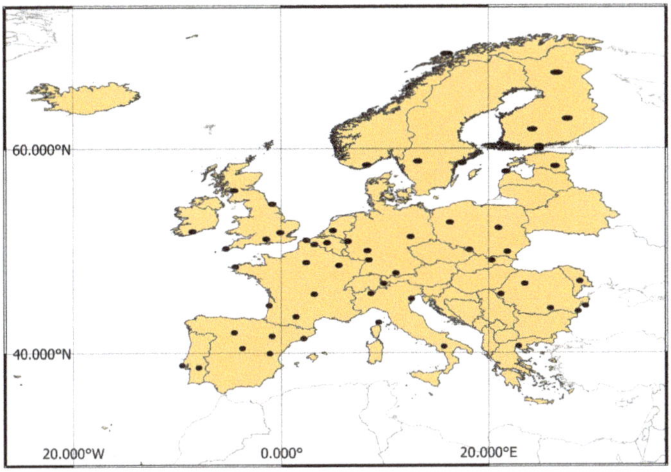

Figure 1. The study area with AERONET stations shown as black dots.

2.2. Data

In this section, we summarize different data used throughout our study.

2.2.1. MODIS Data

MCD19A2 daily product from MODIS collection 6 was released and made publicly available on 30 May 2018. It was generated from both the Aqua and Terra satellites and delivered in Hierarchical Data Format [26]. MCD19A2 has a 1 km spatial resolution and uses the MAIAC algorithm that utilizes time series (TMS) analyses, a set of image-based and pixel processing to enhance the precision of cloud recognition, AOD, and other atmospheric rectification [32,33]. Daily MODIS MCD19A2 data were downloaded, and two science datasets (SDS) were extracted; AOD green band (at 550 nm) and AOD quality assurance layer (AOD_QA), which was used to retrieve only pixels with the best quality. We created daily mosaics that cover the study area.

2.2.2. Copernicus Atmosphere Monitoring Service (CAMS) Data

In this study, modeled AOD at 550 nm data with 80 km spatial resolution produced by the European center for medium-range weather forecasts Atmospheric Composition Reanalysis 4 (EAC4) was used to fill the gaps in the MODIS MCD19A2 product. Reanalysis merges model data with worldwide observations into a compatible dataset generated by an atmospheric model that uses the laws of physics and chemistry. EAC4 estimates modeled AOD every 3 h using the 4D-Var assimilation method [34].

2.2.3. Digital Elevation Model

The elevation of the grid cells was added as a land predictor in our study. The Japan Aerospace Exploration Agency (JAXA) provides a worldwide digital surface model for scientific research and other geospatial services. It provides a horizontal resolution (~30 m) by the Panchromatic Remote-sensing Instrument for Stereo Mapping (PRISM), which was carried on the Advanced Land Observing Satellite "ALOS" [35]. Data was accessed in March 2021 from (https://www.eorc.jaxa.jp/ALOS/).

2.2.4. Ground-Based AOD Data

NASA's Aerosols Robotic Network (AERONET) is considered one of the most reliable aerosol networks [36]. AERONET measures direct solar and sky radiance in various channels every 15 min at the local point to compute columnar AOD at intervals from 350 to 1020 nm with low expected uncertainties ranging between 0.01 to 0.02 under cloud-free conditions [37]. There are several categories of AERONET data: level 1.0 (unscreened), level 1.5 (cloud screened), and level 2.0 (cloud screened and quality assured).

In this study, AERONET level 2.0 quality assurance observations were used from 57 stations over Europe, as shown in Figure 1. Since AERONET stations do not measure AOD at 550 nm, available measurements at the nearest two wavelengths to 550 nm (440 or 500 nm as λ_1 and 675 nm as λ_2) for each station were interpolated to 550 nm using the Ångström's turbidity equation represented in Equation (1) [21,38].

$$\tau_a(\lambda) = \beta\lambda^{-\alpha} \tag{1}$$

where $\tau_a(\lambda)$ is the AOD at λ wavelength in micrometers, β is the Angstrom's turbidity coefficient, and α is the band index represented in Equation (2).

$$\alpha = -\frac{\ln(\tau_a(\lambda_1)/\tau_a(\lambda_2))}{\ln(\lambda_1/\lambda_2)} \tag{2}$$

AOD values at two different wavelengths λ_1, λ_2 are related by Equation (3).

$$\tau_a(\lambda_1) = \tau_a(\lambda_2) * \left(\frac{\lambda_1}{\lambda_2}\right)^{-\alpha} \tag{3}$$

2.2.5. European Centre for Medium-Range Weather Forecasts reanalysis (ECMWF)

ERA-5 is the fifth generation of ECMWF reanalysis for the global climate and weather. Hourly data between 10 a.m. and 2 p.m. of U and V wind components, total precipitation, and 2 m surface temperature for the months of March–June of the years 2015–2020 with 0.1° spatial resolution were extracted from the ERA-5 land hourly data. Relative humidity data between 10 a.m. and 2 p.m. at 0.25° spatial resolution was extracted from the ERA-5 monthly averaged data.

All used data shown in Table 1 were reprojected to the European Terrestrial Reference System 1989 (EPSG:3035), using a 1 km grid cell with bilinear interpolation method for $CAMS_{AOD}$ and ECMWF data and the cubic convolution for the ALOS elevation model. All values of $MODIS_{AOD}$, $CAMS_{AOD}$, and elevations were assigned to the closest grid cell.

Table 1. Summary of data used in this study.

Product	Spatial Resolution	Temporal Resolution	Layer
MODIS MCD19A2	1 km	Daily	AOD-055 Quality Assurance (QA)
CAMS	80 km	3 h	Total aerosol optical depth at 550 nm
ALOS DSM	30 m	-	Elevations
AERONET	-	~15 min	Level 2.0
ECMWF ERA-5	0.1°	Hourly	Wind U and V components Total precipitation 2 m surface temperature
ECMWF ERA-5	0.25°	Monthly	Relative humidity

3. Methodology

In this study, we created a Geo-Harmonized Atmospheric Dataset for Aerosol optical depth (GHADA) that covers the study area. Three stages were applied to generate GHADA: first, we merged the Terra and Aqua datasets of the MODIS MCD19A2 product by applying a simple average for all pixels that passed the quality assurance criteria ($QA_{CloudMask}$ = Clear and $QA_{AdjacencyMask}$ = Clear) of this product. Second, we created a machine learning model for every year of the study period to predict AOD values over the study area. MCD19A2 high-quality retrievals were used as the dependent variable, and since the Terra satellite is passing locally around 10:30 a.m. and the Aqua satellite passes around 1:30 p.m., we used the modeled AOD from CAMS at the closest three times per day to the satellites passing (9 a.m., 12 p.m., and 3 p.m.). In addition, the spatiotemporal information for the grid cells was used as independent variables. Finally, we filled MODIS MCD19A2 gaps with the predicted AOD by merging the outputs from stages one and two. We validated the daily maps of GHADA with ground-based observation, and then we utilized this dataset to analyze how the COVID-19 lockdown has affected AOD levels over Europe during the period of March–June 2020 by comparing AOD levels for this period with the average AOD levels in the last five years (2015–2019) for the same months.

4. Space-Time Models

In this section, we propose a novel approach based on the Extremely Randomized Trees (ET) to predict the missing AOD values in the MODIS MCD19A2 product. First, we illustrate the principles of the ETs and discuss their suitability for the AOD prediction problem. Second, we describe in detail the proposed ET training and parameters setting for AOD prediction.

4.1. Extra Trees Algorithm

ET is a tree-based ensemble learning method used in our study to deal with the supervised regression and create prediction models for AOD. The idea behind ET is to strongly randomize the selection of both attributes and cut points while splitting a tree node. Unlike the widely used random forest algorithm that chooses the optimum split, ET chooses it randomly, which further reduces bias and variance. When needed, the latter algorithm creates independent randomized trees of learning sample output values [38].

The number of attributes that are randomly selected at each node (K) and the minimum sample size for splitting a node (n_{min}) are the two main parameters in the ET splitting process. This procedure is applied several times with the whole learning dataset to create an ensemble model that aggregates the predictions of the decision trees to obtain the final estimation by majority vote in classification problems and arithmetic average in regression problems. In addition to accuracy, ET has high computational efficiency [39], which is required when dealing with big data problems.

4.2. Improved Spatiotemporal Information

To determine the spatial and temporal correlation between $MAIAC_{AOD}$ and $CAMS_{AOD}$, we included the following independent variables. For space, we used both the elevations of the grid cells and the great circle distance (D) between each grid cell and a reference point on a sphere identified by their latitudes and longitudes using the haversine approach (Equations (4)–(6)). For time, we used the day of the year (DOY) to calculate the radian time (Rt) for the grid cells on different days in a year to improve model handling of the seasonal cycle, Equation (7) [40].

$$\theta = f(\lambda_{i,t}, \varphi_{i,t}) = \text{haversin}(\varphi_1 - \varphi_2) + \cos(\varphi_1) * \cos(\varphi_2) * \text{haversin}(\lambda_1 - \lambda_2) \quad (4)$$

$$\text{haversin}(\theta) = \sin^2(\frac{\theta}{2}) = \frac{1 - \cos(\theta)}{2} \quad (5)$$

$$D_{i,t} = r * \text{archaversin}(\theta) = 2 * r * \arcsin\left(\sqrt{\theta}\right) \quad (6)$$

$$Rt_{i,t} = \cos\left(2\pi * \frac{DOYi, t}{T}\right) \quad (7)$$

where θ is the central angle between two points in space, φ_1 and φ_2 denote the geographical latitudes in radians of two points in space, λ_1 and λ_2 denote the geographical longitudes in radians of two points in space, r denotes the earth's radius in km, DOY represents the day of the year, T represents the total number of days in the year, for every grid cell (i) on day (t).

For each year between 2015–2020, the model was built using Equation (8).

$$AOD_{i,t} = f(CAMS\text{-}9_{i,t}, CAMS\text{-}12_{i,t}, CAMS\text{-}15_{i,t}, D_{i,t}, H_{i,t}, Rt_{i,t}) \quad (8)$$

where for each grid cell (i) on day (t): $AOD_{i,t}$ is the target AOD value, CAMS-x represents the AOD value extracted from CAMS at hour x, $D_{i,t}$ represents the great circle distance, $H_{i,t}$ represents the elevation, $Rt_{i,t}$ represents the temporal information identified by the radian time.

5. Results

In this section, we present the results of the space-time ET models when predicting the MAIAC AOD values. Then we utilize these models to generate AOD maps over the study area. The validation process is also stated below. Finally, these maps were used to analyze the effects of COVID-19 lockdowns on AOD levels, as discussed in Section 5.4.

5.1. Models

Due to the great number of $MODIS_{AOD}$ -$CAMS_{AOD}$ pairs over land in the study area (on average 380 million pairs per year), representative subsets consisting of ~10% of the whole population (all $MODIS_{AOD}$ -$CAMS_{AOD}$ pairs per year) were chosen using the Kolmogorov–Smirnov test to be used as learning dataset for a space-time model for each year. Then for each learning dataset, we used the k-fold cross-validation (where k = 5) to train and validate each model. In this method, the learning dataset is divided into 5 folds, which means 80% of the pairs in the learning dataset are used as a training set for the model, and the remaining 20% are used for validation. This procedure was repeated five times to test the model on each fold. Based on learning curve results, we found that increasing the learning dataset size to 15% only increased the accuracy of the models by less than 1%, and the curve reaches a plateau beyond this percentage. Therefore, to decrease the computational complexity, we used ~10% of the whole population as a learning dataset. In other words, a learning dataset size of 10% is enough to reach satisfactory accuracy for each year of the study period. The optimized models (number of trees = 30, maximum depth of the tree = 50) were tested on the remaining ~90% (approximately 340 million pairs) of the population.

The results of the trained models for each year are summarized in Table 2. All models achieved high accuracies when predicting MAIAC AOD with a correlation of determination (R^2) ranging between 92.5% to 95% and root mean squared errors from 0.016 to 0.02. These high achieved accuracies with the relatively small errors show the efficiency of our space-time models in predicting the missing AOD values and emphasize the appropriateness of exploitation modeled AOD with improved spatiotemporal information in improving satellite AOD data.

Table 2. Results of the space-time extremely randomized models used to predict the missing AOD in the MODIS MCD19A2 product for each year of the study period.

Year	R-Squared (%)	RMSE	MAE
2015	95	0.017	0.011
2016	94.3	0.018	0.011
2017	93.8	0.018	0.011
2018	92.5	0.02	0.012
2019	92.9	0.019	0.012
2020	94.1	0.016	0.010

Feature importance was calculated based on the reduction in sum of squared errors whenever a variable is chosen to split. Mean importance scores were calculated for all selected input variables of the models (see Figure 2). $CAMS_{AOD}$ at 12:00 p.m. is the most influential variable, accounting for ~33% of $MODIS_{AOD}$ estimates. The other two modeled AOD at 9:00 a.m. and 3:00 p.m. contributed by 18% and 24%, respectively. The radian time and the great circle distance had almost the same influence (10–10.4%). Finally, the elevation had the lowest influence, with ~5% on $MODIS_{AOD}$ estimates.

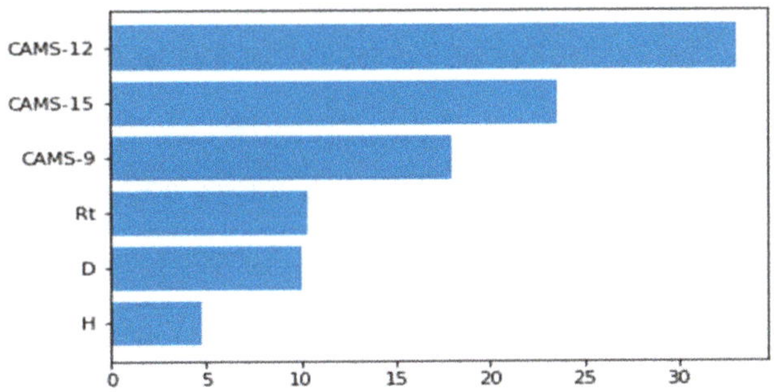

Figure 2. Mean importance scores (%) of independent variables to AOD estimates for the space-time extremely randomized models.

5.2. AOD Maps

We used the optimized space-time models to predict the missing values in the daily MCD19A2 data of the study period. Then we used these predictions to fill the gaps in this product. The outputs of the previous processes were daily AOD maps with 1 km spatial resolution and full coverage over Europe for the period of March–June in the years 2015–2020. To analyze the COVID-19 lockdown effects on AOD levels, we calculated the average AOD levels for the months' March–June of the years 2015–2019 and compared these levels with the same period of the year 2020 (see Figure 3). Moreover, we generated daily AOD maps for the period of January 2018–June 2020 to validate GHADA through all seasons and not solely during the chosen lockdown months.

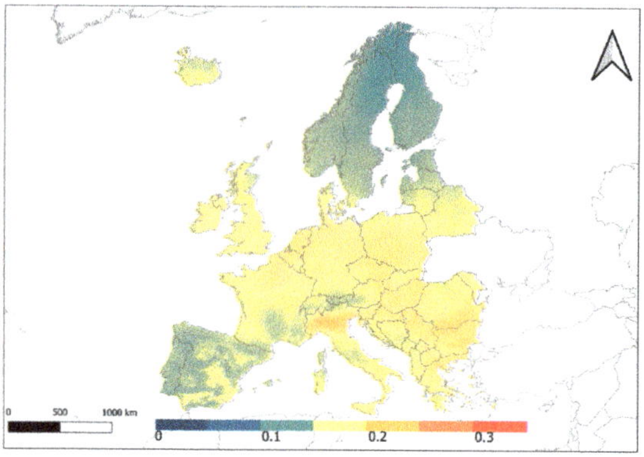

(a)

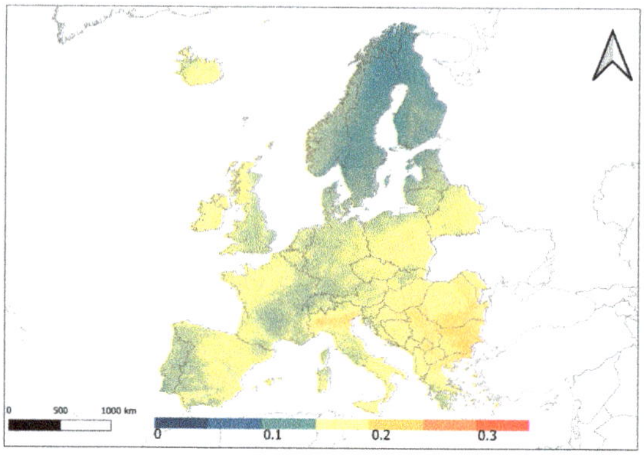

(b)

Figure 3. The average AOD values for the months March–June of (**a**) the years 2015–2019 and (**b**) of the year 2020 during the chosen lockdown period.

5.3. Validation with AERONET

With the assumption that the aerosol column is relatively uniform within a certain time-space boundary [41], the validation of satellite-based AOD products is usually performed between AOD retrievals within the spatiotemporal window and the corresponding AERONET observations [42]. An acceptable accuracy of AOD products can be achieved when 66% of retrievals fall within expected error envelopes (EE) [23,43]. We used for validation the average AERONET level 2.0 quality assurance observations between 10 a.m. and 2 p.m. from 57 stations across Europe during the period of January 2018–June 2020. We chose two spatial diameters, 20 km and 50 km, with AERONET stations in the center for validation and statistical analysis that extensively uses root-mean-square error (RMSE), mean absolute error (MAE), expected error (EE) envelopes, and the fraction of AOD retrievals of the total number (N) falling within EE envelope (Equations (9)–(13)).

$$RMSE = \sqrt{\frac{1}{N}\sum(AOD_{GHADA} - AOD_{AERONET})^2} \tag{9}$$

$$MAE = \frac{1}{N}\sum| AOD_{GHADA} - AOD_{AERONET} | \tag{10}$$

$$Bias = \frac{1}{N}\sum(AOD_{GHADA} - AOD_{AERONET}) \tag{11}$$

$$EE = \pm\ (0.05 + 0.15\ * AOD_{AERONET}) \tag{12}$$

$$AOD_{AERONET} - |EE| \leq AOD_{GHADA} \leq AOD_{AERONET} + |EE| \tag{13}$$

The statistical analysis between daily GHADA maps and AERONET observations has shown similar validation results for the two chosen spatial diameters with ~84% of the samples falling within the EE, good correlations R ~ 76–77%, and relatively small RMSE ~ 0.066–0.067, refer to Table 3.

Table 3. Validation results of GHADA with AERONET at two spatial diameters, where N is the total number of sample points.

D (Km)	N	R	MAE	RMSE	Bias	EE(%)
20	10916	0.762	0.043	0.067	−0.014	83.7
50	12212	0.767	0.043	0.066	−0.014	83.7

Figure 4 represents the density scatter plots for the validation of AOD at 550 nm from GHADA with the AERONET stations at the two chosen spatial diameters.

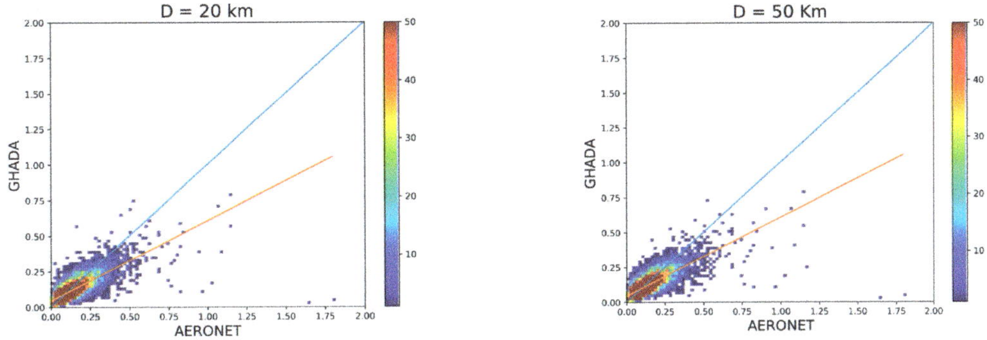

Figure 4. Density scatter plots of validation AOD at 550 nm from GHADA with 57 AERONET stations between 10 a.m. and 2 p.m. at two spatial diameters of 20 km and 50 km. The colored scale bar stands for the frequency of occurrence.

5.4. AOD Relative Percentage Difference

The variations in AOD levels were calculated for each grid cell using the Relative Percentage Difference (RPD) Equation (14).

$$\mathrm{RPD} = \frac{\mathrm{AOD}_{2020} - \mathrm{AOD}_{2015\text{–}2019}}{\mathrm{AOD}_{2015\text{–}2019}} * 100 \quad (14)$$

where AOD_{2020} is the mean AOD value in the study period of 2020 and $\mathrm{AOD}_{2015\text{–}2019}$ is the mean AOD value for the study period covering 2015–2019. The changes are presented in Figure 5.

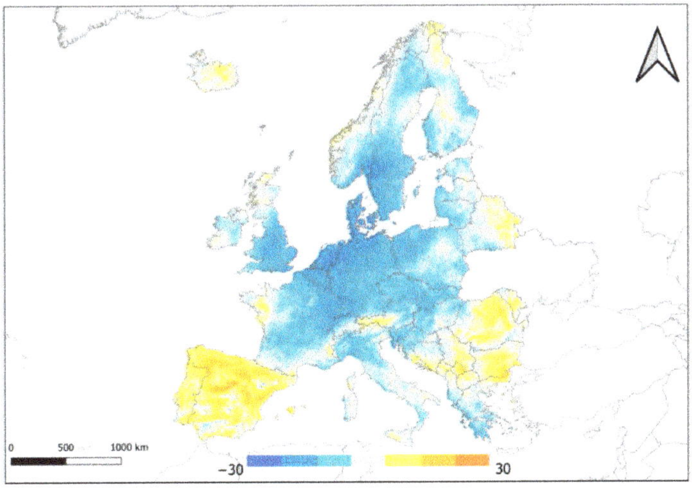

Figure 5. Relative percentage difference of AOD over Europe for the months March–June of the year 2020 and the same months of the previous 5 years.

6. Discussion

In this study, a machine learning-based scheme was used to overcome the limitations in time series analysis concerning AOD. A new dataset for AOD at 550 nm with full coverage over Europe and with 1 km spatial resolution (GHADA) was built. We trained an extra trees model for each year (2015–2020) using the MODIS MCD19A2 as the target variable and CAMS modeled AOD with improved spatiotemporal information as the independent variables. Results showed that the trained models had high accuracies ranging between 92.5–95% when estimating the missing $MAIAC_{AOD}$ retrievals. We compared the AOD_{550} from GHADA and surface observations at 57 AERONET sites over Europe, with two spatial diameters around these AERONET stations within the period of January 2018–June 2020. The overall comparison with ground-based measurements showed a good correlation, with a bias as low as 0.014 and R ~ 0.76. Then we used GHADA to study the influence of COVID-19 pandemic lockdown on AOD levels over Europe in the months March–June by comparing it to AOD levels in the same months for the past five years (2015–2019). The most important advantage of our study when compared to similar work is that we used daily full-coverage AOD maps with high spatial resolution when calculating the average AOD values before and after the lockdown. Such complete coverage reduces bias and uncertainty in such time-series analyses. As shown above, in Figure 5, we have found that AOD levels decreased by 10–30% over most countries of the study area in 2020, mainly the countries located at the center of the analyzed area, while AOD levels increased over the countries that are located on the boundaries of the study area. In the west, AOD increased over Spain and Portugal; in the east, AOD increased over Romania, Bulgaria, Moldova, and Kosovo; in the north, the level slightly increased over Iceland. The decrease in AOD levels was the greatest in the Netherlands, with an average decrease of 20%, while Spain had the highest average increase in AOD levels by 10%. It must be noted that the five AERONET stations in Spain included in this study did not reflect the average increase in AOD over the whole country due to their limited spatial coverage.

As an attempt to justify the findings in areas of increased AOD, we investigated the relationship between the RPD in AOD for the months March–June of the year 2020 and the previous five years and the RPD for four meteorological variables (relative humidity, wind speed, surface temperature, and total precipitation) calculated for the times of MODIS satellites overpassing (10 a.m. to 2 p.m.). We found a close trend between relative humidity and AOD. Spain, Portugal, northern Norway, eastern Belarus, and southern Bulgaria had higher RPD in both AOD and relative humidity. Spain and Portugal had the highest increase of 10–23% in relative humidity. In agreement, areas of decreased humidity had lower RPD of AOD; however, such correlation is to a lower extent than the effect of increased humidity. An exception to this finding is Romania, where RPD in humidity was decreased however AOD was increased. Regarding wind speed, RPD decreased by ~18% in Spain and Portugal, where AOD had a significant increase. Also, the northern part of Italy and the western part of Austria had a clear inverse trend between AOD and wind speed. The average relative humidity over Spain was 65% during the lockdown period of the year 2020. High relative humidity combined with a low average wind speed of less than ~3 m/s play an important role in increasing AOD. Our findings are consistent with [44], where they associated higher humidity and lower wind speed with higher AOD. We found no direct relationship between RPD of neither surface temperature nor total precipitation and RPD of AOD, all of which strengthens the argument that lowering AOD is a consequence of the lockdown. Although we proved that AOD levels increased over Spain, other pollutants such as NO_2 were decreased, which is attributed to the difference in the source of these pollutants as discussed elsewhere [44]. Figure 6 shows the RPD of relative humidity and RPD of wind speed between the lockdown months of the year 2020 and the same period of the previous five years.

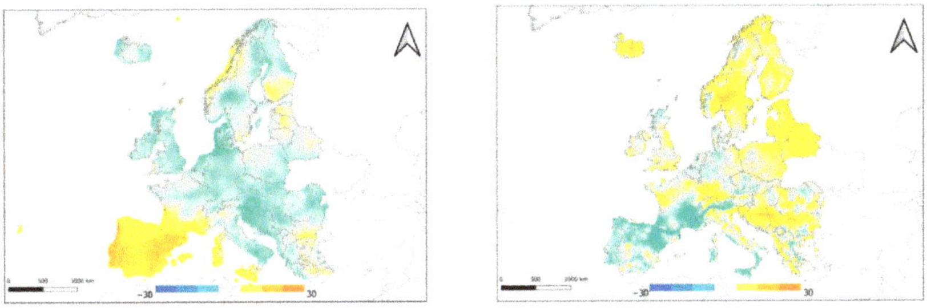

(**a**) (**b**)

Figure 6. Relative percentage difference of (**a**) relative humidity and (**b**) wind speed over Europe between 10 a.m. and 2 p.m. for the months March–June of the year 2020 and the same months of the previous 5 years.

Nevertheless, it must be noted that the average AOD levels over Europe are relatively low (AOD < 0.3) compared to other more polluted regions, where more prominent differences in AOD levels can be observed, for example, as published in [8] where AOD levels over India were investigated. In addition, the extent of restrictions imposed and the adherence to them may contribute to the significance of the change in AOD levels.

7. Conclusions

The advancement of machine learning algorithms provides solutions for AOD satellite-based data drawbacks such as low spatial resolution and gaps caused by persistent clouds, cloud contamination, and high surface reflectance and opens new horizons for studies that can influence decision making. A machine learning-based scheme was used to enhance time series analysis of AOD over the study period. Space-time extremely randomized trees models were built to fill the gaps in the MCD19A2 product of the moderate imaging spectroradiometer (MODIS). The output was a geo-harmonized atmospheric dataset for aerosol optical depth (GHADA) with complete coverage of 1 km spatial resolution over Europe. To the best of our knowledge, GHADA is the first dataset with this coverage and resolution for Europe, and we are the first to analyze how COVID-19 affected AOD levels over Europe with gaps-free AOD maps at high spatial resolution.

We compared AOD levels during the chosen lockdown period to the mean AOD values during the same period in the previous five years. We found a general decrease trend in the countries located at the center of the study area, with the Netherlands scoring the highest average decrease. In contrast, AOD levels increased in the eastern and western European countries as it is distinctly visible in Kosovo and Spain, respectively. We found a correlation between high humidity and low wind speed with AOD increase, which justifies such an increase in countries like Spain and Portugal. We excluded surface temperature and total precipitation as contributing factors to the detected changes in AOD levels, which in return makes COVID-19 lockdown the major cause for the decrease in AOD levels.

Once GHADA is made publicly accessible, it can be used to investigate air quality over Europe with 1 km spatial resolution and improve time series analysis, overcoming the gaps encountered during such studies. The lockdown that happened due to the pandemic generally lowered AOD levels; however, such lockdown is not the ultimate solution to control AOD levels. Cleaner sources of energy and road transport are needed to maintain lower levels of AOD and good air quality. Based on our obtained results, we recommend utilizing machine learning to solve time series analysis limitations and to conduct various applications concerning air quality.

Author Contributions: S.I. and L.H. conceptualized the work. S.I., M.L. and O.P. designed and implemented the workflow and processed the data. M.L., O.P., K.P. and L.H. contributed to the improvement of the draft manuscript. Saleem Ibrahim wrote the paper. All authors have read and agreed to the published version of the manuscript.

Funding: This work is co-financed under Grant Agreement Connecting Europe Facility (CEF) Telecom project 2018-EU-IA-0095 by the European Union and by the Grant Agency of the Czech Technical University in Prague, grant No. SGS21/054/OHK1/1T/11.

Institutional Review Board Statement: Not applicable.

Informed Consent Statement: Not applicable.

Data Availability Statement: The data and data analysis methods are available upon request.

Acknowledgments: The authors sincerely thank NASA EOSDIS for providing the daily MODIS MAIAC AOD product (MCD19A2) available from the Land Processes Distributed Active Archive Center (LPDAAC), AERONET (https://aeronet.gsfc.nasa.gov/) for providing AOD ground-based observation data (was last accessed in May 2021), the European Center for Medium-Range Weather Forecasts (ECMWF) for providing global reanalysis of atmospheric composition, and the Japan Aerospace Exploration Agency (JAXA) for providing the digital surface model used in this study.

Conflicts of Interest: Authors declare no conflict of interest.

References

1. Boffey, D. US Visitors Set to Remain Banned from Entering EU. *Guardian* **2020**. Available online: https://www.theguardian.com/world/2020/jun/29/us-visitors-set-to-remain-banned-from-entering-eu (accessed on 1 May 2021).
2. Henley, J.; Oltermann, P. More Than 250 m People Now in Lockdown in EU as Germany and Belgium Adopt Measures. *Guardian* **2020**. Available online: https://www.theguardian.com/world/2020/mar/18/coronavirus-lockdown-eu-belgium-germany-adopt-measures (accessed on 1 May 2021).
3. Tobias, A.; Carnerero, C.; Reche, C.; Massague, J.; Via, M.; Minguillon, M.C.; Alastuey, A.; Querol, X. Changes in air quality during the lockdown in Barcelona (Spain) one month into the SARS-CoV-2 epidemic. *Sci. Total Environ.* **2020**, *726*, 138540. [CrossRef]
4. Muhammad, S.; Long, X.; Salman, M. COVID-19 pandemic and environmental pollution: A blessing in disguise? *Sci. Total Environ.* **2020**, *728*, 138820. [CrossRef]
5. Filonchyk, M.; Hurynovich, V.; Yan, H. Impact of COVID-19 lockdown on air quality in the Poland, Eastern Europe. *Environ. Res.* **2021**, *198*, 110454. [CrossRef] [PubMed]
6. Soni, M.; Verma, S.; Jethava, H.; Payra, S.; Lamsal, L.; Gupta, P.; Singh, J. Impact of COVID-19 on the Air Quality over China and India Using Long-term (2009–2020) Multi-satellite Data. *Aerosol Air Qual. Res.* **2020**, *21*, e200295. [CrossRef]
7. Filonchyk, M.; Hurynovich, V.; Yan, H.; Gusev, A.; Shpilevskaya, N. Impact Assessment of COVID-19 on Variations of SO_2, NO_2, CO and AOD over East China. *Aerosol Air Qual. Res.* **2020**, *20*, 1530–1540. [CrossRef]
8. Ranjan, A.K.; Patra, A.K.; Gorai, A.K. Effect of lockdown due to SARS COVID-19 on aerosol optical depth (AOD) over urban and mining regions in India. *Sci. Total Environ.* **2020**, *745*, e141024. [CrossRef] [PubMed]
9. Kalluri, R.O.R.; Gugamsetty, B.; Tandule, C.R.; Kotalo, R.G.; Thotli, L.R.; Rajuru, R.R.; Palle, S.N.R. Impact of aerosols on surface ozone during COVID-19 pandemic in southern India: A multi-instrumental approach from ground and satellite observations, and model simulations. *J. Atmos. Sol. Terr. Phys.* **2021**, *212*, 105491. [CrossRef]
10. Liu, Y.; Sarnat, J.A.; Coull, B.A.; Koutrakis, P.; Jacob, D.J. Validation of Multiangle Imaging Spectroradiometer (MISR) aerosol optical thickness measurements using Aerosol Robotic Network (AERONET) observations over the contiguous United States. *J. Geophys. Res.* **2004**, *109*. [CrossRef]
11. Shen, H.; Li, T.; Yuan, Q.; Zhang, H. Estimating Regional Ground-Level PM2.5 Directly from Satellite Top-of-Atmosphere Reflectance Using Deep Belief Networks. *J. Geophys. Res. Atmos.* **2018**, *123*, 13875–13886. [CrossRef]
12. Yang, Q.; Yuan, Q.; Yue, L.; Li, T.; Shen, H.; Zhang, L. The relationships between PM2.5 and aerosol optical depth (AOD) in mainland China: About and behind the spatio-temporal variations. *Environ. Pollut.* **2019**, *248*, 526–535. [CrossRef]
13. Zhang, H.; Hoff, R.M.; Engel-Cox, J.A. The relation between Moderate Resolution Imaging Spectroradiometer (MODIS) aerosol optical depth and PM2.5 over the United States: A geographical comparison by U.S. Environmental Protection Agency regions. *J. Air Waste Manag. Assoc.* **2009**, *59*, 1358–1369. [CrossRef]
14. Ajtai, N.; Mereuta, A.; Stefanie, H.; Radovici, A.; Botezan, C.; Zawadzka-Manko, O.; Stachlewska, I.S.; Stebel, K.; Zehner, C. SEVIRI Aerosol Optical Depth Validation Using AERONET and Intercomparison with MODIS in Central and Eastern Europe. *Remote Sens.* **2021**, *13*, 844. [CrossRef]
15. Torres, O.; Bhartia, P.K.; Sinyuk, A.; Welton, E.J.; Holben, B.N. Total Ozone Mapping Spectrometer measurements of aerosolabsorption from space: Comparison to SAFARI 2000 ground-based observations. *J. Geophys. Res.* **2005**, *110*. [CrossRef]
16. Torres, O.; Tanskanen, A.; Veihelmann, B.; Ahn, C.; Braak, R.; Bhartia, P.K.; Veefkind, P.; Levelt, P. Aerosols and surface UV products from Ozone Monitoring Instrument observations: An overview. *J. Geophys. Res.* **2007**, *112*. [CrossRef]
17. Sayer, A.M.; Hsu, N.C.; Bettenhausen, C.; Ahmad, Z.; Holben, B.N.; Smirnov, A.; Thomas, G.E.; Zhang, H. SeaWiFS Ocean Aerosol Retrieval (SOAR): Algorithm, validation, and comparison with other data sets. *J. Geophys. Res.* **2012**, *117*. [CrossRef]
18. Knapp, K.R.; Frouin, R.; Kondragunta, S.; Prados, A. Toward aerosol optical depth retrievals over land from GOES visible radiances: Determining surface reflectance. *Int. J. Remote Sens.* **2005**, *26*. [CrossRef]

19. Lim, H.; Choi, M.; Kim, J.; Kasai, Y.; Chan, P.W. AHI/Himawari-8 Yonsei Aerosol Retrieval (YAER): Algorithm, Validation and Merged Products. *Remote Sens.* **2018**, *10*, e699. [CrossRef]
20. Kahn, R.A.; Gaitley, B.J.; Garay, M.J.; Diner, D.J.; Eck, T.F.; Smirnov, A.; Holben, B.N. Multiangle Imaging SpectroRadiometer global aerosol product assessment by comparison with the Aerosol Robotic Network. *J. Geophys. Res.* **2010**, *115*. [CrossRef]
21. Sun, L.; Wei, J.; Bilal, M.; Tian, X.; Jia, C.; Guo, Y.; Mi, X. Aerosol Optical Depth Retrieval over Bright Areas Using Landsat 8 OLI Images. *Remote Sens.* **2016**, *8*, 23. [CrossRef]
22. Levy, R.C.; Mattoo, S.; Munchak, L.A.; Remer, L.A.; Sayer, A.M.; Patadia, F.; Hsu, N.C. The Collection 6 MODIS aerosol products over land and ocean. *Atmos. Meas. Tech.* **2013**, *6*, 2989–3034. [CrossRef]
23. Remer, L.A.; Kaufman, Y.J.; Tanré, D.; Mattoo, S.; Chu, D.A.; Martins, J.V.; Li, R.-R.; Ichoku, C.; Levy, R.C.; Kleidman, R.G.; et al. The MODIS Aerosol Algorithm, Products, and Validation. *J. Atmos. Sci.* **2005**, *62*, 947–973. [CrossRef]
24. Hsu, N.C.; Jeong, M.J.; Bettenhausen, C.; Sayer, A.M.; Hansell, R.; Seftor, C.S.; Huang, J.; Tsay, S.C. Enhanced Deep Blue aerosol retrieval algorithm:The second generation. *J. Geophys. Res.* **2013**, *118*, 9296–9315. [CrossRef]
25. Hsu, N.C.; Tsay, S.; King, M.D.; Herman, J.R. Aerosol Properties Over Bright-Reflecting Source Regions. *IEEE Trans. Geosci. Remote Sens.* **2004**, *42*, 557–569. [CrossRef]
26. Lyapustin, A.; Wang, Y.; Laszlo, I.; Kahn, R.; Korkin, S.; Remer, L.A.; Levy, R.; Reid, J.S. Multiangle implementation of atmospheric correction (MAIAC): 2. Aerosol algorithm. *J. Geophys. Res.* **2011**, *116*. [CrossRef]
27. King, M.D.; Platnick, S.; Menzel, W.P.; Ackerman, S.A.; Hubanks, P.A. Spatial and Temporal Distribution of Clouds Observed by MODIS Onboard the Terra and Aqua Satellites. *IEEE Trans. Geosci. Remote Sens.* **2013**, *51*, 3826–3852. [CrossRef]
28. Meraner, A.; Ebel, P.; Zhu, X.X.; Schmitt, M. Cloud removal in Sentinel-2 imagery using a deep residual neural network and SAR-optical data fusion. *ISPRS J. Photogramm. Remote Sens.* **2020**, *166*, 333–346. [CrossRef]
29. Yang, J.; Hu, M. Filling the missing data gaps of daily MODIS AOD using spatiotemporal interpolation. *Sci. Total Environ.* **2018**, *633*, 677–683. [CrossRef]
30. Schneider, R.; Vicedo-Cabrera, A.M.; Sera, F.; Masselot, P.; Stafoggia, M.; de Hoogh, K.; Kloog, I.; Reis, S.; Vieno, M.; Gasparrini, A. A Satellite-Based Spatio-Temporal Machine Learning Model to Reconstruct Daily PM2.5 Concentrations across Great Britain. *Remote Sens.* **2020**, *12*, 3803. [CrossRef]
31. Open Data Science Europe. *Geo-Harmonizer Project Implementation Plan 2020–2022*; Open Data Science Europe: Wageningen, The Netherlands, 2020.
32. Lyapustin, A.; Wang, Y.; Korkin, S.; Huang, W. MODIS Collection 6 MAIAC algorithm. *Atmos. Meas. Tech.* **2018**, *11*, 5741–5765. [CrossRef]
33. Lyapustin, A.; Wang, Y.; Laszlo, I.; Korkin, S. Improved cloud and snow screening in MAIAC aerosol retrievals using spectral and spatial analysis. *Atmos. Meas. Tech.* **2012**, *5*, 843–850. [CrossRef]
34. Inness, A.; Ades, M.; Agustí-Panareda, A.; Barré, J.; Benedictow, A.; Blechschmidt, A.; Dominguez, J.J.; Engelen, R.; Eskes, H.; Flemming, J.; et al. The CAMS reanalysis of atmospheric composition. *Atmos. Chem. Phys.* **2019**, *19*, 3515–3556. [CrossRef]
35. Tadono, T.; Ishida, H.; Oda, F.; Naito, S.; Minakawa, K.; Iwamoto, H. Precise Global DEM Generation by Alos Prism. *ISPRS Ann. Photogramm. Remote Sens. Spat. Inf. Sci.* **2014**, *2*, 71. [CrossRef]
36. Holben, B.N.; Eck, T.F.; Slutsker, I.; Tanré, D.; Buis, J.P.; Setzer, A.; Vermote, E.; Reagan, J.A.; Kaufman, Y.J.; Nakajima, T.; et al. AERONET—A Federated Instrument Network and Data Archive for Aerosol Characterization. *Remote Sens. Environ.* **1998**, *66*, 1–16. [CrossRef]
37. Holben, B.N.; Tanre, D.; Smirnov, A.; Eck, T.F.; Slutsker, I.; Abuhassan, N.; Newcomb, W.W.; Schafer, J.S.; Chatenet, B.; Lavenu, F.; et al. An emerging ground-based aerosol climatology: Aerosol optical depth from AERONET. *J. Geophys. Res.* **2001**, *106*, 12067–12097. [CrossRef]
38. Ångström, A. The parameters of atmospheric turbidity. *Tellus* **1964**, *16*, 64–75. [CrossRef]
39. Geurts, P.; Ernst, D.; Wehenkel, L. Extremely randomized trees. *Mach. Learn.* **2006**, *63*, 3–42. [CrossRef]
40. Wei, J.; Li, Z.; Cribb, M.; Huang, W.; Xue, W.; Sun, L.; Guo, J.; Peng, Y.; Li, J.; Lyapustin, A.; et al. Improved 1 km resolution PM2.5 estimates across China using enhanced space-time extremely randomized trees. *Atmos. Chem. Phys.* **2020**, *20*, 3273–3289. [CrossRef]
41. Anderson, T.L.; Charlson, R.J.; Winker, D.M.; Ogren, J.A.; Holmén, K. Mesoscale Variations of Tropospheric Aerosols. *J. Atmos. Sci.* **2003**, *60*, 119–136. [CrossRef]
42. Martins, V.S.; Lyapustin, A.; de Carvalho, L.A.S.; Barbosa, C.C.F.; Novo, E.M.L.M. Validation of high-resolution MAIAC aerosol product over South America. *J. Geophys. Res. Atmos.* **2017**, *122*. [CrossRef]
43. Levy, R.C.; Remer, L.A.; Kleidman, R.G.; Mattoo, S.; Ichoku, C.; Kahn, R.; Eck, T.F. Global evaluation of the Collection 5 MODIS dark-target aerosol products over land. *Atmos. Chem. Phys.* **2010**, *10*, 10399–10420. [CrossRef]
44. Acharya, P.; Barik, G.; Gayen, B.K.; Bar, S.; Maiti, A.; Sarkar, A.; Ghosh, S.; De, S.K.; Sreekesh, S. Revisiting the levels of Aerosol Optical Depth in south-southeast Asia, Europe and USA amid the COVID-19 pandemic using satellite observations. *Environ. Res.* **2021**, *193*, 110514. [CrossRef] [PubMed]

remote sensing

MDPI

Article

Change of CO Concentration Due to the COVID-19 Lockdown in China Observed by Surface and Satellite Observations

Minqiang Zhou [1], Jingyi Jiang [2,*], Bavo Langerock [1], Bart Dils [1], Mahesh Kumar Sha [1] and Martine De Mazière [1]

1 Royal Belgian Institute for Space Aeronomy (BIRA-IASB), 1180 Brussels, Belgium; minqiang.zhou@aeronomie.be (M.Z.); bavo.langerock@aeronomie.be (B.L.); bart.dils@aeronomie.be (B.D.); mahesh.sha@aeronomie.be (M.K.S.); Martine.DeMaziere@bira-iasb.oma.be (M.D.M.)

2 The College of Forestry, Beijing Forestry University, Beijing 100083, China

* Correspondence: jiangjingyi@bjfu.edu.cn

Abstract: The nationwide lockdown due to the COVID-19 pandemic in 2020 reduced industrial and human activities in China. In this study, we investigate atmospheric carbon monoxide (CO) concentration changes during the lockdown from observations at the surface and from two satellites (TROPOspheric Monitoring Instrument (TROPOMI) and Infrared Atmospheric Sounding Interferometer (IASI)). It is found that the average CO surface concentration in 2020 was close to that in 2019 before the lockdown, and became 18.7% lower as compared to 2019 during the lockdown. The spatial variation of the change in the CO surface concentration is high, with an 8–27% reduction observed for Beijing, Shanghai, Chengdu, Zhengzhou, and Guangzhou, and almost no change in Wuhan. The TROPOMI and IASI satellite observations show that the CO columns decreased by 2–13% during the lockdown in most regions in China. However in South China, there was an 8.8% increase in the CO columns observed by TROPOMI and a 36.7% increase observed by IASI, which is contrary to the 23% decrease in the surface CO concentration. The enhancement of the CO column in South China is strongly affected by the fire emissions transported from Southeast Asia. This study provides an insight into the impact of COVID-19 on CO concentrations both at the surface and in the columns in China, and it can be extended to evaluate other areas using the same approach.

Keywords: carbon monoxide; COVID-19; China; surface concentration; TROPOMI; IASI

Citation: Zhou, M.; Jiang, J.; Langerock, B.; Dils, B.; Sha, M.K.; De Mazière, M. Change of CO Concentration Due to the COVID-19 Lockdown in China Observed by Surface and Satellite Observations. *Remote Sens.* **2021**, *13*, 1129. https://doi.org/10.3390/rs13061129

Academic Editor: Maria João Costa

Received: 24 February 2021
Accepted: 13 March 2021
Published: 16 March 2021

Publisher's Note: MDPI stays neutral with regard to jurisdictional claims in published maps and institutional affiliations.

1. Introduction

The COVID-19 worldwide pandemic has caused millions of deaths, reported by the World Health Organization (WHO) coronavirus disease dashboard. The first COVID-19 patient was detected in Wuhan, Hubei Province, China, in December 2019, and then the disease quickly spread to the whole country before the Chinese New Year 2020 [1]. To prevent the further spread of the outbreak, the Chinese government carried out a nationwide lockdown starting on 23 January 2020 in Wuhan and extending rapidly (in 6 days) to all other provinces [2]. The lockdown outside of Hubei province was eased at the beginning of March, while it continued to 25 March for Hubei province and to 8 April for Wuhan [3].

The strict measures related to COVID-19 had a large impact on economic activities, including energy production, industrial activities, and transportation [4,5]. As a result, the emissions of many atmospheric components were significantly reduced [6–9]. There was a 3.7% decrease in Chinese carbon dioxide (CO_2) emissions in the first half of 2020 related to the COVID-19 pandemic [10]. The reduction mainly occurred in January and February, and the CO_2 emissions in March returned to the emission level of 2019 [11] as the lockdowns were gradually relaxed. Bauwens et al. reported an average 40% decrease in nitrogen dioxide (NO_2) column concentration from satellite measurements over Chinese cities due to measures against the coronavirus outbreak [12]. Based on NO_2 surface observations,

Feng et al. pointed out that nitrogen oxide (NO_X) emissions were reduced by 36% in China due to the COVID-19 lockdown measures [13].

CO is a pollutant that also plays an important role in atmospheric chemistry, e.g., the formation of tropospheric ozone. CO is predominantly removed by OH [14], and the lifetime of CO is relatively long (weeks to months) as compared to other air pollutants [15]. The main atmospheric CO sources are anthropogenic emissions and biomass burning [16], primarily when carbon fuels are not burned completely. According to the Emissions Database for Global Atmospheric Research (EDGAR) v5.0 [17], the anthropogenic CO emissions in China are dominated by residential cooking and heating, and combustion for manufacturing, the power industry, and road transportation.

Previous studies have been carried out to understand the reduction in CO surface concentration due to the COVID-19 lockdown in China on city and regional scales. There was an average 22.7% decrease in the CO surface concentration in Wuhan during the lockdown as compared to the period before lockdown [18]. Shi and Brasseur found that the CO surface concentration during the lockdown decreased from 1.2–1.5 to 0.7–1.0 mg/m^3 before the lockdown in northern China [19]. However, there is a large seasonal variation in CO surface concentrations in eastern Asia, with a maximum in winter and a minimum in summer [20], which has not been taken into account in the these studies. The atmospheric compositions can also be observed by the satellite remote sensing technique using their absorption or emission spectra, which has been applied to understand the CO column changes due to the COVID-19 lockdown in China [21,22]. It is important to compare the CO concentration changes observed by the surface and satellite measurements. However, to our knowledge, few studies have been performed to investigate this. Here, we aim at looking into the changes in CO concentration due to the COVID-19 lockdown in China using both surface and satellite observations, and investigating whether CO reduction can be observed by both surface and satellite observations. The data and method are presented in Section 2. To reduce the impact from the seasonal variation of CO, the observations in 2020 are compared to similar observations in 2019. In Section 3, the changes in CO surface concentrations in China and the variations at six megacities are discussed. In addition, the column-averaged dry-air mole fraction of CO (XCO) observed from the TROPOspheric Monitoring Instrument (TROPOMI) onboard the Sentinel 5 Precursor (S5P) satellite and the CO column observed from the Infrared Atmospheric Sounding Interferometer (IASI) onboard the Meteorological Operational (Metop)-B satellite are analyzed and compared to the surface measurements. The discussions about the results as well as the limitations of this study are carried out in Section 4 and the conclusions are drawn in Section 5.

2. Materials and Methods

2.1. Data

Hourly CO surface observations are carried out at air pollution monitoring sites by the Ministry of Ecology and Environment of China (http://www.mee.gov.cn/, accessed on 10 March 2021). The CO concentration is reported in units of mg/m^3. In this study, we used the sites where observations were available in both 2019 and 2020: we found 1375 sites in China (Figure 1), including 12 sites at Beijing, 10 sites at Shanghai, 11 sites at Wuhan, 10 sites at Chengdu, 9 sites at Zhengzhou, and 12 sites at Guangzhou. Note that few sites were in western China, and most sites were located in highly polluted regions with large CO anthropogenic emissions.

The offline level 2 CO product from the TROPOMI was used in this study, which was downloaded from https://scihub.copernicus.eu/ accessed on 10 March 2021. The XCO product was retrieved from the 2.3 μm spectral range of the shortwave infrared solar radiance measurements under clear-sky conditions; it is sensitive to the tropospheric boundary layer [23]. The spatial resolution of the TROPOMI XCO observations was 7.2×7.2 km^2 for the footprint at nadir before 6 August 2019 and changed to 7.2×5.6 km^2 afterwards. The overpass time was about 13:00. The TROPOMI CO level 2 measurements were filtered out with the qa_value less than 0.5, which is recommended by the user guide (https://sentinel.esa.int/documents/247904/3541451/Sentinel-5P-Carbon-Monoxide-Level-2-Product-Readme-File, accessed on 10 March 2021). After that, the daily TROPOMI level 2 observations were binned to 0.05° × 0.05° (latitude by longitude) grids as the level 3 data, and we studied the CO changes based on these level 3 daily products.

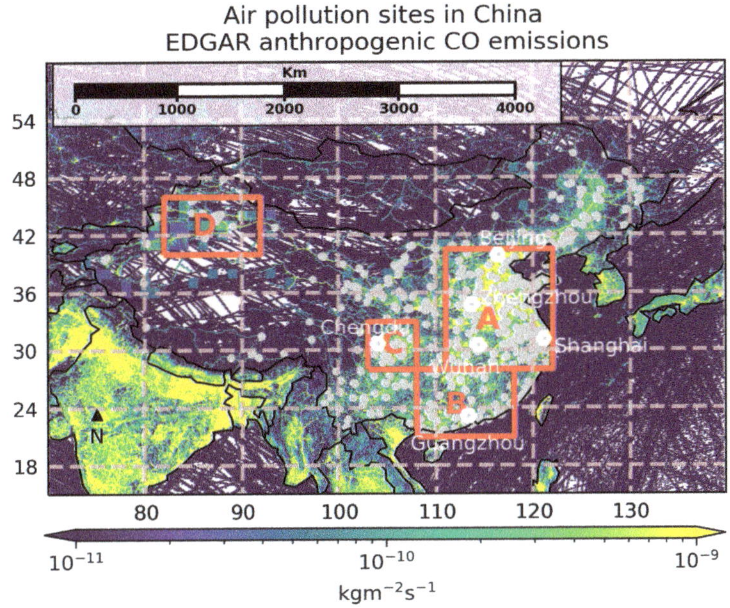

Figure 1. The location of the air pollution sites (light gray dots), six megacities (white hexagons) and regions (red boxes), together with the CO anthropogenic emission annual mean in 2015 from the Emissions Database for Global Atmospheric Research (EDGAR) v5.0 inventory.

The IASI level 2 CO column dataset was processed using the Fast Optimal Retrievals on Layers for IASI (FORLI) software [24] by the Université Libre de Bruxelles, Laboratoire Atmosphères, Milieux, Observations Spatiales (ULB-LATMOS) before 14 May 2019 (v20140922) and by the European Organisation for the Exploitation of Meteorological Satellites (EUMETSAT) afterward (v6.5.0), which was downloaded from https://iasi.aeris-data.fr/cos_iasi_b_arch/ accessed on 10 March 2021. The field of view at nadir of the IASI instrument is about 12 km. The CO is retrieved from the thermal infrared spectra in the spectral range 4.58 to 4.69 μm, so that IASI CO product is more sensitive to the mid- and upper-troposphere, and less sensitive to the lower-troposphere [25]. IASI provides both daytime and nighttime CO measurements (9:30 and 21:30). As the diurnal variation in CO at the mid- and upper-troposphere is much weaker than for the surface, we used both daytime and nighttime IASI CO observations to generate the 0.5° × 0.5° daily product.

As fire emissions are an important source of CO, we used the Visible Infrared Imaging Radiometer Suite (VIIRS) 375 m data [26] onboard the Suomi National Polar-Orbiting Partnership (Suomi NPP) satellite to understand the fire impacts in 2019 and 2020. The VIIRS sensor has a swath width of 3060 km, which is able to provide complete coverage of the Earth everyday. There are 22 spectral channels, between 0.412 μm and 12.01 μm: 16 channels are moderate resolution bands (M-bands), which have a spatial resolution of 750 m at the nadir; 5 channels are imaging resolution bands (I-bands), which have a spatial resolution of 375 m at the nadir; 1 channel is a one day/night panchromatic band with a spatial resolution of 750 m [27]. The VIIRS fire data were download from https://firms.modaps.eosdis.nasa.gov/ accessed on 10 March 2021.

Apart from the measurements, four emission datasets were used to understand the CO anthropogenic and wildfire fluxes in China. The EDGAR v5.0 and the Regional Emission Inventory in Asia (REAS) v3.2 [28] were used to estimate the CO anthropogenic emissions in China. Note that both the EDGAR v5.0 and the REAS v3.2 only hold data up to 2015 for CO, and there is no information about the CO anthropogenic emissions in 2019 and 2020. The Global Fire Assimilation System (GFAS) [29] and the Fire Inventory from NCAR (FINN) [30] were used to understand the CO wildfire emissions. The GFAS and FINN data are up to date and available for both 2019 and 2020, as they use satellite measurements as the inputs. For both the anthropogenic and wildfire emissions, two datasets were compared to each other to assess the uncertainty.

2.2. Method

The surface and satellite CO data in 2020 were compared to similar observations in 2019 during four periods: the month before the Chinese New Year (BCNY; before lockdown), the month after the Chinese New Year (ACNY; lockdown), the month between 11 March and 10 April (3/11–4/10), and the month between 11 April and 10 May (4/11–5/10). To reduce the impact of the Spring Festival, the national holidays in 2019 (4 February to 10 February) and 2020 (24 January to 2 February) were not considered in our study. We considered that in 2019, BCNY was between 1 January and 3 February, and ACNY was between 11 February and 10 March, and that in 2020, BCNY was between 1 January and 23 January, and ACNY was between 3 February and 10 March. From 3/11–4/10, the lockdown was relaxed at most places in China except Hubei Province, and from 4/10–5/10, the lockdown was officially ended throughout the whole of China. The four periods in 2019 and 2020 are summarized in Figure 2.

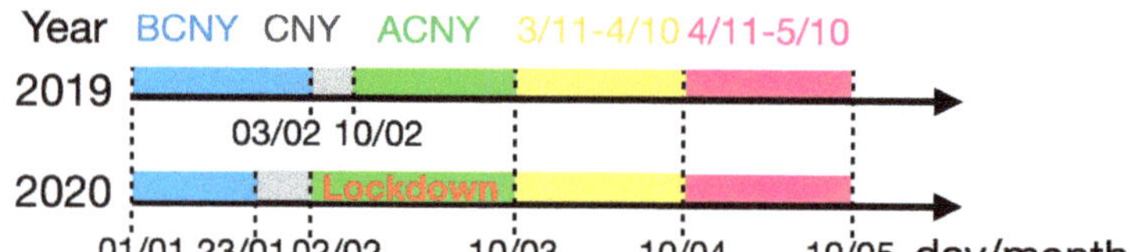

Figure 2. The four periods (before the Chinese New Year (BCNY), after the Chinese New Year (ACNY), 3/11–4/10, and 4/11–5/10) in 2019 and 2020. Note that the Chinese New Year (CNY) national holiday was not considered in this study.

According to the European Centre for Medium-Range Weather Forecasts (ECMWF) ERA5 reanalysis data, the winds at 850 hPa above China during these four periods, especially for the first three periods, were similar in 2019 and 2020 (Figure S1). The layer at 850 hPa (about 1.5 km a.s.l.) is between the lower troposphere and the free atmosphere, as it is close to the Planetary Boundary Layer (PBL) height. On a small scale, such as in a city, the winds in 2019 and 2020 could be very different, but the winds in 2019 and 2020 were generally similar in both wind speed and wind direction on a large scale, such as for the whole of China. Therefore, it is indicated that the changes in CO concentration during the COVID-19 lockdown on the national scale had limited influence from meteorological conditions.

The relative difference in CO concentration at the surface observed by the air pollution sites or in the column observed by the satellite measurements during these periods between 2020 and 2019 was calculated as ($\Delta CO = (2020 - 2019)/2019 \times 100\%$). Then, the mean and standard deviation (std) of the differences were derived from all measurement locations (sites or grids) within a city, a region, or the whole of China:

$$\Delta CO_m = \frac{\sum(\Delta CO_i)}{N}, \quad (1)$$

$$\Delta CO_{std} = \sqrt{\frac{\sum(\Delta CO_i - \Delta CO_m)^2}{N}}, \quad (2)$$

where N is the total number of locations and i is the index of the location. To reduce the impact from outliers, we also used the median when comparing CO changes at the surface with those in the column.

3. Results and Discussions

3.1. CO Surface Concentration

The CO surface concentrations during the four periods in 2019, together with the relative differences between 2020 and 2019, are shown in Figure 2. The mean CO concentrations were 1.21, 1.00, 0.74, 0.71 mg/m^3 during BCNY, ACNY, 3/11–4/10, and 4/11–5/10 in 2019, respectively. There was a large month-to-month variation, and the average CO concentration during ACNY was about 17% less than that during BCNY in 2019.

The mean and std of the relative difference between 2020 and 2019 at all sites are 1.1 ± 24.3%, −18.7 ± 22.2%, −6.2 ± 20.2%, and −4.8 ± 23.6% during BCNY, ACNY, 3/11–4/10, and 4/11–5/10, respectively (Table 1). The CO concentrations during BCNY in 2019 and 2020 were at the same level. The mean difference during ACNY indicates that there was an 18.7% reduction in CO surface concentration due to the COVID-19 lockdown. The reduction in CO surface concentration is also observed for 3/11–4/10 and 4/11–5/10, but the amplitudes become much weaker as compared to that during ACNY. The large std (20–24%) suggests that the spatial variability of CO surface concentration changes across China is high, as CO is affected by local as well as transported emissions from hundreds and thousands of kilometers away due to its lifetime of weeks to months.

Table 1. The mean and standard deviation (std) of the relative change in CO surface concentrations.

	BCNY	ACNY	3/11–4/10	4/11–5/10
China (1375 sites)	1.1 ± 24.3%	−18.7 ± 22.2%	−6.2 ± 20.2%	−4.8 ± 23.6%
Beijing (12 sites)	−12.5 ± 5.6%	−8.0 ± 11.3%	−15.6 ± 14.4%	13.1 ± 5.2%
Shanghai (10 sites)	8.9 ± 2.6%	−20.3 ± 3.1%	−25.4 ± 2.1%	7.1 ± 3.1%
Wuhan (11 sites)	−20.0 ± 2.4%	0.4 ± 2.7%	−6.5 ± 4.8%	−23.1 ± 2.7%
Chengdu (10 sites)	1.1 ± 2.8%	−27.0 ± 3.4%	−16.2 ± 5.1%	−15.8 ± 6.1%
Zhengzhou (9 sites)	−3.1 ± 5.3%	−25.0 ± 6.0%	−12.4 ± 4.1%	−17.2 ± 2.1%
Guangzhou (12 sites)	−13.7 ± 1.9%	−25.9 ± 2.3%	−6.9 ± 4.0%	−10.5 ± 3.7%

As the change in CO surface concentration varied with location (Figure 3), we investigated in detail six megacities (Beijing, Shanghai, Wuhan, Chengdu, Zhengzhou, and Guangzhou). The hourly means and stds of CO surface concentrations in these cities during the four periods in 2019 and 2020, together with their relative changes between 2020 and 2019, are shown in Figure 4. The diurnal variations of CO surface concentrations in these cities are similar, with two peaks around 10:00 and 24:00 local hours. During BCNY, the phase and amplitude of the diurnal variations in 2020 were close to those in 2019. During ACNY, except in Wuhan, the peak-to-peak amplitudes of the diurnal variations became smaller in 2020 as compared to 2019 despite the large stds. Large reductions of the CO surface concentrations of 20–27% are observed during ACNY at Shanghai, Chengdu, Zhengzhou, and Guangzhou. Reductions by 6–25% in the CO surface concentration are also observed from 3/11–4/10 in these cities. A reduction during BCNY is also observed in Guangzhou, but it is less significant as compared to that during ACNY. However in Beijing, the reduction of the CO surface concentration during ACNY was only 8%, which is less than the observed 12% reduction during BCNY and 16% reduction from 3/11–4/10. The relatively low reduction in CO during ACNY in Beijing was affected by the meteorological background. Previous studies found that the wind speed was decreased by

20% and the PBL heights were generally lower during the lockdown period as compared to the climatology for Beijing, leading to higher surface concentrations of atmospheric pollutants [31,32].

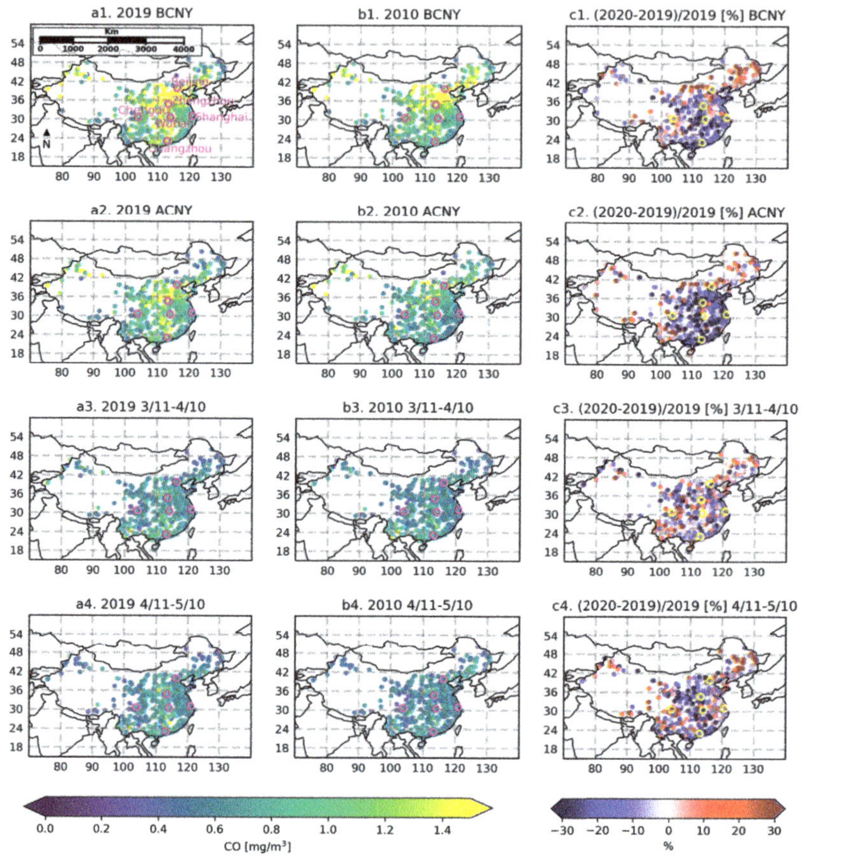

Figure 3. The mean CO surface concentrations in units of mg/m^3 observed at all sites in China during BCNY, ACNY, 3/11–4/10, and 4/11–5/10 in 2019 (**first column**) and 2020 (**second column**), together with their percentage differences between 2020 and 2019 ((2020–2019)/2019 × 100%) (**third column**). The six megacities are marked as the purple (**a**,**b**) and yellow (**c**) circles.

The city of Wuhan shows a behavior that is different from the five other cities: the CO surface concentration in 2020 was even slightly larger than that in 2019 during ACNY but was about 20% less than that in 2019 during BCNY. As the city was hit heavily by the virus, the most strict measures were carried out in Wuhan. More than a 50% reduction in atmospheric NO_2 concentrations was observed from both satellite measurements of column abundances [12] and surface in-situ observations [18] during the lockdown period. Apart from anthropogenic emissions, biomass burning is also an important CO source [33]. The VIIRS satellite observed many fires (burning or combustion at places giving out bright light, heat, and smoke) in Wuhan and in the northern area of Wuhan during BCNY, and the fires were almost extinguished during ACNY in 2019. In contrast to 2019, there was almost no fire observed during BCNY, but more fires existed during ACNY in 2020 (Figure S2). First, we looked at the CO wildfire emissions from the GFAS during BCNY and ACNY in 2019 and 2020. Consistent with VIIRS fire measurements, the CO wildfire emissions during BCNY in 2019 were higher than those in 2020, and the CO wildfire emissions during ACNY in 2019 were lower than those in 2020. However, the CO wildfire emissions were much lower as compared to the CO anthropogenic emissions from the EDGAR v5.0 and the REAS v3.2 around Wuhan (Figure S3). There were two things to be addressed there: (1) to assess the uncertainty of the CO wildfire emission, we compared the GFAS with the FINN. It was found that the difference between GFAS and FINN CO wildfire emissions around Wuhan was within 20%; (2) the anthropogenic CO emission from EDGAR v5.0 or REAS v3.2 was only available for 2015, and it was decreasing during the last decade in China with an annual change of about 3–4% [34]. Even though we took the 4%/year decrease in the CO anthropogenic emissions into account, the contributions from the CO wildfire emissions were still less than 1.0% of the CO anthropogenic emissions during ACNY and BCNY in 2019 and 2020 within the $1.0^{\circ} \times 1.0^{\circ}$ box around Wuhan. In summary, the change in CO surface concentration in Wuhan cannot be explained by the local wildfire emissions (biomass burning). Second, we looked at the concentrations of other air pollutants (NO_2, SO_2, $PM_{2.5}$, PM_{10}) in 2019 and 2020 in Wuhan (Figure S4). The averaged NO_2, SO_2, $PM_{2.5}$, PM_{10} concentrations during BCNY in 2020 were 17%, 9%, 34%, and 31% less than those in 2019. The decreases of those four air pollutants are consistent with the 20% decrease in CO during BCNY in 2020 as compared to 2019. The averaged NO_2, $PM_{2.5}$, PM_{10} concentrations in Wuhan during ACNY in 2020 were 51%, 43% and 42% less than those in 2019. However, SO_2 and CO increased slightly during ACNY in 2020 as compared to 2019. The similar behavior of CO and SO_2 suggests that these two gases come from common sources, e.g., the burning of fossil fuels by power plants and other industrial facilities. Finally, we looked at the VIIRS and MODIS fire observations inside Wuhan, where more fires were observed above a large coke factory (Wuhan Pingmei Wugang Joint Coking Company) during ACNY in 2020 as compared to 2019 (Figure S5). According to the sources of SO_2, CO, and NO_X in Asia [35], it is inferred that the CO and SO_2 emissions from industry (such as the coke factory) during ACNY in 2020 were larger than the reduced emissions from road transportation.

3.2. CO Column Observed from Satellites

The TROPOMI XCO and IASI CO column measurements in 2019, together with the relative differences between 2020 and 2019 during the four periods, are shown in Figure 5. In general, the TROPOMI and IASI measurements have a similar spatial distribution in China. The means and stds of XCO observed by TROPOMI in 2019 in China are 110.1 ± 24.1, 109.7 ± 24.9, 113.3 ± 26.9, and 112.9 ± 22.8 ppb during the BCNY, ACNY, 3/11–4/10, and 4/11–5/10 periods, respectively. There is almost no change in the mean XCO in China during these four periods, which is different from the large month-to-month variation of CO surface concentration. The means and stds of CO columns observed by IASI in 2019 in China are $1.83 \pm 0.47 \times 10^{18}$, $1.98 \pm 0.50 \times 10^{18}$, $2.18 \pm 0.54 \times 10^{18}$ and $2.25 \pm 0.53 \times 10^{18}$ molecules/cm^2 during BCNY, ACNY, 3/11–4/10, and 4/11–5/10 periods, respectively. The month-to-month change of the CO column is opposite to that observed at the surface.

As satellite measurements are contaminated by cloud, the variability in them is relatively high. To reduce random uncertainty, the satellite measurements (both TROPOMI and IASI) were averaged on regional scales, and we focused on the CO changes in four regions with high values (Figure 5a2,c2; Figure 1): North, East, and Central China (A); South China (B), Sichuan basin (C), and Urumqi region (D). The quantitative estimates of the CO changes are shown in Figure 6 and Table 2. The medians of the XCO relative changes during ACNY in 2020 relative to 2019 observed by the TROPOMI satellite are −10.5%, 8.8%, −1.9%, and −4.6% in regions A, B, C, and D, respectively. The medians of the CO column relative changes during ACNY in 2020 relative to 2019 observed by the IASI satellite are −13.3%, 36.7%, −1.8%, and −3.6% in regions A, B, C, and D, respectively. The largest reduction in CO concentration was found by both satellites in Region A during ACNY in 2020, with a minimum in the region between Zhengzhou and Beijing. The reductions in the CO column during ACNY were also significant in Regions C and D, especially when we compare the CO changes during ACNY to the changes during BCNY, 3/11–4/10, and 4/11–5/10. However, there was an 8.8% increase in XCO observed by TROPOMI and a 36.7% increase in CO columns observed by IASI for Region B, which was related to the fires in Southeast Asia, and will be discussed later. To compare the satellite with the surface observations, the relative changes in CO surface concentrations for the same regions are also shown in Figure 6. The medians of the relative changes in CO surface concentrations during ACNY in 2020 as compared to that in 2019 are −25.1%, −23.1%, −15.8%, and −18.2%, for Regions A, B, C, and D, respectively. At these regions, the CO surface concentrations decreased dramatically during the lockdown and then increased afterward, with Region A being the most prominent.

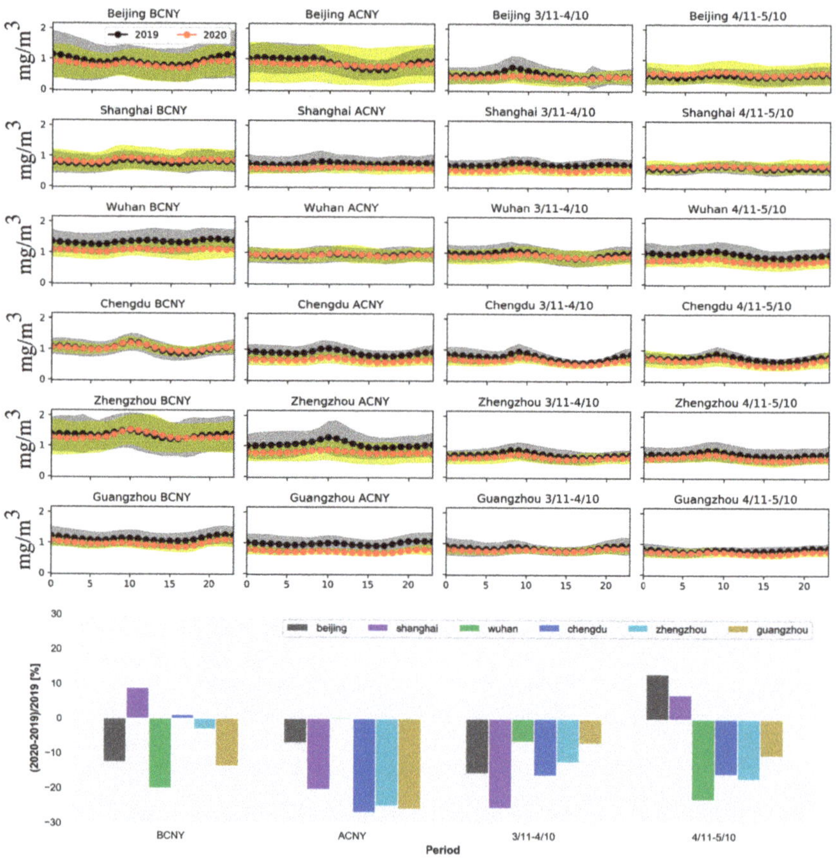

Figure 4. Upper: the hourly means (solid line) and standard deviations (shadow) of CO surface concentrations observed in Beijing, Shanghai, Wuhan, Chengdu, Zhengzhou, and Guangzhou during BCNY (**first column**), ACNY (**second column**), 3/11–4/10 (**third column**), and 4/11–5/10 (**last column**) in 2019 and 2020. Lower: the relative changes in CO surface concentrations between 2020 and 2019 in these six megacities during BCNY, ACNY, 3/11–4/10, and 4/11–5/10.

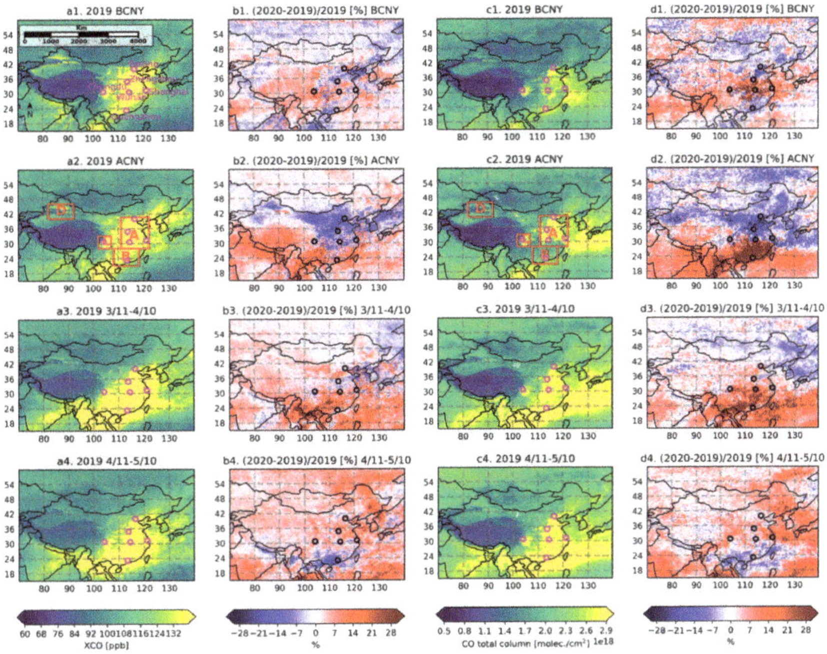

Figure 5. The TROPOspheric Monitoring Instrument (TROPOMI) satellite XCO observations in units of ppb (**a1–a4**) and the Infrared Atmospheric Sounding Interferometer (IASI) CO column observations in unit of molec./cm^2 (**c1–c4**) over China during BCNY, ACNY, 3/11–4/10, and 4/11–5/10 in 2019, together with the relative differences between 2020 and 2019 ((2020–2019)/2019 × 100%) (TROPOMI: **b1–b4**, IASI: **d1–d4**). The six megacities are marked as the purple and black circles. The four regions are marked in (**a2**,**c2**).

The surface and satellite observations both showed reductions during ACNY in Regions A, C, and D, but the reduction in CO columns was less significant as compared to the reduction in the CO surface concentrations. The satellites observe the column CO abundance. The CO partial columns in the PBL only account for 20–40% of the total columns in these regions according to the Copernicus Atmosphere Monitoring Service (CAMS) operational model [36]. Assuming that there is no change in CO partial columns above the PBL, the magnitude of the CO total column reduction is expected to be 2.5–5 times less than that at the surface. In this case, the relative changes in CO during ACNY in 2020 observed by the satellite and surface observations are generally in good agreement for Regions A, C, and D.

A large disagreement between the satellite and surface observations was found in Region B, where the CO surface concentrations were significantly reduced (>20%) during the lockdown in 2020, while the TROPOMI and IASI observations show that the CO during ACNY and 3/11–4/10 in 2020 was much larger than that in 2019. As the weather conditions between January and March (cool and dry) are favorable for burning, there are vast numbers of fires that emerge across the countryside in Southeast Asia (Myanmar, Laos, Thailand, and Cambodia). The VIIRS satellite detected more fires in Southeast Asia during ACNY and 3/11–4/10 in 2020 as compared to 2019 (Figure 7). The CO columns in Southeast Asia observed by TROPOMI and IASI during ACNY and 3/11–4/10 in 2020 were also increased as compared to 2019 (Figure 5). The CO wildfire emissions from GFAS in March 2019 and March 2020 in Southeast Asia (blue box in Figure 7) were 1.75×10^{-10} and 9.88×10^{-10} kg/m^2/s, respectively. Both the absolute values and the variation of CO wildfire emissions in Southeast Asia are comparable to the CO anthropogenic emission annual means in 2015 in Region B of 1.05×10^{-9} kg/m^2/s from REAS v3.2 and of 7.13×10^{-10} kg/m^2/s from EDGAR v5.0. As CO has a lifetime of about weeks to months, CO observed in Region B could be transported from the surrounding areas. The 3-day backward trajectories of 2m-height air at local noon for each day during ACNY and 3/11–4/10 in 2020 were simulated by the Hybrid Single-Particle Lagrangian Integrated Trajectory (HYSPLIT) model driven by the National Centers for Environmental/Prediction Global Data Assimilation System (NCEP/GDAS) meteorological data with a $1.0^\circ \times 1.0^\circ$ (latitude by longitude) spatial resolution. Note that we only plotted the backward trajectories at the center of Region B, as the wind is generally harmonized in this region. The backward trajectories from 2 m height at the center of Region B suggest that the CO surface concentration in this region has little influence from the fire in Southeast Asia. The 3-day backward trajectories from a 2 km height at the center of Region B show that the fire emission in Southeast Asia can be transported to South China, which is consistent with the winds at 750 hPa from the ERA5 reanalysis data. As a result, the CO column in Region B is strongly affected by the fire in Southeast Asia, and more fires in 2020 led to a CO enhancement in the free troposphere in South China during the lockdown observed by the satellite. The CO increase during ACNY and 3/11–4/10 in 2020 observed by IASI was even larger than that observed by TROPOMI, as the IASI retrieval is more sensitive to the mid- and upper-troposphere.

Table 2. The median of the relative changes in CO surface concentration observed by surface measurements, and in CO columns observed by TROPOMI and IASI satellite measurements during four periods in each region.

		BCNY	ACNY	3/11–4/10	4/11–5/10
Surface	Region A	−5.9%	−25.1%	−9.4%	−7.0%
	Region B	−16.7%	−23.1%	−7.3%	−16.4%
	Region C	−2.4%	−15.8%	−7.0%	−2.6%
	Region D	−1.7%	−18.2%	−19.8%	−14.0%
TROPOMI	Region A	−1.3%	−10.5%	0.6%	5.5%
	Region B	0.2%	8.8%	11.7%	−6.4%
	Region C	7.0%	−1.9%	3.0%	3.7%
	Region D	0.1%	−4.6%	1.8%	3.9%
IASI	Region A	3.5%	−13.3%	2.8%	4.1%
	Region B	−2.9%	36.7%	20.6%	−3.2%
	Region C	9.4%	−1.8%	16.6%	3.8%
	Region D	−0.3%	−3.6%	0.8%	0.6%

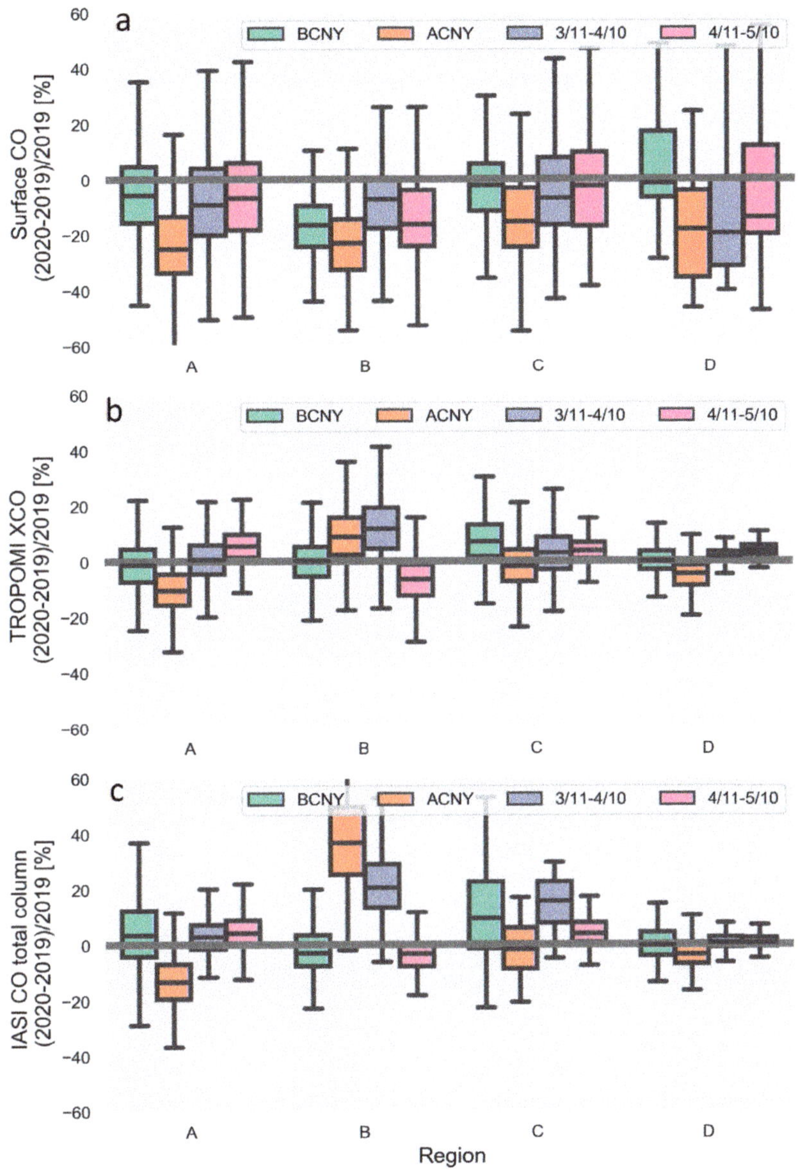

Figure 6. Box plots of the CO changes from the surface (**a**), TROPOMI (**b**), and IASI observations (**c**) during 4 periods in 2020 against those in 2019. Each box plot shows the values of relative difference for the maximum (top of solid line), 75th percentile (top of box), median (line through middle of box), 25th percentile (bottom of box) and minimum (bottom of solid line) of the distribution.

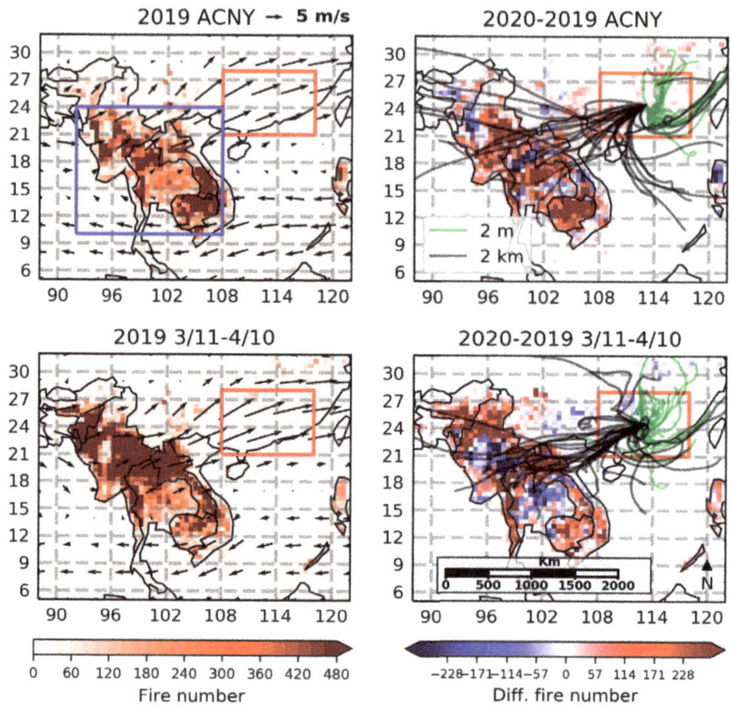

Figure 7. The number of fires observed by the Visible Infrared Imaging Radiometer Suite (VIIRS) satellite in 0.5° × 0.5° (latitude by longitude) grids over Southeast Asia (blue box) together with the wind at the 750 hPa from the ERA5 reanalysis data during ACNY and 3/11–4/10 in 2019 (**left**) and the difference in the number of fires between 2020 and 2019 (**right**). The green and black lines in the right panels are 3-day backward trajectories at 12:00 (local time) from 2 m and 2 km heights at the center of Region B (red box) for each day during ACNY and 3/11–4/10 2020 simulated by the Hybrid Single-Particle Lagrangian Integrated Trajectory (HYSPLIT) model.

4. Discussions

In this study, we investigated CO changes based on both CO surface measurements and satellite column measurements. CO reduction was observed by both surface and satellite observations during the COVID-19 lockdown at most places in China. We have highlighted the importance of seasonal variations of CO surface concentration, which must be taken into account when looking at the CO changes during the COVID-19 lockdown, but have not been done in several previous studies [18,19]. In addition, we found that the specific changes in the industrial emissions at the city scale are important to the changes in CO surface concentration at Wuhan, which are suggested by the simultaneous SO_2 measurements and VIIRS/MODIS fire measurements. However, the limitation of this study is that the impact of the industrial emission on the CO change was only discussed qualitatively, because up-to-date CO anthropogenic emissions for the year 2020 are not currently available. A further study could focus on the application of the inverse modeling approach with the surface CO measurements as the inputs to optimize each anthropogenic component.

Different from the CO surface concentration, the changes in CO columns during the COVID-19 lockdown in China observed by TROPOMI satellite measurements using the difference between 2019 and 2020 in this study are similar to the results using only 2020 measurements before and after the lockdown [21], because the XCO means from TROPOMI were almost the same during these four periods. However, the month-to-month variation in CO columns observed by IASI cannot be ignored. In order to reduce the uncertainty, the satellite measurements were only discussed with the median values during each period on the regional scale. The changes in the CO columns observed by satellites are generally consistent with those at the surface in most regions in China under the assumption that the CO concentration above the PBL is not greatly changed. The assumption works well for NO_x [12], as it has a short lifetime of several hours in the atmosphere. However, due to the relatively long lifetime of CO, the assumption does not work for CO in Region B, where the CO concentration above the PBL was strongly affected by the fire emissions transported from Southeast Asia. We addressed the fact that the CO changes in the free atmosphere are important when comparing the surface and satellite measurements.

5. Conclusions

Surface observations have shown that CO concentrations were at the same level during BCNY in 2019 and 2020, and there was a mean reduction of 18.7% during ACNY in 2020 as compared to 2019, from 1375 sites in China due to the COVID-19 lockdown. Reductions in CO surface concentration were also observed from 3/11–4/10 and 4/11–5/10 in 2020, but they were smaller than the reduction during ACNY. As the spatial variability of CO surface concentration changes across China is high, we investigated the CO changes at six megacities specifically. Large reductions in CO concentration between 20% and 27% during ACNY in 2020 were found in Shanghai, Chengdu, Zhengzhou, and Guangzhou. The CO surface reduction during ACNY in Beijing was only 8%, which may be explained by the exceptional meteorological conditions in that period in 2020. The most strict measures related to COVID-19 were carried out at Wuhan, but there was no decrease in the CO surface concentration observed during the lockdown in 2020 as compared to 2019. By looking at other air pollutants in Wuhan, we found that NO_2, $PM_{2.5}$ and PM_{10} were significantly reduced (>40%) during ACNY in 2020 as compared to 2019, and SO_2 and CO were both slightly increased. The similar behavior of CO and SO_2 suggests that they came from common sources, e.g., the burning of fossil fuels by industrial facilities. The TROPOMI and IASI CO column observations captured the reduction in CO columns (by 2 to 13%) during ACNY in Regions A, C, and D, but the reductions in CO columns were less significant than the reductions in the surface CO concentrations. However, the TROPOMI and IASI observations show that there were 8.8% and 36.7% CO column enhancements during ACNY in 2020 in Region B, which is contrary to the significant reduction (>20%) observed in CO surface concentrations.

Supplementary Materials: The following are available online at https://www.mdpi.com/2072-4292/13//1129/s1, Figure S1: the wind above China during four periods in 2019 and 2020. Figure S2: VIIRS fire map around Wuhan during BCNY and ACNY in 2019 and 2020. Figure S3: GFAS CO wildfire emissions around Wuhan during BCNY and ACNY in 2019 and 2020, together with the CO anthropogenic emissions from EDGAR v5.0 and REAS v3.2 around Wuhan in 2015. Figure S4: The time series of CO, SO_2, NO_x, $PM_{2.5}$ and PM_{10} in Wuhan. Figure S5: VIIRS fire map inside Wuhan during BCNY and ACNY in 2019 and 2020.

Author Contributions: Conceptualization, M.Z. and M.D.M.; methodology, M.Z. and J.J.; writing—original draft preparation, M.Z. and J.J.; writing—review and editing, J.J., B.L., B.D., M.K.S., and M.D.M.; visualization, M.Z. All authors have read and agreed to the published version of the manuscript.

Funding: This research was funded by S5P-MPC (No. 4000117151/16/I-LG) and CAMS84.

Data Availability Statement: The TROPOMI CO data are publicly available at ESA Copernicus Open Access Hub https://scihub.copernicus.eu/. The IASI satellite data are publicly available at https://iasi.aeris-data.fr/cos_iasi_b_arch/. The surface CO measurements are publicly available at https://quotsoft.net/air/. The VIIRS fire observations are publicly available at https://firms.modaps.eosdis.nasa.gov/.

Acknowledgments: The authors would like to thank the Ministry of Ecology and Environment of China for providing the CO surface observations, ESA for making the TROPOMI satellite data publicly available, ULB-LATMOS and EUMETSAT for providing the IASI data, and NASA for providing the VIIRS fire products.

Conflicts of Interest: The authors declare no conflict of interest.

References

1. Wu, Z.; McGoogan, J. Characteristics of and Important Lessons From the Coronavirus Disease 2019 (COVID-19) Outbreak in China. *JAMA* **2020**, *323*, 1239–1242. [CrossRef] [PubMed]
2. Tian, H.; Liu, Y.; Li, Y.; Wu, C.H.; Chen, B.; Kraemer, M.U.G.; Li, B.; Cai, J.; Xu, B.; Yang, Q.; et al. An investigation of transmission control measures during the first 50 days of the COVID-19 epidemic in China. *Science* **2020**, *368*, 638–642. [CrossRef] [PubMed]
3. Zhou, T.; Nguyen, T.v.T.; Zhong, J.; Liu, J. A COVID-19 descriptive study of life after lockdown in Wuhan, China. *R. Soc. Open Sci.* **2020**, *7*, 200705. [CrossRef] [PubMed]
4. Ali, G.; Abbas, S.; Qamer, F.M.; Wong, M.S.; Rasul, G.; Irteza, S.M.; Shahzad, N. Environmental impacts of shifts in energy, emissions, and urban heat island during the COVID-19 lockdown across Pakistan. *J. Clean. Prod.* **2021**, *291*, 125806. [CrossRef]
5. Sharifi, A.; Khavarian-Garmsir, A.R. The COVID-19 pandemic: Impacts on cities and major lessons for urban planning, design, and management. *Sci. Total Environ.* **2020**, *749*, 142391. [CrossRef] [PubMed]
6. Petetin, H.; Bowdalo, D.; Soret, A.; Guevara, M.; Jorba, O.; Serradell, K.; Pérez García-Pando, C. Meteorology-normalized impact of the COVID-19 lockdown upon NO2 pollution in Spain. *Atmos. Chem. Phys.* **2020**, *20*, 11119–11141. [CrossRef]
7. Alqasemi, A.S.; Hereher, M.E.; Kaplan, G.; Al-Quraishi, A.M.F.; Saibi, H. Impact of COVID-19 lockdown upon the air quality and surface urban heat island intensity over the United Arab Emirates. *Sci. Total Environ.* **2021**, *767*, 144330. [CrossRef]
8. Rahman, M.S.; Azad, M.A.K.; Hasanuzzaman, M.; Salam, R.; Islam, A.R.M.T.; Rahman, M.M.; Hoque, M.M.M. How air quality and COVID-19 transmission change under different lockdown scenarios? A case from Dhaka city, Bangladesh. *Sci. Total Environ.* **2021**, *762*, 143161. [CrossRef]
9. Bekbulat, B.; Apte, J.S.; Millet, D.B.; Robinson, A.L.; Wells, K.C.; Presto, A.A.; Marshall, J.D. Changes in criteria air pollution levels in the US before, during, and after Covid-19 stay-at-home orders: Evidence from regulatory monitors. *Sci. Total Environ.* **2021**, *769*, 144693. [CrossRef]
10. Liu, Z.; Ciais, P.; Deng, Z.; Lei, R.; Davis, S.J.; Feng, S.; Zheng, B.; Cui, D.; Dou, X.; Zhu, B.; et al. Near-real-time monitoring of global CO2 emissions reveals the effects of the COVID-19 pandemic. *Nat. Commun.* **2020**, *11*, 5172. [CrossRef]
11. Zheng, B.; Geng, G.; Ciais, P.; Davis, S.J.; Martin, R.V.; Meng, J.; Wu, N.; Chevallier, F.; Broquet, G.; Boersma, F.; et al. Satellite-based estimates of decline and rebound in China's CO2 emissions during COVID-19 pandemic. *Sci. Adv.* **2020**, *6*. [CrossRef]
12. Bauwens, M.; Compernolle, S.; Stavrakou, T.; Müller, J.F.; van Gent, J.; Eskes, H.; Levelt, P.F.; van der, A.R.; Veefkind, J.P.; Vlietinck, J.; et al. Impact of Coronavirus Outbreak on NO2 Pollution Assessed Using TROPOMI and OMI Observations. *Geophys. Res. Lett.* **2020**, *47*, e2020GL087978. [CrossRef]
13. Feng, S.; Jiang, F.; Wang, H.; Wang, H.; Ju, W.; Shen, Y.; Zheng, Y.; Wu, Z.; Ding, A. NOx Emission Changes Over China During the COVID-19 Epidemic Inferred From Surface NO2 Observations. *Geophys. Res. Lett.* **2020**, *47*, e2020GL090080. [CrossRef]
14. Aschi, M.; Largo, A. Reactivity of gaseous protonated ozone: a computational investigation on the carbon monoxide oxidation reaction. *Int. J. Mass Spectrom.* **2003**, *228*, 613–627. [CrossRef]
15. Novelli, P.C.; Masarie, K.A.; Lang, P.M. Distributions and recent changes of carbon monoxide in the lower troposphere. *J. Geophys. Res. Atmos.* **1998**, *103*, 19015–19033. [CrossRef]
16. Granier, C.; Bessagnet, B.; Bond, T.; D'Angiola, A.; Denier van der Gon, H.; Frost, G.J.; Heil, A.; Kaiser, J.W.; Kinne, S.; Klimont, Z.; et al. Evolution of anthropogenic and biomass burning emissions of air pollutants at global and regional scales during the 1980–2010 period. *Clim. Change* **2011**, *109*, 163. [CrossRef]
17. Crippa, M.; Solazzo, E.; Huang, G.; Guizzardi, D.; Koffi, E.; Muntean, M.; Schieberle, C.; Friedrich, R.; Janssens-Maenhout, G. High resolution temporal profiles in the Emissions Database for Global Atmospheric Research. *Sci. Data* **2020**, *7*, 121. [CrossRef] [PubMed]
18. Lian, X.; Huang, J.; Huang, R.; Liu, C.; Wang, L.; Zhang, T. Impact of city lockdown on the air quality of COVID-19-hit of Wuhan city. *Sci. Total Environ.* **2020**, *742*, 140556. [CrossRef]
19. Shi, X.; Brasseur, G.P. The Response in Air Quality to the Reduction of Chinese Economic Activities During the COVID-19 Outbreak. *Geophys. Res. Lett.* **2020**, *47*, e2020GL088070. [CrossRef] [PubMed]
20. Verma, R.L.; Kondo, Y.; Oshima, N.; Matsui, H.; Kita, K.; Sahu, L.K.; Kato, S.; Kajii, Y.; Takami, A.; Miyakawa, T. Seasonal variations of the transport of black carbon and carbon monoxide from the Asian continent to the western Pacific in the boundary layer. *J. Geophys. Res. Atmos.* **2011**, *116*. [CrossRef]
21. Fan, C.; Li, Y.; Guang, J.; Li, Z.; Elnashar, A.; Allam, M.; de Leeuw, G. The Impact of the Control Measures during the COVID-19 Outbreak on Air Pollution in China. *Remote Sens.* **2020**, *12*, 1613. [CrossRef]
22. Miyazaki, K.; Bowman, K.; Sekiya, T.; Jiang, Z.; Chen, X.; Eskes, H.; Ru, M.; Zhang, Y.; Shindell, D. Air Quality Response in China Linked to the 2019 Novel Coronavirus (COVID-19) Lockdown. *Geophys. Res. Lett.* **2020**, *47*, e2020GL089252. [CrossRef] [PubMed]
23. Landgraf, J.; aan de Brugh, J.; Scheepmaker, R.; Borsdorff, T.; Hu, H.; Houweling, S.; Butz, A.; Aben, I.; Hasekamp, O. Carbon monoxide total column retrievals from TROPOMI shortwave infrared measurements. *Atmos. Meas. Tech.* **2016**, *9*, 4955–4975. [CrossRef]

24. Hurtmans, D.; Coheur, P.F.; Wespes, C.; Clarisse, L.; Scharf, O.; Clerbaux, C.; Hadji-Lazaro, J.; George, M.; Turquety, S. FORLI radiative transfer and retrieval code for IASI. *J. Quant. Spectrosc. Radiat. Transf.* **2012**, *113*, 1391–1408. [CrossRef]
25. Kerzenmacher, T.; Dils, B.; Kumps, N.; Blumenstock, T.; Clerbaux, C.; Coheur, P.F.; Demoulin, P.; García, O.; George, M.; Griffith, D.W.T.; et al. Validation of IASI FORLI carbon monoxide retrievals using FTIR data from NDACC. *Atmos. Meas. Tech.* **2012**, *5*, 2751–2761. [CrossRef]
26. Schroeder, W.; Oliva, P.; Giglio, L.; Csiszar, I.A. The New VIIRS 375m active fire detection data product: Algorithm description and initial assessment. *Remote Sens. Environ.* **2014**, *143*, 85–96. [CrossRef]
27. Cao, C.; De Luccia, F.J.; Xiong, X.; Wolfe, R.; Weng, F. Early On-Orbit Performance of the Visible Infrared Imaging Radiometer Suite Onboard the Suomi National Polar-Orbiting Partnership (S-NPP) Satellite. *IEEE Trans. Geosci. Remote. Sens.* **2014**, *52*, 1142–1156. [CrossRef]
28. Kurokawa, J.; Ohara, T. Long-term historical trends in air pollutant emissions in Asia: Regional Emission inventory in ASia (REAS) version 3. *Atmos. Chem. Phys.* **2020**, *20*, 12761–12793. [CrossRef]
29. Kaiser, J.W.; Heil, A.; Andreae, M.O.; Benedetti, A.; Chubarova, N.; Jones, L.; Morcrette, J.J.; Razinger, M.; Schultz, M.G.; Suttie, M.; et al. Biomass burning emissions estimated with a global fire assimilation system based on observed fire radiative power. *Biogeosciences* **2012**, *9*, 527–554. [CrossRef]
30. Wiedinmyer, C.; Akagi, S.K.; Yokelson, R.J.; Emmons, L.K.; Al-Saadi, J.A.; Orlando, J.J.; Soja, A.J. The Fire INventory from NCAR (FINN): A high resolution global model to estimate the emissions from open burning. *Geosci. Model Dev.* **2011**, *4*, 625–641. [CrossRef]
31. Huang, X.; Ding, A.; Gao, J.; Zheng, B.; Zhou, D.; Qi, X.; Tang, R.; Wang, J.; Ren, C.; Nie, W.; et al. Enhanced secondary pollution offset reduction of primary emissions during COVID-19 lockdown in China. *Natl. Sci. Rev.* **2020**. [CrossRef]
32. Le, T.; Wang, Y.; Liu, L.; Yang, J.; Yung, Y.L.; Li, G.; Seinfeld, J.H. Unexpected air pollution with marked emission reductions during the COVID-19 outbreak in China. *Science* **2020**, *369*, 702–706. [CrossRef] [PubMed]
33. Van der Werf, G.R.; Randerson, J.T.; Giglio, L.; Collatz, G.J.; Kasibhatla, P.S.; Arellano, A.F., Jr. Interannual variability in global biomass burning emissions from 1997 to 2004. *Atmos. Chem. Phys.* **2006**, *6*, 3423–3441. [CrossRef]
34. Zheng, B.; Chevallier, F.; Ciais, P.; Yin, Y.; Deeter, M.N.; Worden, H.M.; Wang, Y.; Zhang, Q.; He, K. Rapid decline in carbon monoxide emissions and export from East Asia between years 2005 and 2016. *Environ. Res. Lett.* **2018**, *13*, 044007. [CrossRef]
35. Li, M.; Zhang, Q.; Kurokawa, J.I.; Woo, J.H.; He, K.; Lu, Z.; Ohara, T.; Song, Y.; Streets, D.G.; Carmichael, G.R.; et al. MIX: A mosaic Asian anthropogenic emission inventory under the international collaboration framework of the MICS-Asia and HTAP. *Atmos. Chem. Phys.* **2017**, *17*, 935–963. [CrossRef]
36. Inness, A.; Blechschmidt, A.M.; Bouarar, I.; Chabrillat, S.; Crepulja, M.; Engelen, R.J.; Eskes, H.; Flemming, J.; Gaudel, A.; Hendrick, F.; et al. Data assimilation of satellite-retrieved ozone, carbon monoxide and nitrogen dioxide with ECMWF's Composition-IFS. *Atmos. Chem. Phys.* **2015**, *15*, 5275–5303. [CrossRef]

remote sensing

MDPI

Article

A Satellite-Based Land Use Regression Model of Ambient NO_2 with High Spatial Resolution in a Chinese City

Lina Zhang [1], Changyuan Yang [1], Qingyang Xiao [2], Guannan Geng [2], Jing Cai [1], Renjie Chen [1], Xia Meng [1,3,*] and Haidong Kan [1]

1 Key Lab of Public Health Safety of the Ministry of Education and NHC Key Lab of Health Technology Assessment, School of Public Health, Fudan University, Shanghai 200032, China; 19111020024@fudan.edu.cn (L.Z.); bslk1980@163.com (C.Y.); jingcai@fudan.edu.cn (J.C.); chenrenjie@fudan.edu.cn (R.C.); kanh@fudan.edu.cn (H.K.)

2 State Key Joint Laboratory of Environment Simulation and Pollution Control, School of Environment, Tsinghua University, Beijing 100084, China; qingyang_xiao18@163.com (Q.X.); guannangeng@tsinghua.edu.cn (G.G.)

3 Shanghai Key Laboratory of Meteorology and Health, Shanghai Typhoon Institute/CMA, Shanghai 200030, China

* Correspondence: mengxia@fudan.edu.cn

Abstract: Previous studies have reported that intra-urban variability of NO_2 concentrations is even higher than inter-urban variability. In recent years, an increasing number of studies have developed satellite-derived land use regression (LUR) models to predict ground-level NO_2 concentrations, though only a few have been conducted at a city scale. In this study, we developed a satellite-derived LUR model to predict seasonal NO_2 concentrations at a city scale by including satellite-retrieved NO_2 tropospheric column density, population density, traffic indicators, and NO_x emission data. The R^2 of model fitting and 10-fold cross validation were 0.70 and 0.61 for the satellite-derived seasonal LUR model, respectively. The satellite-based LUR model captured seasonal patterns and fine gradients of NO_2 variations at a 100 m × 100 m resolution and demonstrated that NO_2 pollution in winter is 1.46 times higher than that in summer. NO_2 concentrations declined significantly with increasing distance from roads and with increasing distance from the city center. In Suzhou, 84% of the total population lived in areas with NO_2 concentrations exceeding the annual-mean standard at 40 $\mu g/m^3$ in 2014. This study demonstrated that satellite-retrieved data could help increase the accuracy and temporal resolution of the traditional LUR models at a city scale. This application could support exposure assessment at a high resolution for future epidemiological studies and policy development pertaining to air quality control.

Keywords: satellite-based; NO_2; land use regression; exposure assessment

Citation: Zhang, L.; Yang, C.; Xiao, Q.; Geng, G.; Cai, J.; Chen, R.; Meng, X.; Kan, H. A Satellite-Based Land Use Regression Model of Ambient NO_2 with High Spatial Resolution in a Chinese City. *Remote Sens.* **2021**, *13*, 397. https://doi.org/10.3390/rs13030397

Academic Editor: Vijay Natraj

Received: 9 December 2020

Accepted: 21 January 2021

Published: 24 January 2021

1. Introduction

Nitrogen dioxide (NO_2) is not only a primary pollutant mainly from fossil fuel emissions but also a secondary pollutant arising in large part from a photochemical conversion combining NO with O_3 [1,2]. It is a common indicator for traffic-related air pollution and proven to be associated with a myriad of adverse health effects. NO_2 has been positively linked to lung cancer mortality in California by the American Cancer Society Cancer Prevention II Study [3]. In China, short-term exposure to NO_2 was significantly associated with total natural causes mortality and cardiorespiratory disease mortality across 272 cities [4]. Even at or below the current European Air quality limit values, the associations between NO_2 exposure and adverse effects have been found for both short-term and long-term exposure in Europe [5]. In previous epidemiological studies, exposure to NO_2 was mostly evaluated using ground-based fixed monitoring data, interpolation methods, or land use regression (LUR) models [6,7].

The concentrations of NO_2 may decline at a distance of several hundred meters from emission sources [8], and the spatial distributions of NO_2 differ significantly between, and especially within, cities [9,10]. In Canada, variations in NO_2 concentrations within a city further showed a stronger association with cause-specific mortality than that between cities [11]. Thus, it is an essential issue to evaluate intra-urban NO_2 concentrations with a high spatial resolution for epidemiological studies. The LUR models are one of the most common assessment methods used to capture spatial variability of NO_2 with a high spatial resolution, and have been applied in NO_2-related cohort studies in Europe and the United States [9,12–15]. Land use regression models also have been developed for predicting NO_2 concentrations in Chinese cities, including Shanghai, Tianjin, and Wuhan [16–18]. Traditional LUR models highly depend on land use data and have lower temporal resolution, but these do not satisfy the flexible requirements of exposure assessment in epidemiological studies.

Satellite data have been proven to be one of the key predictors for estimating ambient NO_2 concentrations with a high temporal resolution [19–21]. Specifically, a study in Western Europe indicated that the adjusted R^2 of LUR models with satellite data was increased by 0.02–0.06 compared to the models without satellite data with the R^2 of 0.48–0.56 [22]. Other studies showed that the satellite-based LUR models could expand the temporal resolution of traditional LUR models for predicting air pollutants' concentrations, from annual level to monthly or seasonal scales [19,23–25]. NO_2 column density from the Ozone Monitoring Instrument (OMI) aboard satellite Aura is the most commonly used dataset for establishing satellite-based LUR or machine learning models [26–28]. The satellite-based LUR models not only expanded the temporal resolution of traditional ones [19], but also simultaneously helped improve model performance [22,29,30]. However, in China, most of these studies were conducted at regional or national scales [21,31]; whether satellite data can improve the resolution and model performance of LUR models at a city scale, has not been fully evaluated. In addition, the row anomaly of OMI led to a large amount of missing data at the daily level [32], hence OMI NO_2 column density data might be inappropriate to be directly used to assess NO_2 exposure levels within a city at a daily scale, and some studies resampled the data at a seasonal scale [33].

Therefore, in this study, we developed a satellite-derived LUR model, in a Chinese metropolis, to capture intra-urban NO_2 temporal variations at a seasonal level with a high spatial resolution. This model with a high spatial resolution is expected to capture the finer gradients of NO_2 variations within a city at a higher temporal resolution than that of the traditional LUR model, which could provide more accurate exposure assessment for epidemiological studies.

2. Materials and Methods

2.1. Study Area

Suzhou is a city located in southeastern Jiangsu Province of East China (Figure 1). It includes five urban districts (Gusu, Huqiu, Wuzhong, Xiangcheng, and Wujiang) and four satellite cities (Changshu, Taicang, Kunshan, and Zhangjiagang). Suzhou is one of five urban locations in the China Kadoorie Biobank (CKB) cohort that have focused on common chronic diseases since 2004 [34]. We developed a satellite-derived LUR model in Suzhou as a case study to establish the methodology for the assessment of exposure to NO_2 of the CKB cohort study to support the next phase of air pollution-related epidemiological studies. Suzhou covered 8488.42 km^2 in 2018 and about 42.5% of the total area was covered by waterbody. The total registered population in Suzhou reached 7.04 million by the end of 2018 (http://tjj.suzhou.gov.cn/sztjj/tjnj/2019/zk/indexce.htm). Suzhou is located in a subtropical monsoon climate zone with four distinct seasons.

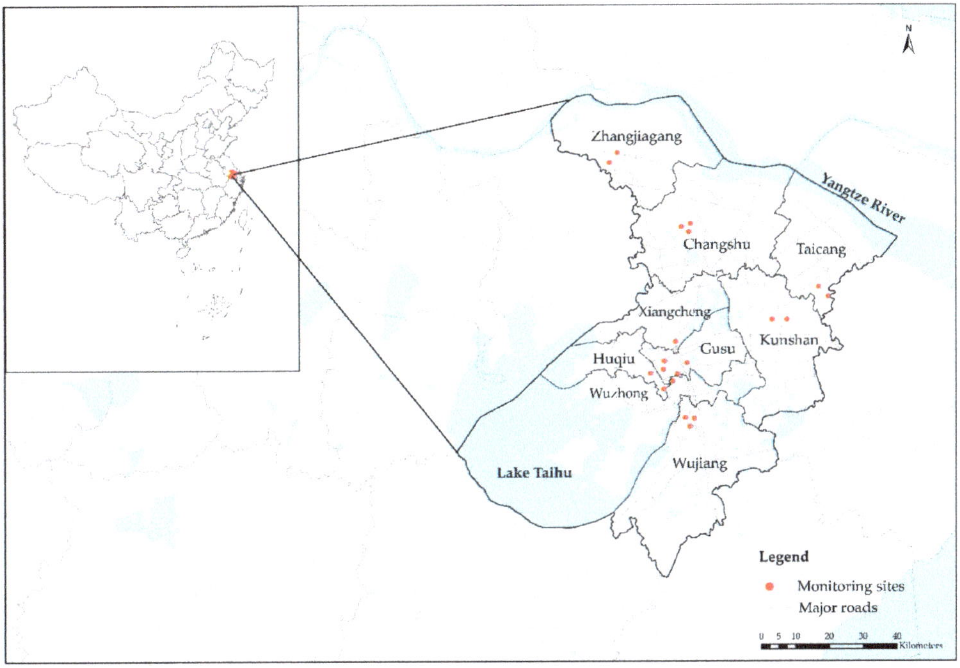

Figure 1. The location of Suzhou in China and the NO_2 monitoring sites in Suzhou that were used in this study.

2.2. Data

The database included data on NO_2 monitoring, NO_2 tropospheric column density from the OMI instrument, population density, road network, land use parameters, and NO_x emissions.

2.2.1. Monitoring Data

Daily NO_2 monitoring data of 20 fixed air quality stations were obtained from the National Environmental Monitoring Network, and the locations of the stations are shown in Figure 1. In accordance with the Chinese Ambient Air Quality Standard (GB3095-2012), at least 20 hourly measurements were included to calculate the daily NO_2 concentration; at least 27 daily values were needed to calculate monthly concentrations (25 daily values for February); at least 324 daily values were needed to calculate the annual concentration. Most of the fixed stations were located in areas with a relatively high population density to represent the averaged exposure levels for public health.

2.2.2. Satellite Data

The OMI instrument is on board the National Aeronautics and Space Administration (NASA) Aura satellite that was launched in 2004. It measures radiances across 270–500 nm of the ultraviolet and visible waveband. Global tropospheric vertical column NO_2 density data of OMI level 2 ($OMNO_2$) product, with a spatial resolution of 13 km × 24 km at nadir [35], are available online at a daily time step and were downloaded from NASA Goddard Earth Sciences Data and Information Services Center (https://earthdata.nasa.gov/). Cloud cover and a dynamic row anomaly problem of OMI were responsible for a significantly high rate of missing values of daily data. The "row anomaly" occurred due to the technical issues of the OMI, which has produced invalid data in the center-right part of each swath of observations since 2008 [32]. Within a city, the high missing rate might cause low availability of OMI NO_2 tropospheric column density data at a daily level. Therefore, seasonal resampling was done by averaging all daily OMI NO_2 tropospheric column density data falling inside a 40 km × 40 km grid to fill the gap caused by missing data and smooth the noise [33]. The satellite data were then interpolated to the fixed monitoring stations using an inverse distance weighted (IDW) method.

2.2.3. Other Predictors

Land Use Parameters

Land use data (agricultural, forest, grassland, waterbody, urban and built up, and unused land) from 2014 were interpreted from the Landsat TM5 dataset (https://earthexplorer.usgs.gov/) with a 30 m spatial resolution (Figure 2). Specifically, agricultural land included dry land and paddy fields; forest land included dense forests, shrub forests, loose forests, and other forests; grassland included highly-covered grassland; waterbody included rivers, lakes, beaches, bottomlands, and reservoirs; urban and built up land included urban and rural settlements and other built-up land; unused land included bare rock and sand. In Suzhou, the major land use types were urban and built-up land, agricultural land, and waterbody; and agricultural land mainly consisted of paddy fields. To optimize the correlation between NO_2 measurements and land use predictors, different buffer distances were applied, from 100 m to 5000 m, at 100-m intervals, around the 20 fixed monitoring sites [10,17,36]. The areas of each land use type were then calculated within these buffer zones separately.

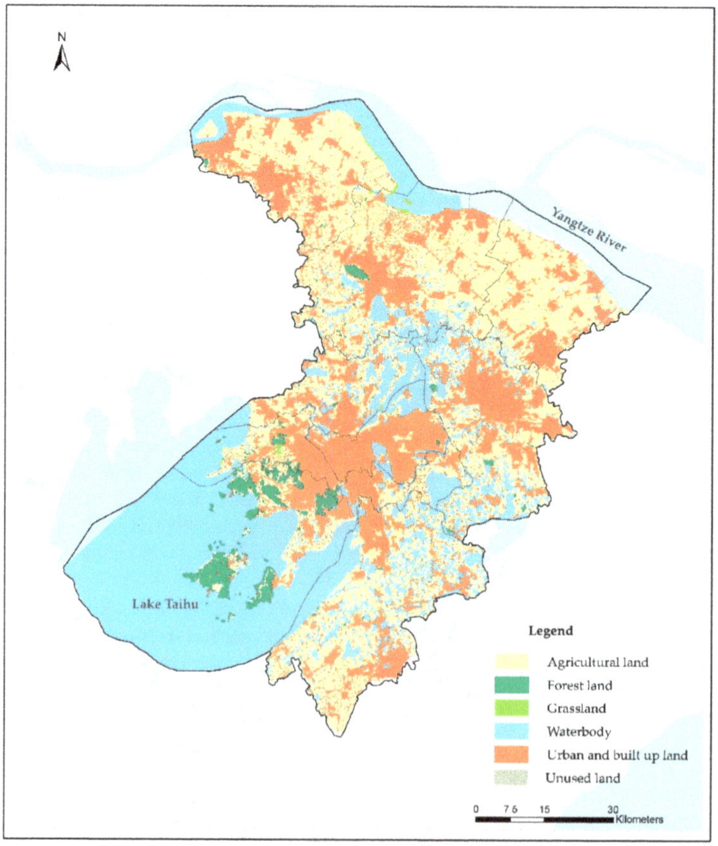

Figure 2. The spatial distribution of types of land use in this study in Suzhou in 2014.

Road Network

Lengths of major roads and distances to the nearest major road were calculated as indicators of traffic emissions. Types of roads included expressways, national roads, provincial roads, urban expressways, county roads, town roads, and other roads. Then, expressways, national roads, provincial roads, and urban expressways were merged as major roads. Within the buffers from 100 m to 5000 m (at 100 m intervals) around the 20 fixed monitoring sites, the lengths of major roads were then calculated [6,17]. Distance from monitoring sites to the nearest major road, inverse of the distance, and logarithmic transformation of the inverse distance were also calculated as indicators of traffic emissions [6,10].

Population Density

Population density data were obtained from the Oak Ridge National Laboratory (ORNL)'s LandScan 2014 global database at 30″ × 30″ resolution in raster format (http://www.ornl.gov/sci/landscan/), which were then interpolated to the NO_2 monitoring stations using the IDW method. The population data, with an ESRI binary raster format, is approximately at a 1 km × 1 km resolution and each grid represents an average population number within the grid at an annual level (https://landscan.ornl.gov/documentation). Figure 3 shows the spatial distribution of the population in Suzhou in 2014, suggesting that more people tended to live in the center of five urban districts and four satellite cities in Suzhou.

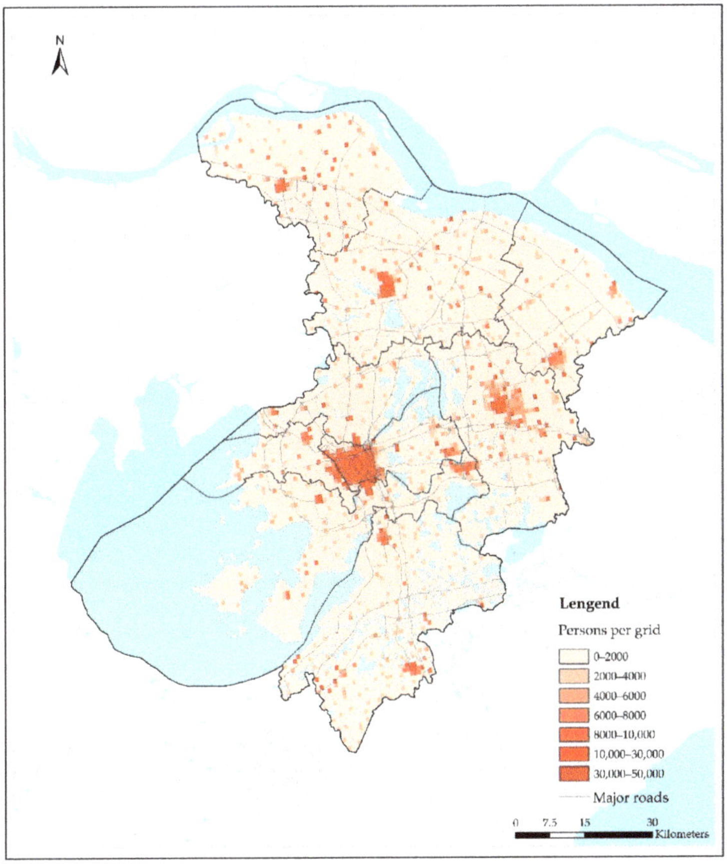

Figure 3. The distribution of population and major roads in Suzhou.

NO_x Emissions

NO_x emission inventory data were collected from the Multiresolution Emission Inventory of China (MEIC, http://www.meicmodel.org) at a spatial resolution of 1 km × 1 km. The industrial NO_2 emissions from power plants and non-power plants were computed separately within buffer zones of 1 km to 10 km, at 1-km intervals, around each monitoring site.

2.3. Model Development and Evaluation

A traditional LUR model was developed, as the first step, to select the most optimized predictors from all parameters with a linear regression model [6,10,20,36]. Since the OMI NO_2 tropospheric column density was aggregated at a seasonal level to fill the gap caused by the high missing rate of the satellite data [32], this model was developed at a seasonal level [37,38]. First, we set every potential variable a prior direction. Second, manual backward supervised regression was conducted based on NO_2 seasonal concentrations to select the most optimized predictor variables. Predictors were kept in the model if they satisfied the criteria proposed by previous studies [6,10,17]: (1) the variables improved the model R^2 by at least 1%; (2) the effect directions of the variables were consistent with the prior directions; (3) the variables that were already in the model did not change their effect directions; (4) the variable would be excluded from the model if the p value was less than 0.1. This process continued until there were no more variables meeting the criteria. Variance inflation factors (VIFs) were calculated as an indicator of multicollinearity. Variables with VIF values greater than three were removed from the satellite-based LUR model and this step was repeated.

In the second step, a linear mixed effects model was developed (see Equation (1)) by involving random effects of OMI NO_2 tropospheric column density [23,37]. The advantage of employing this model was to include the variability of associations between NO_2 concentrations and OMI NO_2 tropospheric column density over time. Similar satellite-based models had been developed for predicting $PM_{2.5}$ concentrations in a national assessment [37] and PM_{10} concentrations within a city in Shanghai [23]. In this model, the OMI NO_2 tropospheric column density had both random effect and fixed effect coefficients, which represented seasonal variability in the association between NO_2 measurements and OMI NO_2 tropospheric column density and the average effect of satellite measurements on the ground NO_2 measurements for the whole year, respectively [23,37]. The model structure can be summarized as:

$$NO_{2,st} = (\beta_0 + \beta_0') + (\beta_1 + \beta_1')\, OMI_{st} + \beta_{is} X_{is} + \varepsilon_{st} \quad (1)$$

where $NO_{2,st}$ indicates the mean observed NO_2 concentrations (μg/m^3) at the fixed station s in season t; OMI_{st} is the only independent variable with both fixed and random effects, which represents OMI NO_2 tropospheric column density data at the fixed station s in season t; β_0 and β_0' are the intercepts of the fixed and season-specific random effects for the model, respectively; β_1 and β_1'indicate the fixed and season-specific random slopes for OMI_{st}, respectively; X_{is} represents a series of predictors, which are selected by satisfying the criteria from the first step; and β_{is} represents the fixed slope for predictor i at the fixed station s; and ε_{st} is the error term at the fixed station s in season t.

In the third step, 10-fold cross validation (CV) was applied to evaluate the model performance [17,37]: 90% of the data were randomly selected for model development, which was used to predict NO_2 concentrations of the remaining 10% of the data; and this process was repeated 10 times. Root mean squared error (RMSE) was calculate as the standard deviation of the residuals. RMSE and R^2 were used to evaluate the model's performance by comparing measured and predicted NO_2 concentrations during model development and 10-fold CV, respectively. The relative prediction error (RPE, defined as RMSE divided by the mean NO_2 measurements) from 10-fold CV was then calculated to evaluate prediction accuracy.

In the fourth step, seasonal prediction maps of NO_2 concentrations in Suzhou were produced based on the satellite-derived LUR models, at a 100 m × 100 m resolution at a seasonal timescale. In addition, we further calculated annual-mean and seasonal-mean population-weighted NO_2 concentrations in Suzhou [39] (see Equation (2)).

$$C_{Pop} = \sum Pop_i \times C_i / \sum Pop_i \tag{2}$$

where C_{Pop} indicates the annual-mean or seasonal-mean population-weighted NO_2 exposure concentrations in Suzhou; Pop_i represents the population density of grid i; and C_i indicates the estimated annual-mean or seasonal-mean NO_2 concentrations of grid i.

Figure 4 shows the workflow for the development of the satellite-derived LUR model in our study. Statistical analyses were performed with nlme packages (https://www.rdocumentation.org/packages/nlme/versions/3.1-151/topics/nlme) of R3.6.1.

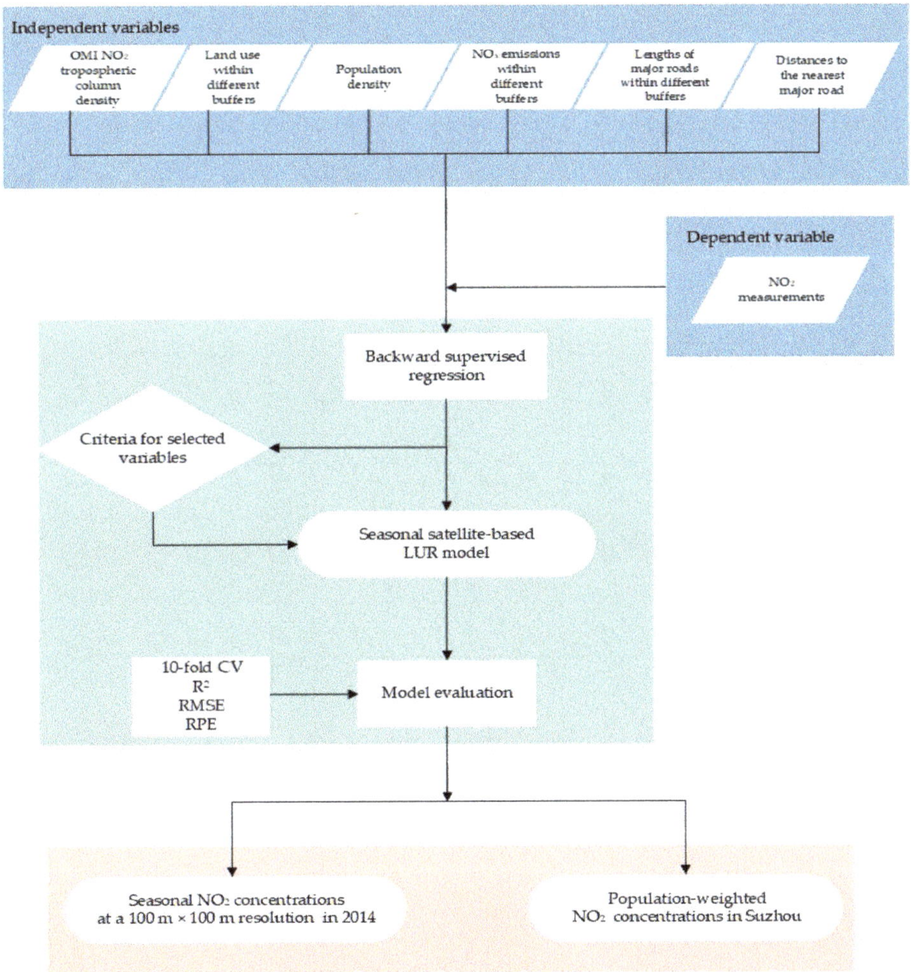

Figure 4. Workflow for the development of the satellite-derived LUR model.

3. Results

3.1. Descriptive Statistics Analyses

In 2014, the annual-mean NO_2 was 46.23 μg/m^3 in Suzhou, with the lowest concentration of 36.52 μg/m^3 recorded in summer and the highest concentration of 53.22 μg/m^3 in winter, as measured at fixed monitoring sites. Among all predictors, the Pearson's correlation coefficient between seasonal OMI NO_2 tropospheric column density and seasonal NO_2 measurements was highest with the value of 0.65.

3.2. Model Development and Evaluation

After variable selection, as the results of the first step, the satellite-derived LUR model included four predictors: NO_2 tropospheric column density from OMI, population density, log transformed inverse of nearest distances to major roads (Log_distance), and NO_2 non-power plants emissions within a 10-km buffer zone (Table 1). The R^2 and RMSE of this model were 0.63 and 5.76 μg/m^3, respectively. The R^2 and RMSE of the 10-fold CV were 0.59 and 6.09 μg/m^3, respectively. The VIFs of the four variables were all less than 2, showing weak multicollinearity among them.

Table 1. The traditional land use regression (LUR) model for predicting NO_2 concentrations.

Variables	β	SE	*p* Value
Intercept	33.57	5.13	<0.001
NO_2 tropospheric column density	0.85	0.11	<0.001
Population density	0.00016	0.0001	0.043
Log_distance	2.92	1.38	0.038
Non-power emissions within 10 km buffer zone	0.0001	0.00003	0.002

The results of the second step, including the estimated coefficients of fixed effects of the four predictor variables, are shown in Table 2. All predictors were positively and significantly associated with measured NO_2 concentrations, with *p* values less than 0.05. The absolute contribution (IQR × β), for each influencing predictor, was calculated as the regression coefficient (β) of fixed effects multiplied by the inter-quartile range (IQR) of the corresponding predictor. The results indicated that the non-power emissions within a 10-km buffer zone and OMI NO_2 tropospheric column density contributed most to NO_2 concentrations, because they had higher IQR × β values (Table 2).

Table 2. The fixed effects of the satellite-derived LUR model for predicting NO_2 concentrations.

Variables	β	SE	*p* Value	IQR × β [1]
Intercept	39.617	7.348	<0.001	
NO_2 tropospheric column density	0.618	0.293	0.039	4.389
Population density	0.00016	0.0001	0.029	1.976
Log_distance	3.240	1.272	0.013	1.546
Non-power emissions within 10-km buffer zone	0.0001	0.00003	<0.001	4.792

[1] represents the regression coefficient (β) of fixed effects multiplied by the inter-quartile range (IQR) for each predictor at 20 monitoring sites.

The R^2 and RMSE of the seasonal satellite-derived LUR model were 0.70 and 5.24 μg/m^3, respectively. The R^2 and RMSE of the 10-fold CV were 0.61 and 5.91 μg/m^3, respectively, for the seasonal model (Figure 5). The RPE from 10-fold CV was 12.78%, which indicated a relatively high predicting accuracy at the seasonal level. The linear mixed effects model performed better than the traditional linear regression model, suggesting the importance of considering the seasonal variability of the association between ground NO_2 measurements and OMI NO_2 tropospheric column density.

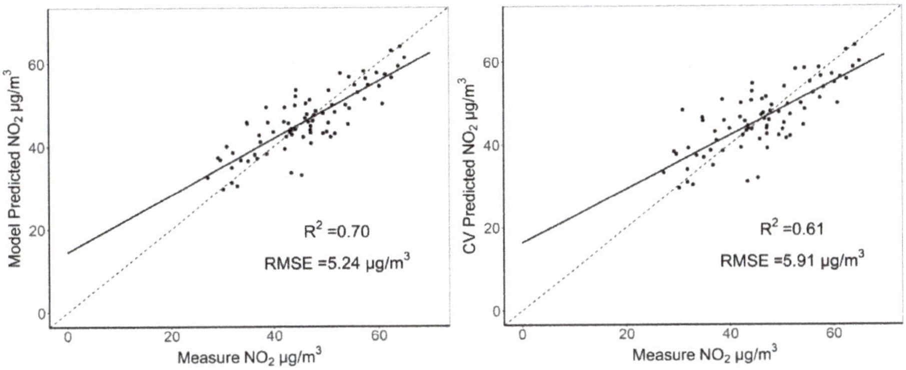

Figure 5. Scatter plots of measured and predicted NO_2 concentrations from model fitting (**left**), and 10-fold cross validation (**right**), respectively, for the satellite-derived linear mixed effects model at a seasonal timescale.

3.3. Spatiotemporal Trends of Predicting NO_2 Concentrations

Predictive maps of NO_2 concentrations with a spatial resolution of 100 m × 100 m were produced at a seasonal timescale (Figure 6). The seasonal pattern of predicted NO_2 concentrations agreed well with field measurements. Mean NO_2 concentration was highest in winter (47.3 µg/m^3) in Suzhou, which was 1.46 times higher than that in summer. The spatial patterns of NO_2 predictions were similar at different seasons throughout the year. Maps with high spatial resolution showed that severe NO_2 pollution occurred along the major roads and declined significantly with increasing distance from the road. Urban centers with high population density and an intensive road network also experienced higher NO_2 concentrations than that of the rural areas (Figure 6). For example, in summer, the maximum NO_2 concentration (58.99 µg/m^3) that occurred in urban areas was 2.77 times higher than the minimum value (21.33 µg/m^3) in rural areas; and in winter, the maximum concentration (76.93 µg/m^3) was 2.03 times higher compared to the lowest value (37.91 µg/m^3) in rural areas. The results indicated that the NO_2 concentration was generally higher in urban areas than that in rural areas both in winter and summer.

The population-weighted annual mean NO_2 concentration in 2014 was 44.94 µg/m^3 in Suzhou, higher than the annual-mean predicted concentration of 41.4 µg/m^3 and also higher than the annual-mean NO_2 standard of 40 µg/m^3 defined in the Chinese National Ambient Air Quality Standards (GB 3095-2012). In winter, 99% of the total population lived in areas with NO_2 concentrations exceeding 40 µg/m^3 in Suzhou (Table 3).

Table 3. Population-weighted NO_2 exposure concentrations.

Parameter	Annual	Spring	Summer	Autumn	Winter
Population-weighted concentration (µg/m^3)	44.94	46.33	35.64	46.59	51.21
Proportion (%) *	84	92	22	96	99

* Proportion: Proportion of population living in areas with NO_2 concentrations exceeding 40 µg/m^3.

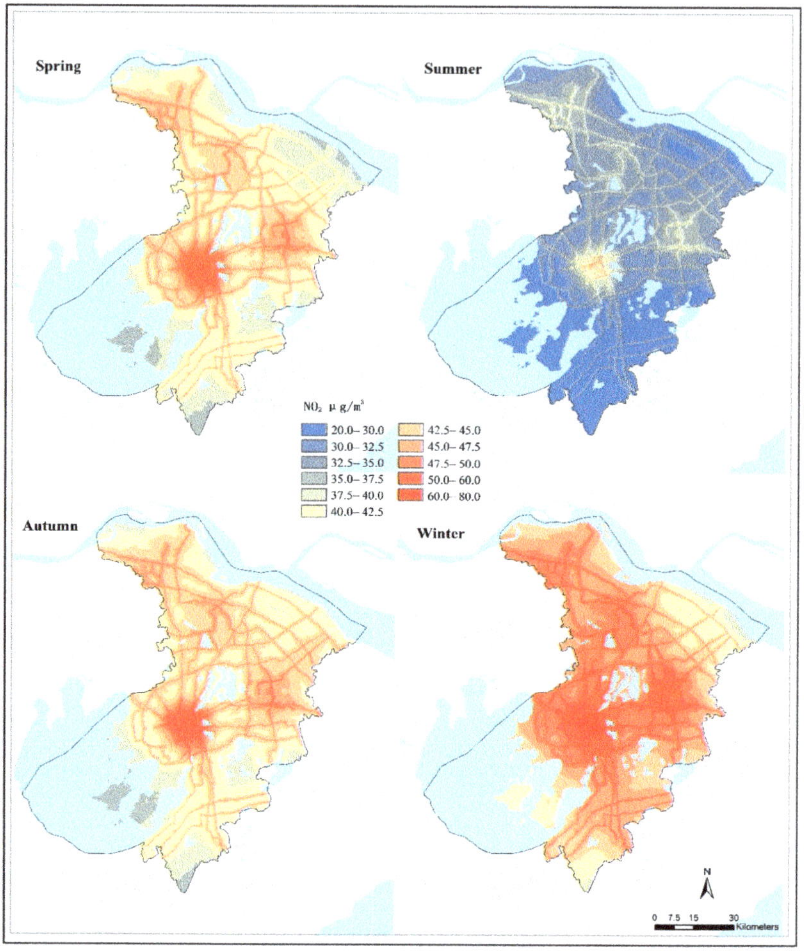

Figure 6. NO_2 spatial distribution at the seasonal level in Suzhou, 2014.

4. Discussion

Our study built a satellite-derived LUR model with OMI NO_2 tropospheric column density data to predict NO_2 concentrations at seasonal timescales with a high spatial resolution (100 m × 100 m) in Suzhou. The R^2 values of model fitting and 10-fold CV were 0.70 and 0.61 at seasonal timescales, respectively, reflecting the relatively high stability of the model.

Our seasonal satellite-derived LUR model performance was comparable with previous satellite-based LUR models on NO_2 concentration assessment at global, national, and regional scales. For the global satellite-based LUR model, the R^2 and MAE (mean absolute error) for the model were 0.54 and 3.7 ppb at a 100 m×100 m resolution, respectively [20]. The adjusted R^2 values of models with satellite data were 0.48–0.58 in 17 contiguous countries of Western Europe [22]. The R^2 of the model fitting and CV were 0.79 and 0.77 of the national satellite-derived LUR in the United States, respectively [19]. Similarly, in China, Xu et al. and Yang et al. developed satellite-derived LUR models at national and regional scales, respectively [21,31]. The R^2 of 10-fold cross-validation (CV) was 0.78 for the national model in 2015 [31], and the R^2 of model fitting was 0.61 for the regional model [21]. Although increasing studies have used machine learning methods with satellite data to evaluate NO_2 concentrations based on a large number of measurements from fixed monitors at regional or national scales [40–43], the training data may be insufficient to develop machine learning models within a city because of the limited number of fixed stations in this study. The comparison suggested that our satellite-derived LUR model, including satellite-retrieved NO_2 tropospheric column density, population density, traffic indicators, and NO_x emission data, predicted ground NO_2 concentrations with relatively high accuracy based on the fixed stations in Suzhou.

In terms of NO_2 concentration, our results exhibited significant spatial variability within a city at a fine spatial resolution (100 m $\times$ 100 m), and found a distinctive decline with increasing distance from the roads and significant differences between urban and rural areas. The high variability within a city suggested that exposure assessments of NO_2 might be inaccurate if they just depended on measurements of a limited number of fixed monitoring sites. This high spatial heterogeneity may be mainly dependent on NO_2 pollution-related sources, such as traffic and industrial emissions. Traffic and industrial emissions are known as the main sources of NO_2, contributing to the high spatial heterogeneity of NO_2 concentrations along roads and within a city. On one hand, NO_2 is emitted as a primary pollutant from these sources. On the other hand, NO_2 is also a secondary pollutant [1,2]. In our study, NO_2 concentrations were significantly higher along roads and declined gradually with increased distance from roads in Suzhou, consistent with previous results of NO_2 spatial heterogeneity along roads [8]. The variables indicating traffic-related sources in our study were also frequently used in the previous LUR models for NO_2 concentrations assessment [6,17,36]. Additionally, industrial emissions, an important influencing predictor for NO_2 assessment in our model, had also been found to be an important variable in the previous LUR models to predict ground NO_2 concentrations within cities such as in Shanghai and Tianjin [16,17]. A recent study observed a notable decrease of NO_2 concentrations during the Chinese New Year holiday in 2020 led by the novel coronavirus (COVID-19) lockdown compared to those before or after this period in Suzhou [44]. A sharp decline in traffic emissions and a slight reduction in industry emissions caused by the shut-down policies might be the main contributors to the decrease of NO_2 concentrations during the lockdown period in Suzhou [44], suggesting that both traffic and industrial emissions are crucial sources of NO_2 in Suzhou. Additionally, our results found that mean NO_2 concentrations were higher in winter compared to that in summer. This was consistent with the previous studies on the seasonal pattern of NO_2 concentrations in China [24,45]. In winter, NO_2-related emissions are stronger due to more emissions from coal combustion for heating; while meteorological conditions are less favorable and could impede the dispersion and transportation of NO_2 pollution [44,46,47]. Both of these might be contributors to the higher NO_2 concentrations in winter [44,46,47]. Our results in Figure 6 showed an approximately lower ratio between urban and rural NO_2 concentrations in winter compared to those in summer. This might be due to more coal combustion for the heating of houses in rural areas in winter compared to that in urban areas [48].

As another influencing factor for NO_2 spatial heterogeneity, the spatial pattern of population density was highly consistent with that of NO_2 predictions in Suzhou, suggesting that population density can be used as an indicator of anthropogenic emissions that reflects a series of emissions including traffic, industrial process, and heating sources [6]. High population density not only intensified the NO_2 pollution, but also resulted in an increased exposure of populations to high NO_2 levels. In this study, 84% of the population were exposed to higher NO_2 levels than the national annual-mean NO_2 standards (40 $\mu g/m^3$) in Suzhou in 2014; while the proportion of the population exposed to concentrations exceeding the World Health Organization (WHO) annual NO_2 standards (40 $\mu g/m^3$) was only 8% in Western Europe [39], which was much smaller than that in Suzhou. This might be because a high population density and high concentrations of air pollution coexist in Chinese cities. For example, many residential buildings are located along major roads for the convenience of transportation, and residents living in these buildings might be both influenced by the traffic-related emissions and housing heating emissions, especially during winter in the rural areas. Our results suggested that policy makers should take effective interventions for these areas of higher NO_2 concentrations, especially for urban regions with the higher population density, which is an urgent need for the public health.

The satellite-based LUR model also expanded the temporal resolution and improved the accuracy of seasonal NO_2 predictions. Land use data, including land cover, road network, and population data, used in traditional LUR models commonly have lower temporal resolution, whereas the NO_2 tropospheric column density data could represent temporal variability of NO_2 concentration with a strong correlation with ground NO_2 concentration. Previous studies mostly employed satellite data to expand the temporal resolution of the LUR model for the assessment of NO_2 concentrations to seasonal or monthly timescales at national or regional scales [19,21,30]; however, few satellite-based LUR models on NO_2 concentrations assessment have been developed at a city scale considering the local influencing factors with a flexible timescale in China. In this study, we developed a satellite-based LUR model in Suzhou to capture the fine gradients of NO_2 concentrations at a spatial resolution of 100 m $\times$ 100 m. More importantly, our predictions captured the significant seasonal variability of NO_2 concentrations within a city, which could not be achieved by traditional LUR models. These findings suggested that the satellite-derived model could provide exposure assessment of NO_2 concentrations at a flexible timescale for epidemiological studies and scientific evidence for protecting residents from NO_2 pollution.

Our study has several limitations. First, the OMI NO_2 tropospheric column density for spatial prediction was relatively coarse (13 km $\times$ 24 km). Satellite-based NO_2 data with a higher spatial resolution could help improve the model performance in the future when they are available. Second, our model was developed at a seasonal level rather than a daily level. The cloud cover and row anomaly problem of OMI lead to missing data at a daily level within a city; therefore, we resampled OMI data at a seasonal level to fill the gap. Satellite-based NO_2 data with a lower missing rate might help improve the temporal resolution of our model in the future. Third, traffic counts are an ideal predictor to identify the traffic emissions, but these were not accessible for this study. We used major road lengths and distance to the nearest major road as surrogates of traffic counts to indicate the influence of traffic emissions on NO_2 concentrations. This was also applied as a traffic variable in NO_2 LUR models in the European Study of Cohorts for Air Pollution Effects (ESCAPE) project and other studies of the development of NO_2 LUR models [6,36].

5. Conclusions

In summary, the satellite-derived LUR model could predict seasonal NO_2 concentrations at a 100 m $\times$ 100 m resolution with relatively high accuracy, at a city scale. This model could capture the fine gradients both along the road and within the urban-rural areas for each season based on the satellite data. According to the predictions, we found that 84% of the city's total population lived in areas with NO_2 concentrations exceeding the national annual standard of NO_2 of 40 $\mu g/m^3$ in Suzhou in 2014. Hence, reducing NO_2 concentrations is urgently needed, especially for urban areas with a higher population density. This model and its predictions could support policy developments in the control of air quality and accurate exposure assessment for future epidemiological studies.

Author Contributions: Conceptualization, L.Z., Q.X., G.G., J.C., R.C., X.M., and H.K.; methodology, C.Y., Q.X., G.G., J.C., R.C., and X.M.; software, L.Z., and X.M.; validation, L.Z., and X.M.; formal analysis, L.Z., and X.M.; investigation, C.Y.; resources, C.Y., and X.M.; data curation, C.Y.; writing—original draft preparation, L.Z.; writing—review and editing, L.Z., and X.M.; visualization, L.Z.; supervision, L.Z., and X.M.; project administration, H.K.; funding acquisition, H.K. All authors have read and agreed to the published version of the manuscript.

Funding: This research was funded by the National Natural Science Foundation of China (91843302, 91643205 and 82003413) and the National Key Research and Development Program of China (2016YFC0206504).

Institutional Review Board Statement: Not applicable.

Informed Consent Statement: Not applicable.

Data Availability Statement: Publicly available datasets were analyzed in this study. These data can be found here: https://earthexplorer.usgs.gov/ (OMI NO_2 tropospheric column density data and Landsat TM5 dataset), http://www.ornl.gov/sci/landscan/ (population density data).

Acknowledgments: Thanks to Qiang Zhang, University of Tsinghua, for providing the NO_x emission inventory data.

Conflicts of Interest: The authors declare no conflict of interest.

References

1. Casquero-Vera, J.A.; Lyamani, H.; Titos, G.; Borras, E.; Olmo, F.J.; Alados-Arboledas, L. Impact of primary NO_2 emissions at different urban sites exceeding the European NO_2 standard limit. *Sci. Total Environ.* **2019**, *646*, 1117–1125. [CrossRef]
2. Anttila, P.; Tuovinen, J.-P.; Niemi, J.V. Primary NO_2 emissions and their role in the development of NO_2 concentrations in a traffic environment. *Atmos. Environ.* **2011**, *45*, 986–992. [CrossRef]
3. Jerrett, M.; Burnett, R.T.; Beckerman, B.S.; Turner, M.C.; Krewski, D.; Thurston, G.; Martin, R.V.; van Donkelaar, A.; Hughes, E.; Shi, Y.; et al. Spatial analysis of air pollution and mortality in California. *Am. J. Respir. Crit. Care Med.* **2013**, *188*, 593–599. [CrossRef] [PubMed]
4. Chen, R.; Yin, P.; Meng, X.; Wang, L.; Liu, C.; Niu, Y.; Lin, Z.; Liu, Y.; Liu, J.; Qi, J.; et al. Associations Between Ambient Nitrogen Dioxide and Daily Cause-specific Mortality: Evidence from 272 Chinese Cities. *Epidemiology* **2018**, *29*, 482–489. [CrossRef] [PubMed]
5. WHO Regional Office for Europe. *Review of Evidence on Health Aspects of Air Pollution—REVIHAAP Project: Technical Report*; WHO Regional Office for Europe: Copenhagen, Denmark, 2013. Available online: https://www.ncbi.nlm.nih.gov/books/NBK361805/ (accessed on 13 November 2020).
6. Beelen, R.; Hoek, G.; Vienneau, D.; Eeftens, M.; Dimakopoulou, K.; Pedeli, X.; Tsai, M.-Y.; Künzli, N.; Schikowski, T.; Marcon, A.; et al. Development of NO_2 and NO_x land use regression models for estimating air pollution exposure in 36 study areas in Europe—The ESCAPE project. *Atmos. Environ.* **2013**, *72*, 10–23. [CrossRef]
7. Son, J.Y.; Bell, M.L.; Lee, J.T. Individual exposure to air pollution and lung function in Korea: Spatial analysis using multiple exposure approaches. *Environ. Res.* **2010**, *110*, 739–749. [CrossRef] [PubMed]
8. Karner, A.A.; Eisinger, D.S.; Niemeier, D.A. Near-roadway air quality: Synthesizing the findings from real-world data. *Environ. Sci. Technol.* **2010**, *44*, 5334–5344. [CrossRef] [PubMed]
9. Cyrys, J.; Eeftens, M.; Heinrich, J.; Ampe, C.; Armengaud, A.; Beelen, R.; Bellander, T.; Beregszaszi, T.; Birk, M.; Cesaroni, G.; et al. Variation of NO_2 and NO_x concentrations between and within 36 European study areas: Results from the ESCAPE study. *Atmos. Environ.* **2012**, *62*, 374–390. [CrossRef]
10. Lee, J.H.; Wu, C.F.; Hoek, G.; de Hoogh, K.; Beelen, R.; Brunekreef, B.; Chan, C.C. Land use regression models for estimating individual NO_x and NO_2 exposures in a metropolis with a high density of traffic roads and population. *Sci. Total Environ.* **2014**, *472*, 1163–1171. [CrossRef]
11. Crouse, D.L.; Peters, P.A.; Villeneuve, P.J.; Proux, M.O.; Shin, H.H.; Goldberg, M.S.; Johnson, M.; Wheeler, A.J.; Allen, R.W.; Atari, D.O.; et al. Within- and between-city contrasts in nitrogen dioxide and mortality in 10 Canadian cities; a subset of the Canadian Census Health and Environment Cohort (CanCHEC). *J. Expo. Sci. Environ. Epidemiol.* **2015**, *25*, 482–489. [CrossRef]
12. de Hoogh, K.; Gulliver, J.; Donkelaar, A.v.; Martin, R.V.; Marshall, J.D.; Bechle, M.J.; Cesaroni, G.; Pradas, M.C.; Dedele, A.; Eeftens, M.; et al. Development of West-European $PM_{2.5}$ and NO_2 land use regression models incorporating satellite-derived and chemical transport modelling data. *Environ. Res.* **2016**, *151*, 1–10. [CrossRef] [PubMed]
13. Fischer, P.H.; Marra, M.; Ameling, C.B.; Hoek, G.; Beelen, R.; de Hoogh, K.; Breugelmans, O.; Kruize, H.; Janssen, N.A.; Houthuijs, D. Air Pollution and Mortality in Seven Million Adults: The Dutch Environmental Longitudinal Study (DUELS). *Environ. Health Perspect.* **2015**, *123*, 697–704. [CrossRef] [PubMed]
14. Foraster, M.; Basagana, X.; Aguilera, I.; Rivera, M.; Agis, D.; Bouso, L.; Deltell, A.; Marrugat, J.; Ramos, R.; Sunyer, J.; et al. Association of long-term exposure to traffic-related air pollution with blood pressure and hypertension in an adult population-based cohort in Spain (the REGICOR study). *Environ. Health Perspect.* **2014**, *122*, 404–411. [CrossRef] [PubMed]
15. Ohlwein, S.; Klumper, C.; Vossoughi, M.; Sugiri, D.; Stolz, S.; Vierkotter, A.; Schikowski, T.; Kara, K.; Germing, A.; Quass, U.; et al. Air pollution and diastolic function in elderly women—Results from the SALIA study cohort. *Int. J. Hyg. Environ. Health* **2016**, *219*, 356–363. [CrossRef] [PubMed]
16. Chen, L.; Wang, Y.; Li, P.; Ji, Y.; Kong, S.; Li, Z.; Bai, Z. A land use regression model incorporating data on industrial point source pollution. *J. Environ. Sci.* **2012**, *24*, 1251–1258. [CrossRef]
17. Meng, X.; Chen, L.; Cai, J.; Zou, B.; Wu, C.-F.; Fu, Q.; Zhang, Y.; Liu, Y.; Kan, H. A land use regression model for estimating the NO_2 concentration in Shanghai, China. *Environ. Res.* **2015**, *137*, 308–315. [CrossRef]
18. Xu, G.; Jiao, L.M.; Zhao, S.L.; Yuan, M.; Li, X.M.; Han, Y.Y.; Zhang, B.; Dong, T. Examining the Impacts of Land Use on Air Quality from a Spatio-Temporal Perspective in Wuhan, China. *Atmosphere* **2016**, *7*, 62. [CrossRef]

19. Bechle, M.J.; Millet, D.B.; Marshall, J.D. National Spatiotemporal Exposure Surface for NO_2: Monthly Scaling of a Satellite-Derived Land-Use Regression, 2000–2010. *Environ. Sci. Technol.* **2015**, *49*, 12297–12305. [CrossRef]
20. Larkin, A.; Geddes, J.A.; Martin, R.V.; Xiao, Q.; Liu, Y.; Marshall, J.D.; Brauer, M.; Hystad, P. Global Land Use Regression Model for Nitrogen Dioxide Air Pollution. *Environ. Sci. Technol.* **2017**, *51*, 6957–6964. [CrossRef]
21. Yang, X.; Zheng, Y.; Geng, G.; Liu, H.; Man, H.; Lv, Z.; He, K.; de Hoogh, K. Development of $PM_{2.5}$ and NO_2 models in a LUR framework incorporating satellite remote sensing and air quality model data in Pearl River Delta region, China. *Environ. Pollut.* **2017**, *226*, 143–153. [CrossRef]
22. Vienneau, D.; de Hoogh, K.; Bechle, M.J.; Beelen, R.; van Donkelaar, A.; Martin, R.V.; Millet, D.B.; Hoek, G.; Marshall, J.D. Western European land use regression incorporating satellite- and ground-based measurements of NO_2 and PM_{10}. *Environ. Sci. Technol.* **2013**, *47*, 13555–13564. [CrossRef] [PubMed]
23. Meng, X.; Fu, Q.; Ma, Z.; Chen, L.; Zou, B.; Zhang, Y.; Xue, W.; Wang, J.; Wang, D.; Kan, H.; et al. Estimating ground-level PM_{10} in a Chinese city by combining satellite data, meteorological information and a land use regression model. *Environ. Pollut.* **2016**, *208*, 177–184. [CrossRef] [PubMed]
24. Ma, X.; Jia, H.; Sha, T.; An, J.; Tian, R. Spatial and seasonal characteristics of particulate matter and gaseous pollution in China: Implications for control policy. *Environ. Pollut.* **2019**, *248*, 421–428. [CrossRef] [PubMed]
25. Zhang, Z.; Wang, J.; Hart, J.E.; Laden, F.; Zhao, C.; Li, T.; Zheng, P.; Li, D.; Ye, Z.; Chen, K. National scale spatiotemporal land-use regression model for $PM_{2.5}$, PM_{10} and NO_2 concentration in China. *Atmos. Environ.* **2018**, *192*, 48–54. [CrossRef]
26. Hoek, G.; Eeftens, M.; Beelen, R.; Fischer, P.; Brunekreef, B.; Boersma, K.F.; Veefkind, P. Satellite NO_2 data improve national land use regression models for ambient NO_2 in a small densely populated country. *Atmos. Environ.* **2015**, *105*, 173–180. [CrossRef]
27. Novotny, E.V.; Bechle, M.J.; Millet, D.B.; Marshall, J.D. National Satellite-Based Land-Use Regression: NO_2 in the United States. *Environ. Sci. Technol.* **2011**, *45*, 4407–4414. [CrossRef]
28. Zhan, Y.; Luo, Y.; Deng, X.; Zhang, K.; Zhang, M.; Grieneisen, M.L.; Di, B. Satellite-Based Estimates of Daily NO_2 Exposure in China Using Hybrid Random Forest and Spatiotemporal Kriging Model. *Environ. Sci. Technol.* **2018**, *52*, 4180–4189. [CrossRef]
29. Knibbs, L.D.; Coorey, C.P.; Bechle, M.J.; Marshall, J.D.; Hewson, M.G.; Jalaludin, B.; Morgan, G.G.; Barnett, A.G. Long-term nitrogen dioxide exposure assessment using back-extrapolation of satellite-based land-use regression models for Australia. *Environ. Res.* **2018**, *163*, 16–25. [CrossRef]
30. Knibbs, L.D.; Hewson, M.G.; Bechle, M.J.; Marshall, J.D.; Barnett, A.G. A national satellite-based land-use regression model for air pollution exposure assessment in Australia. *Environ. Res.* **2014**, *135*, 204–211. [CrossRef]
31. Xu, H.; Bechle, M.J.; Wang, M.; Szpiro, A.A.; Vedal, S.; Bai, Y.; Marshall, J.D. National $PM_{2.5}$ and NO_2 exposure models for China based on land use regression, satellite measurements, and universal kriging. *Sci. Total Environ.* **2019**, *655*, 423–433. [CrossRef]
32. McPeters, R.D.; Frith, S.; Labow, G.J. OMI total column ozone: Extending the long-term data record. *Atmos. Meas. Tech.* **2015**, *8*, 4845–4850. [CrossRef]
33. Xiao, Q.; Chang, H.H.; Geng, G.; Liu, Y. An Ensemble Machine-Learning Model to Predict Historical $PM_{2.5}$ Concentrations in China from Satellite Data. *Environ. Sci. Technol.* **2018**, *52*, 13260–13269. [CrossRef] [PubMed]
34. Chen, Z.; Chen, J.; Collins, R.; Guo, Y.; Peto, R.; Wu, F.; Li, L. China Kadoorie Biobank of 0.5 million people: Survey methods, baseline characteristics and long-term follow-up. *Int. J. Epidemiol.* **2011**, *40*, 1652–1666. [CrossRef] [PubMed]
35. Levelt, P.F.; van den Oord, G.H.J.; Dobber, M.R.; Malkki, A.; Huib, V.; Johan de, V.; Stammes, P.; Lundell, J.O.V.; Saari, H. The ozone monitoring instrument. *IEEE Trans. Geosci. Remote Sens.* **2006**, *44*, 1093–1101. [CrossRef]
36. Hoek, G.; Beelen, R.; de Hoogh, K.; Vienneau, D.; Gulliver, J.; Fischer, P.; Briggs, D. A review of land-use regression models to assess spatial variation of outdoor air pollution. *Atmos. Environ.* **2008**, *42*, 7561–7578. [CrossRef]
37. Ma, Z.; Hu, X.; Sayer, A.M.; Levy, R.; Zhang, Q.; Xue, Y.; Tong, S.; Bi, J.; Huang, L.; Liu, Y. Satellite-Based Spatiotemporal Trends in $PM_{2.5}$ Concentrations: China, 2004–2013. *Environ. Health Perspect.* **2016**, *124*, 184–192. [CrossRef]
38. Saucy, A.; Roosli, M.; Kunzli, N.; Tsai, M.-Y.; Sieber, C.; Olaniyan, T.; Baatjies, R.; Jeebhay, M.; Davey, M.; Fluckiger, B.; et al. Land Use Regression Modelling of Outdoor NO_2 and $PM_{2.5}$ Concentrations in Three Low Income Areas in the Western Cape Province, South Africa. *Int. J. Environ. Res. Public Health* **2018**, *15*, 1452. [CrossRef]
39. de Hoogh, K.; Chen, J.; Gulliver, J.; Hoffmann, B.; Hertel, O.; Ketzel, M.; Bauwelinck, M.; van Donkelaar, A.; Hvidtfeldt, U.A.; Katsouyanni, K.; et al. Spatial $PM_{2.5}$, NO_2, O_3 and BC models for Western Europe—Evaluation of spatiotemporal stability. *Environ. Int.* **2018**, *120*, 81–92. [CrossRef]
40. Chen, Z.Y.; Zhang, R.; Zhang, T.H.; Ou, C.Q.; Guo, Y. A kriging-calibrated machine learning method for estimating daily ground-level NO_2 in mainland China. *Sci. Total Environ.* **2019**, *690*, 556–564. [CrossRef]
41. Di, Q.; Amini, H.; Shi, L.; Kloog, I.; Silvern, R.; Kelly, J.; Sabath, M.B.; Choirat, C.; Koutrakis, P.; Lyapustin, A.; et al. Assessing NO_2 Concentration and Model Uncertainty with High Spatiotemporal Resolution across the Contiguous United States Using Ensemble Model Averaging. *Environ. Sci. Technol.* **2020**, *54*, 1372–1384. [CrossRef]
42. Li, T.; Wang, Y.; Yuan, Q. Remote Sensing Estimation of Regional NO_2 via Space-Time Neural Networks. *Remote Sens.* **2020**, *12*, 2514. [CrossRef]
43. Zhang, T.; He, W.; Zheng, H.; Cui, Y.; Song, H.; Fu, S. Satellite-based ground $PM_{2.5}$ estimation using a gradient boosting decision tree. *Chemosphere* **2020**, 128801. [CrossRef] [PubMed]

44. Wang, H.; Miao, Q.; Shen, L.; Yang, Q.; Wu, Y.; Wei, H. Air pollutant variations in Suzhou during the 2019 novel coronavirus (COVID-19) lockdown of 2020: High time-resolution measurements of aerosol chemical compositions and source apportionment. *Environ. Pollut.* **2020**, *271*, 116298. [CrossRef] [PubMed]
45. Kuerban, M.; Waili, Y.; Fan, F.; Liu, Y.; Qin, W.; Dore, A.J.; Peng, J.; Xu, W.; Zhang, F. Spatio-temporal patterns of air pollution in China from 2015 to 2018 and implications for health risks. *Environ. Pollut.* **2020**, *258*, 113659. [CrossRef] [PubMed]
46. Ye, S.; Ma, T.; Duan, F.; Li, H.; He, K.; Xia, J.; Yang, S.; Zhu, L.; Ma, Y.; Huang, T.; et al. Characteristics and formation mechanisms of winter haze in Changzhou, a highly polluted industrial city in the Yangtze River Delta, China. *Environ. Pollut.* **2019**, *253*, 377–383. [CrossRef]
47. Shen, Y.; Jiang, F.; Feng, S.; Zheng, Y.; Cai, Z.; Lyu, X. Impact of weather and emission changes on NO_2 concentrations in China during 2014–2019. *Environ. Pollut.* **2021**, *269*, 116163. [CrossRef]
48. Shen, G.; Ru, M.; Du, W.; Zhu, X.; Zhong, Q.; Chen, Y.; Shen, H.; Yun, X.; Meng, W.; Liu, J.; et al. Impacts of air pollutants from rural Chinese households under the rapid residential energy transition. *Nat. Commun.* **2019**, *10*, 3405. [CrossRef]

remote sensing

MDPI

Article

Global-Scale Patterns and Trends in Tropospheric NO_2 Concentrations, 2005–2018

Sadegh Jamali [1,*], Daniel Klingmyr [2] and Torbern Tagesson [3,4]

1 Department of Technology and Society, Lund University, Box 118, 221 00 Lund, Sweden
2 Golder Environmental Services, Golder Associates AB, P.O. Box 20127, 104 60 Stockholm, Sweden; daniel_klingmyr@golder.se
3 Department of Physical Geography and Ecosystem Science, Lund University, Sölvegatan 12, SE-223 62 Lund, Sweden; torbern.tagesson@nateko.lu.se
4 Department of Geosciences and Natural Resource Management, University of Copenhagen, Øster Voldgade 10, DK-1350 Copenhagen, Denmark
* Correspondence: sadegh.jamali@tft.lth.se; Tel.: +46-46-222-9139

Received: 9 October 2020; Accepted: 26 October 2020; Published: 28 October 2020

Abstract: Nitrogen dioxide (NO_2) is an important air pollutant with both environmental and epidemiological effects. The main aim of this study is to analyze spatial patterns and temporal trends in tropospheric NO_2 concentrations globally using data from the satellite-based Ozone Monitoring Instrument (OMI). Additional aims are to compare the satellite data with ground-based observations, and to find the timing and magnitude of greatest breakpoints in tropospheric NO_2 concentrations for the time period 2005–2018. The OMI NO_2 concentrations showed strong relationships with the ground-based observations, and inter-annual patterns were especially well reproduced. Eastern USA, Western Europe, India, China and Japan were identified as hotspot areas with high concentrations of NO_2. The global average trend indicated slightly increasing NO_2 concentrations (0.004×10^{15} molecules cm^{-2} y^{-1}) in 2005–2018. The contribution of different regions to this global trend showed substantial regional differences. Negative trends were observed for most of Eastern USA, Western Europe, Japan and for parts of China, whereas strong, positive trends were seen in India, parts of China and in the Middle East. The years 2005 and 2007 had the highest occurrence of negative breakpoints, but the trends thereafter in general reversed, and the highest tropospheric NO_2 concentrations were observed for the years 2017–2018. This indicates that the anthropogenic contribution to air pollution is still a major issue and that further actions are necessary to reduce this contribution, having a substantial impact on human and environmental health.

Keywords: tropospheric NO_2 concentrations; nitrogen dioxide; OMI; spatio-temporal trends; DBEST; PolyTrend; time-series analysis; breakpoint detection

1. Introduction

Air pollution is one of the main threats to human health, ecosystems and climate on a global scale [1,2]. The global population is growing substantially, and more than half of the world's population now live in urban areas. Large urban areas and high population densities are hotspots for air pollution [1,3]. According to the World Health Organization (WHO), about 3 million people die annually due to ambient air pollution, mainly in low- and middle-income countries, and about 90% of the world's population are exposed to air that exceeds the WHO air quality guidelines [4].

Nitrogen dioxide (NO_2) is one of the most important air pollutants in the atmosphere [5] and linked to a number of both environmental and epidemiological effects [2,6]. It is formed in processes where nitrogen reacts with oxygen in high temperatures, e.g., through lightning and the combustion of fuels [7]. The main anthropogenic sources of NO_2 emissions are transport, industry processes and energy production [8]. Some of the main environmental effects linked to high NO_2 concentrations are acidification, eutrophication and photochemical formation of ozone (O_3) [6,7,9]. NO_2 also modifies the radiative balance in the atmosphere and influences the atmospheric lifetime of greenhouse gases [10,11]. NO_2 is toxic at high concentrations, and the epidemiological effects include respiratory illnesses such as lung cancer, asthma exacerbations and cardiopulmonary mortality [2,5,7,12]. NO_2 has a short atmospheric lifetime, on average 3.8 ± 1.0 h (mean ± 1 standard deviation) [8] as it reacts with sunlight, which triggers the production of hydroxyl radical OH [13]. Therefore, high concentrations of tropospheric NO_2 are mainly confined to its emission sources, which in general are urban and industrialized areas [2,5].

Monitoring of NO_2 concentrations can be done with ground-based monitoring stations. However, monitoring stations tend to be clustered in city centers, have a small spatial coverage and are often lacking in developing countries [2,14]. Ground-based air quality monitoring is thereby unevenly distributed, and large areas are under-represented [14,15]. An alternative approach to monitor air pollution is the usage of remotely sensed satellite data that increase the spatial coverage. Major advances have been made over the past decades to use satellite sensors to monitor atmospheric pollutants [1]. Satellite monitoring of NO_2 started in 1995 with the Global Ozone Monitoring Experiment (GOME) instrument [3]. Since then, other satellite instruments have been used to monitor tropospheric NO_2, such as GOME-2, the SCanning Imaging Absorption spectroMeter for Atmospheric CHartographY (SCIAMACHY), the Ozone Monitoring Instrument (OMI) and the recent TROPOspheric Monitoring Instrument (TROPOMI) aboard Sentinel-5 Precursor. Out of these instruments, OMI offers the longest continuous monitoring record (ongoing since 2004) and has a relatively high spatial resolution (13 × 24 km^2 at nadir) [6,7]. Potential errors in estimating NO_2 concentrations from satellite data include uncertainties in surface albedo, aerosols, cloud parameters, slant column density and air mass factor calculations [2,6,16]. Therefore, for satellite-based products to be trustworthy, the data need to be compared against other observations of NO_2 concentrations, such as from ground-based monitoring stations [17].

Studies of long-term trends in air pollution provide information about likely changes and distribution patterns that are useful for assessing the effects of emission mitigation efforts [18–20]. Such studies investigating NO_2 trends using OMI data and validating derived results against ground-based measurements have been performed previously. For instance, there are studies on NO_2 trends over USA [2,15,21], over China [22], Russia [23], in eight European cities [1] and in cities around the globe [24]. These studies have reported declining NO_2 trends in their respective study areas and relationship between OMI and ground-based measurements with correlation coefficients ranging between 0.3 and 0.93. NO_2 trend studies on a global scale have also been performed previously using various satellite sensors, but these studies have overall found both negative and positive trends [3,5,19,25,26].

For trend analysis, one of the most widely used methods is the ordinary least-squares (OLS) linear regression, as performed in most of the above-mentioned studies. These simple linear models only provide partial insights on the mechanism essential for an appropriate attribution of drivers of changes. Actual changes can abruptly occur caused by climatic extreme events, anthropogenic mitigations efforts or changes in contributing factors to air pollution. These changes may only be visible for a short period in time, despite having long-lasting effects, and will therefore remain undetected using such traditional linear trend models [27–29].

Recent advances in time-series and breakpoint analysis open new possibilities for studying tropospheric NO_2 concentrations observed by Earth observation satellites, as they allow for the detection of nonlinear trends and turning points in the concentrations. Nonlinear trend models (e.g., PolyTrend) can separate trends into linear and nonlinear trend types [30]. Piecewise linear models, such as Break For Additive Season and Trend (BFAST) [27] or Detecting Breakpoints and Estimating Segments in Trend (DBEST) [29], allow for separating time-series into individual segments, capturing dynamics in specific explanatory variables [28,31–33]. By using these methods, dynamics in tropospheric NO_2 concentrations may be better characterized by capturing specific atmospheric conditions and stages of pollution development through time.

Hence, the main aim of this paper is to analyze global and regional patterns and trends in tropospheric NO_2 concentrations using a continuous time-series of tropospheric NO_2 concentrations from the OMI instrument from 2005 to 2018 with novel methods within time-series and breakpoint analyses. Specifically, we aim at (1) comparing the OMI data against NO_2 concentrations from ground-based monitoring stations, (2) analyzing spatial patterns and temporal (nonlinear) trends, (3) investigating whether regional differences can be found in global NO_2 concentrations and (4) spatially explicitly detecting major breakpoints in NO_2 concentrations and estimating their timing and magnitude at global scale.

2. Materials and Methods

2.1. Satellite-Based NO_2 Dataset

Aura is one of the National Aeronautics and Space Administration's (NASA) Earth Observing System (EOS) satellites. It was launched in 2004 with the mission to collect data of global air pollution and to monitor the chemistry and dynamics of Earth's atmosphere on a daily basis [34]. Aboard Aura there are four instruments, one of which is OMI [34,35]. OMI is a nadir-looking push broom hyperspectral imaging spectrometer that measures reflected solar radiation in the ultraviolet and the visible light (UV/VIS) channels of the electromagnetic spectrum (wavelength range of 264–504 nm) with a spectral resolution of 0.42–0.63 nm [36,37].

We used the OMI/Aura level 3 NO_2 (OMNO2d) standard product (the cloud screened subset 4) downloaded from NASA's Earth Observation data collection [38]. The OMNO2d product contains composites of daily total tropospheric column NO_2 data with a spatial resolution of 0.25° × 0.25°. In this study, we used OMI data from 1 January 2005 until 31 December 2018 (in total, 5092 daily OMNO2d files considering 21 gaps in the daily data files). We also excluded all pixels with less than 50 days of data per year, in order to minimize influences of errors in the retrieval process.

2.2. Ground-Based NO_2 Dataset

The ground-based data are annual averages (n = 6093) of daily observations of atmospheric NO_2 concentrations (n = 1,706,830) from monitoring stations in the USA between the years 2005 to 2018, provided by the United States Environmental Protection Agency (US EPA) [39]. The reference method used by the US EPA for collection of ambient NO_2 is chemiluminescence analysis [40] based on the reaction of nitric oxide (NO) with ozone (O_3). The principle of the method is that a sample of ambient gas enters a reaction chamber where NO molecules react with O_3 to form NO_2. The reaction produces a quantity of light, a phenomenon known as chemiluminescence. The intensity of the light, which is proportional to the concentration of NO_2, is then measured to determine the concentration of NO_2 [40,41].

2.3. Comparison against Ground Observations

The daily OMI NO_2 data were first averaged monthly, and thereafter annually. Annual averages were used since this study focuses on long-term trends, and it is therefore the inter-annual variability that must be validated. The annual averages were then compared to corresponding ground-based NO_2 data in order to verify the validity of OMI NO_2 product. Since the two datasets use different units (10^{15} molecules cm^{-2} for the satellite-based data and part per billion (ppb) for the ground-based data), we calculated z-scores using the z-statistic ((data value − average)/standard deviation) for both datasets. The relationships between the two datasets were quantified using the root-mean-square error (RMSE), and by goodness-of-fit when fitting the ordinary least-squares linear regression on the z-scores for the two datasets.

2.4. Analysis of Spatial Patterns and Temporal Trends

The spatial patterns were analyzed by averaging all OMI NO_2 data pixel-wise over the study period. For analyzing the temporal trends, time-series of annual mean NO_2 concentration were first calculated. Then we applied PolyTrend to analyze and classify trends in the annual NO_2 time-series 2005–2018. We also applied the DBEST program to detect the greatest significant breakpoints in the annual NO_2 time-series and estimate their timing and magnitude. The PolyTrend and DBEST analyses were both performed at pixel level having a statistical significance threshold (α) of 0.05. Pixels with an absolute value of annual average tropospheric NO_2 concentration below the OMI detection limit (0.5×10^{15} molec.cm^{-2}) [42] were excluded from the trend analyses.

2.4.1. Nonlinear Trend Analysis with PolyTrend

PolyTrend is an automated method with an algorithm that accounts for nonlinear change in a trend [30]. It uses a polynomial fitting-based scheme that divides trends into linear and nonlinear trend behaviors and then subdivides the nonlinear trends into classes of cubic, quadratic, and concealed trend types. The linear trend type means that the trend line has a uniform direction over the study period (either increasing or decreasing). The quadratic trend type is a trend line with one bend in its curve, implying that the cell has experienced one direction-change in its trend line over the study period (i.e., first positive and then negative trend, or vice versa). The cubic trend type means that the trend line has two bends, implying that corresponding cell has experienced more than one change in the trend direction over the study period (i.e., first decreasing followed by increasing and then again decreasing change, or vice versa). The concealed trend type consists of cells with either cubic or quadratic trend types, but with no significant net change in tropospheric NO_2 concentrations over the study period. We refer to Jamali et al. [30] for more details.

2.4.2. Breakpoint Analysis with DBEST

DBEST was developed for analyzing time-series of satellite sensor data, and it uses a segmentation method for two main algorithms of trend generalization and change detection [29]. We used its change detection algorithm in order to detect breakpoints with greatest change in tropospheric NO_2 concentrations. Our input data in DBEST were the pixel-wise time-series of the annual average NO_2 concentrations data.

First, DBEST tests for the occurrence of discontinuities, in this case of tropospheric NO_2 concentrations, by analyzing the absolute differences between consecutive data points and comparing this to the first level-shift-threshold set by the user (Table 1). If the difference is greater than the first level-shift-threshold, then it tests whether or not this difference caused a significant shift in the mean level of tropospheric NO_2 concentrations and persisted over the duration-threshold. If the mean level before and after this identified discontinuity is greater than the second level-shift-threshold, DBEST considers this a level-shift point. DBEST then repeats this process for all data points, sorts them into descending order based on the absolute value of tropospheric NO_2 concentrations difference, and tests if the spacing between a data point and an identified level-shift point is at least the duration-threshold. The trend component of the time-series is then segmented using a peak/valley detector function and a method that draws a straight line through detected peak/valley points and compares perpendicular distances to the non-peak and non-valley points between them with the distance-threshold parameter. If the distance is greater than this threshold, these points are added to the set of detected peak/valley points and level-shift points, all of which are called turning points. Detected turning points are then fit to the tropospheric NO_2 concentrations trend using piecewise linear modelling, and those turning points that minimize the Bayesian Information Criterion (BIC) [43] are considered breakpoints. Here, we used the change detection algorithm of DBEST with a set value (2) for the number of significant breakpoints of interest for detection (Table 1), and as such, DBEST identifies a final set of greatest significant breakpoints as requested by user. The results of the change detection algorithm include the starting time of the breakpoints (break date); the change duration, or the temporal period over which this change occurred; the change value, or the amount of change that occurred over this time period; the change type, whether the change is abrupt (level-shift) or non-abrupt; the change significance, based on the statistical significance level ($\alpha = 0.05$).

Table 1. DBEST setting parameters, description and the threshold values used in this study.

Parameter	Description	Set Value
Algorithm	The algorithm used by DBEST (either generalization or change detection)	change detection
Data type	Cyclical for time-series with seasonal cycle, and non-cyclical for time-series without seasonal cycle	non-cyclical
Seasonality	The seasonality period for cyclical data, and empty for non-cyclical data	empty
First level-shift-threshold	The lowest absolute difference allowed in input data before and after a breakpoint	0.1×10^{15} molecules cm^{-2}
Duration-threshold	The lowest time period (time steps) within which the shift in the mean level before and after the breakpoint persists	2 years
Second level-shift-threshold	The lowest absolute difference allowed in the means of the data calculated over the duration-threshold before and after the breakpoint	0.5×10^{15} molecules cm^{-2}
Distance-threshold	An internal fitting parameter computed by DBEST	default
Breakpoint number	The number of greatest breakpoints of interest for detection	2
Alpha (α)	Statistical significance level used for testing significance of detected breakpoints	0.05

Here, the annual average tropospheric NO_2 concentrations time-series data were set as non-cyclical type (Table 1). The first level-shift-threshold was set to 0.1×10^{15} molecules cm^{-2} and the second level-shift-threshold to 0.5×10^{15} molecules cm-^{2}. It is recommended that the first level-shift-threshold be set to a smaller value than that for the second level-shift-threshold [29]. Therefore, if a detected change was quick (between two consecutive observations/years) and large enough (0.1×10^{15} molecules cm^{-2}) to shift the mean over the user-set duration (2 years) by 0.5×10^{15} molecules cm^{-2} before and after the point, it was characterized as an abrupt change, otherwise it was considered a non-abrupt change, provided that it was a significant breakpoint. The distance-threshold is normally set to be a default that is derived internally by DBEST.

3. Results

3.1. Data Comparison against Ground Observations

The comparison of OMI data against ground-based observations showed a strong relationship (Pearson's correlation coefficient R = 0.65) that was statistically significant (p-value < 0.01) (Figure 1). The relationship was equally strong (R = 0.65) when separating the analysis into a comparison of how well OMI captured the spatial variability (data averaged site-wise; Figure 1b). The OMI data were most successful at reproducing the inter-annual variability (data averaged annually), for which the observations were in a very close relationship with the ground-based observations (R = 0.99) (Figure 1c). The ordinary least-squares linear trend in annual averages of the z-scores in OMI NO_2 concentrations (-0.220 ± 0.027 z-scores y^{-1}; $R^2 = 0.85$) was very similar to the corresponding trend in the ground-based NO_2 concentrations (-0.218 ± 0.022 z-scores y^{-1}; $R^2 = 0.83$).

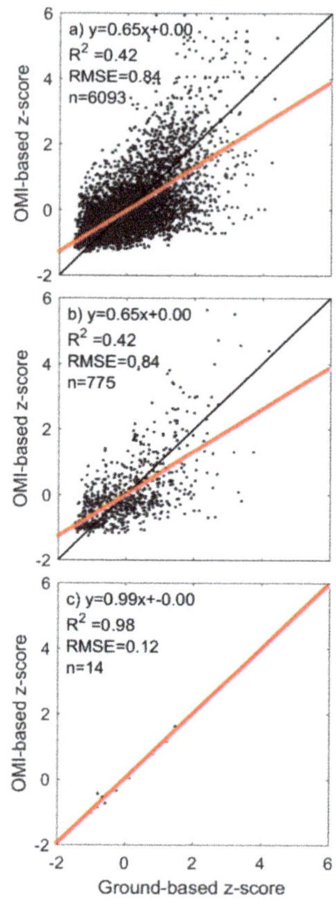

Figure 1. Comparison between z-scores from Ozone Monitoring Instrument (OMI)-based and ground-based tropospheric NO_2 concentrations. (**a**) All annual averages of the ground-based stations against the annual averages of the corresponding OMI pixels. (**b**) The site-wise average for each ground-based station against the corresponding OMI-based pixels. (**c**) The annual averages of all ground-based stations against the annual averages of all corresponding OMI-based pixels. Included are also the ordinary least-squares linear regression (red) with corresponding regression equation and coefficient of determination (R^2), the root-mean-square errors (RMSE) and the number of data points (n). Slope of the linear regression fit indicates Pearson's correlation coefficient (R). The black lines are the one-to-one lines.

3.2. *Spatial Patterns*

There is a distinct difference in the NO_2 concentration distribution between the northern and southern hemispheres, where the higher concentrations are almost exclusively found in the northern hemisphere (Figure 2a). The primary hotspot areas are USA (Figure 2b), Western Europe (Figure 2c), and India, China and Japan (Figure 2d). While the mean global NO_2 concentration was 0.2×10^{15} molecules cm^{-2}, The Netherlands, Belgium, Germany, France, UK, Italy and Spain had the highest average NO_2 concentration (on average 1.91×10^{15} molecules cm^{-2}), followed by Japan (0.91×10^{15} molecules cm^{-2}), India (0.43×10^{15} molecules cm^{-2}), USA (0.38×10^{15} molecules cm^{-2}) and China (0.36×10^{15} molecules cm^{-2}) (Table 2). The maximum NO_2 concentration was for China (28.24×10^{15} molecules cm^{-2}) followed by Japan (14.28×10^{15} molecules cm^{-2}), Italy (11.84×10^{15} molecules cm^{-2}), Germany (11.34×10^{15} molecules cm^{-2}), USA (11.25×10^{15} molecules cm^{-2}) and India (9.22×10^{15} molecules cm^{-2}). Due to their high concentrations in tropospheric NO_2, we selected these areas as focus areas used for further analysis in the remaining part of the study.

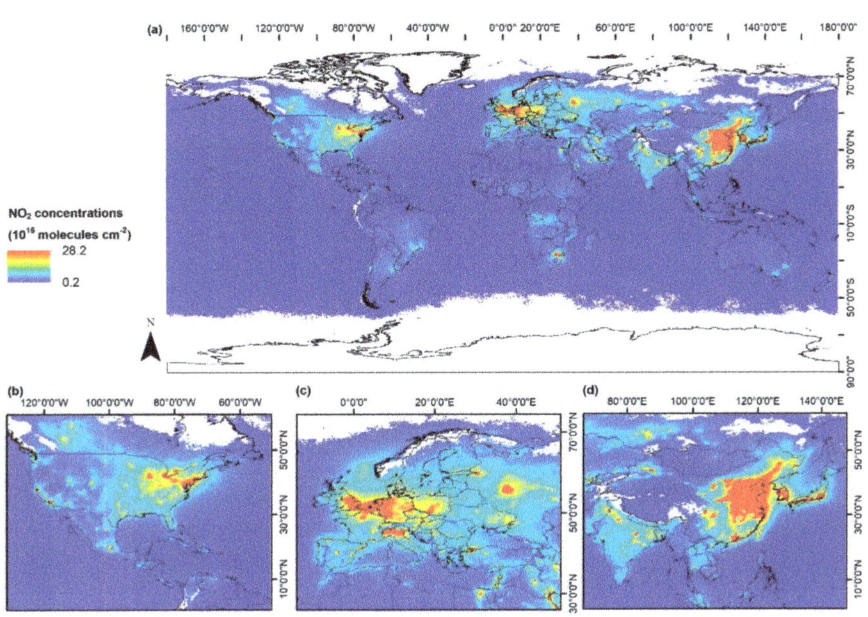

Figure 2. Spatial distribution of tropospheric NO_2 concentrations (10^{15} molecules cm^{-2}) averaged over the years 2005–2018: (**a**) globally; (**b**) USA; (**c**) Europe; (**d**) India, China, Japan. Pixels with less than 50 days of data per year were excluded.

Table 2. The average, maximum and range of tropospheric NO_2 concentrations (10^{15} molecules cm^{-2}), 2005–2018, for the focus areas. Included are the trends in tropospheric NO_2 concentrations averaged country-wise, as well as their strongest positive and negative trend slope (10^{15} molecules cm^{-2} y^{-1}).

Country	Average NO2 Concentration	Max NO2 Concentration	Average Range	Average Trend	Strongest Trend Slope	
					+	−
USA	0.38	11.25	10.87	−0.033	0.055	−0.732
The Netherlands	4.63	9.34	4.70	−0.132	0.000	−0.298
Belgium	3.43	9.26	5.83	−0.143	0.000	−0.285
Germany	1.67	11.34	9.72	−0.035	0.096	−0.361
UK	0.93	7.87	6.94	−0.089	0.016	−0.348
Spain	0.60	5.66	5.06	−0.044	0.012	−0.336
Italy	1.00	11.84	10.84	−0.070	0.047	−0.527
France	1.12	7.42	6.30	−0.042	0.015	−0.309
India	0.43	9.22	8.79	0.040	0.302	−0.031
China	0.36	28.24	27.88	0.014	0.363	−0.946
Japan	0.91	14.28	13.37	−0.049	0.036	−0.671
Global	0.20	28.24	28.04	0.004	0.363	−0.969

3.3. *Temporal Trends*

Significant trends in NO_2 concentrations were observed largely over land and to a much lower degree over oceans along boundaries with lands (Figure 3). With the insignificant no-trends masked out, 79.55% of the remaining cells had positive trend whereas 20.45% had negative trend. The increasing trends were distributed over large parts of land, but the decreasing trends were generally observed over USA (Figure 3a), Western Europe (Figure 3b), Japan and the eastern parts of China (Figure 3c). The global average trend in 2005–2018 was slightly increasing (0.004×10^{15} molecules cm^{-2} y^{-1}); however, the regional negative trends were strong enough to compensate for the global rising trend of NO_2 concentrations over larger areas. Globally, the strongest negative trend was -0.969×10^{15} molecules cm^{-2} y^{-1} while the strongest positive trend was only 0.363×10^{15} molecules cm^{-2} y^{-1} (Table 2).

Areas with high average NO_2 concentrations, except India and western parts of China (Figure 2), generally showed negative trends (Figure 3; Table 2). On average, the strongest negative trends were found in Europe (Belgium: -0.143×10^{15} molecules cm^{-2} y^{-1}; Netherlands: -0.132×10^{15} molecules cm^{-2} y^{-1}; U.K.: -0.089×10^{15} molecules cm^{-2} y^{-1}; Italy: -0.070×10^{15} molecules cm^{-2} y^{-1}) followed by Japan (-0.049×10^{15} molecules cm^{-2} y^{-1}) and USA (-0.033×10^{15} molecules cm^{-2} y^{-1}). The average trend was positive over India and the Middle East. The strongest positive average trend (0.040×10^{15} molecules cm^{-2} y^{-1}) was for India. Although the strongest negative trend in the focus areas (-0.946×10^{15} molecules cm^{-2} y^{-1}) was for China, the average trend for the entire country was just slightly increasing (0.014×10^{15} molecules cm^{-2} y^{-1}) because strong increasing trends (0.363×10^{15} molecules cm^{-2} y^{-1}) were observed over large parts of the country as well (Figure 3c).

3.3.1. Trend Types

In a global context, the linear trend was the dominant trend type with a spatial coverage of 61.98%, out of which 54.47% were positive and 7.51% negative (Figure 4a, Table 3). The concealed trend was the second trend type with 21.89% spatial coverage and mainly found over east of China and Southwestern Europe (Figure 4c,d). For the remaining trends, 9.77% were quadratic and 6.36% were cubic, out of which the majority was found over the eastern parts of USA (Figure 4a) and west of Europe (Spain and Portugal) (Figure 4b,c). In the focus areas, the dominant trend type was different for different areas. In the USA, the nonlinear trends (67.59%) were spatially more than the linear trends (32.41%) (Figure 4a, Table 3). In the focus areas in Europe, the most common trend type was linear (negative), except for Spain where the nonlinear trends, particularly the quadratic negative trends (57.96%), were dominant (Figure 4c, Table 3). The most common trend type over India was linear (increasing) (84.36%), and over Japan was linear (decreasing) (43.03%). China was the country with the largest proportion of nonlinear concealed trends in NO_2 concentration (45.81%); it was also the second country with the highest proportion of linearly increasing trends (39.19%) after India (84.36%) (Figure 4c,d, Table 3).

3.3.2. Breakpoints in Tropospheric NO_2 Concentrations

The global tropospheric NO_2 concentrations showed a slightly decreasing trend from 2005 to 2008, followed by a small, positive change (0.03×10^{15} molecules cm^{-2}) starting in 2008, and then a gradual increasing trend between 2011 and 2018 (Figure 5a). The annual average reached its highest values towards the end of the period in 2017–2018 (0.66×10^{15} and 0.67×10^{15} molecules cm^{-2}). Among the focus areas, only India showed a similar trend behavior but at a higher NO_2 level and with a much greater positive change (0.20×10^{15} molecules cm^{-2}) in 2015 (Figure 5d). Japan was also similar in showing a linear long-term trend with only one breakpoint change but different in that the detected breakpoint was a great negative change (-0.47×10^{15} molecules cm^{-2}), thus resulting in an overall decreasing trend (Figure 5f). In contrast, the number of the greatest changes detected in NO_2 concentrations over USA, Europe and China was two. The two greatest changes of USA (-0.50×10^{15} molecules cm^{-2} and -0.08×10^{15} molecules cm^{-2}) as well as Europe (-0.08×10^{15} molecules cm^{-2} and -0.16×10^{15} molecules cm^{-2}) were both negative and started either in the beginning (2004–2005) or towards the end of the studied period (2013–2016) (Figure 5b,c). The first greatest change detected over China was positive (0.78×10^{15} molecules cm^{-2}) and started in 2008, but then a second big reverse change (-0.81×10^{15} molecules cm^{-2}) happened in 2011 (Figure 5e). These two almost equally big but opposite changes (upward and then downward) with no relax time in between caused the overall NO_2 trend being insignificant with no net-change in NO_2 concentrations throughout the time period over China. This type of significant nonlinear trend was identified as concealed trend type (Figure 5e).

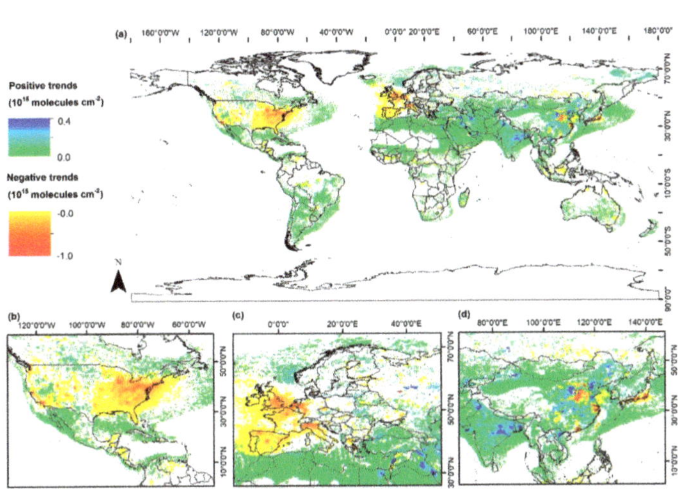

Figure 3. The slope of trend in tropospheric NO_2 concentrations obtained by using the annual average tropospheric NO_2 concentrations data series, 2005–2018, in PolyTrend: (**a**) globally; (**b**) USA; (**c**) Europe; (**d**) India, China, Japan. Insignificant no-trends were masked out ($\alpha = 0.05$).

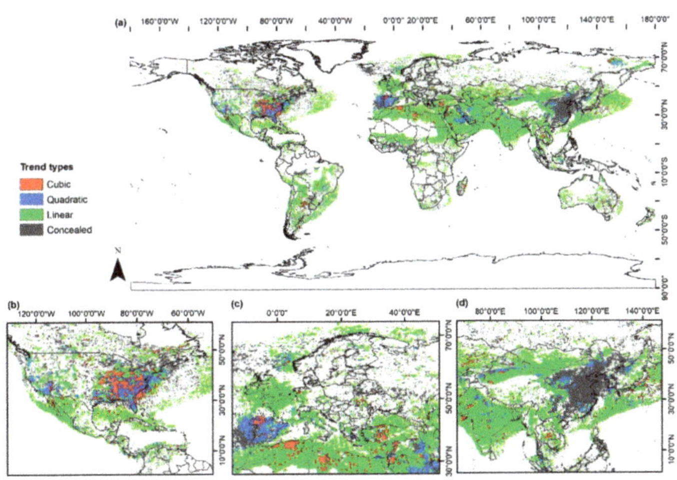

Figure 4. The type of trend in tropospheric NO_2 concentrations obtained by using the annual average tropospheric NO_2 concentration data series, 2005–2018, in PolyTrend: (**a**) globally; (**b**) USA; (**c**) Europe; (**d**) India, China, Japan. Insignificant no-trends were masked out ($\alpha = 0.05$).

Table 3. Spatial coverage (%) of the significant increasing and decreasing trend types globally and in the focus areas with hotspots in average NO_2 concentration. Insignificant no-trends were masked out ($\alpha = 0.05$).

	Trend Types [1]								
	Lin. +	**Lin. −**	**Quad. +**	**Quad. −**	**Cub. +**	**Cub. −**	**Conc. +**	**Conc. −**	**Cell Count**
USA	7.51	24.90	1.17	25.98	1.20	16.82	8.15	14.27	8052
The Netherlands	0.00	82.35	0.00	0.00	0.00	13.73	0.00	3.92	51
Belgium	0.00	98.04	0.00	0.00	0.00	1.96	0.00	0.00	51
Germany	4.51	68.44	0.41	2.46	0.41	5.74	2.87	15.16	244
UK	0.00	94.23	0.00	2.41	0.00	0.96	0.96	1.44	416
Spain	0.13	6.44	0.00	57.96	0.13	10.10	2.28	22.98	792
Italy	0.59	75.81	0.00	14.45	0.00	2.07	3.83	3.25	339
France	0.00	87.31	0.00	7.02	0.00	0.90	1.34	3.43	670
India	84.36	0.03	9.64	0.03	4.53	0.03	1.07	0.34	3840
China	39.19	0.85	10.89	0.53	2.64	0.09	33.46	12.35	10,259
Japan	10.09	43.03	0.00	11.87	0.89	13.06	9.19	11.87	337
Global	54.47	7.51	6.19	3.58	4.56	1.80	14.33	7.56	123,256

[1] Lin = linear, Quad = quadratic, Cub = cubic, Conc = concealed.

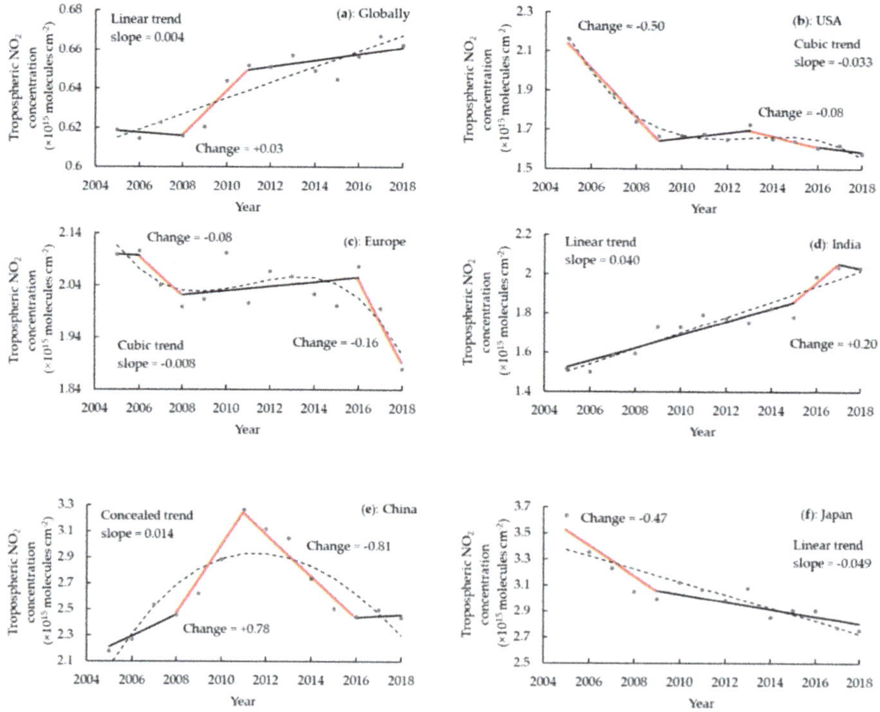

Figure 5. Time-series of annual average tropospheric NO_2 concentrations, 2005–2018, with a segmented trend estimated by Detecting Breakpoints and Estimating Segments in Trend (DBEST): (**a**) globally; (**b**) USA; (**c**) Europe; (**d**) India; (**e**) China; (**f**) Japan. The line segments in red denote breakpoints with greatest change (10^{15} molecules cm^{-2}), and the dashed curves denote the type of trend estimated by PolyTrend.

Figure 6a shows the greatest breakpoint change detected in the annual average NO_2 concentrations at pixel level. The spatial patterns of the detected short-term changes were similar to the long-term overall trends observed over lands (Figure 3): positive breakpoints were found over large areas in all continents (79.4%) and negative breakpoints mainly over the focus areas (20.6%). The greatest negative drop was for China (-12.41×10^{15} molecules cm^{-2}), followed by USA (-5.60×10^{15} molecules cm^{-2}), Italy (-3.81×10^{15} molecules cm^{-2}) and Japan (-3.78×10^{15} molecules cm^{-2}) and then the other focus countries in Europe (Figure 6a, Table 4). The greatest positive change was also for China (6.65×10^{15} molecules cm^{-2}) followed by India (2.13×10^{15} molecules cm^{-2}). Range of the change values was therefore the highest for China (19.06×10^{15} molecules cm^{-2}) and the least for Netherlands (1.59×10^{15} molecules cm^{-2}) and Belgium (1.75×10^{15} molecules cm^{-2}), where the average changes were high and no positive change was detected at all (Table 4). The type of majority of the detected greatest changes was non-abrupt, indicating that most of the changes occurred gradually over time, except for Belgium where the changes mainly happened abruptly (56.86%).

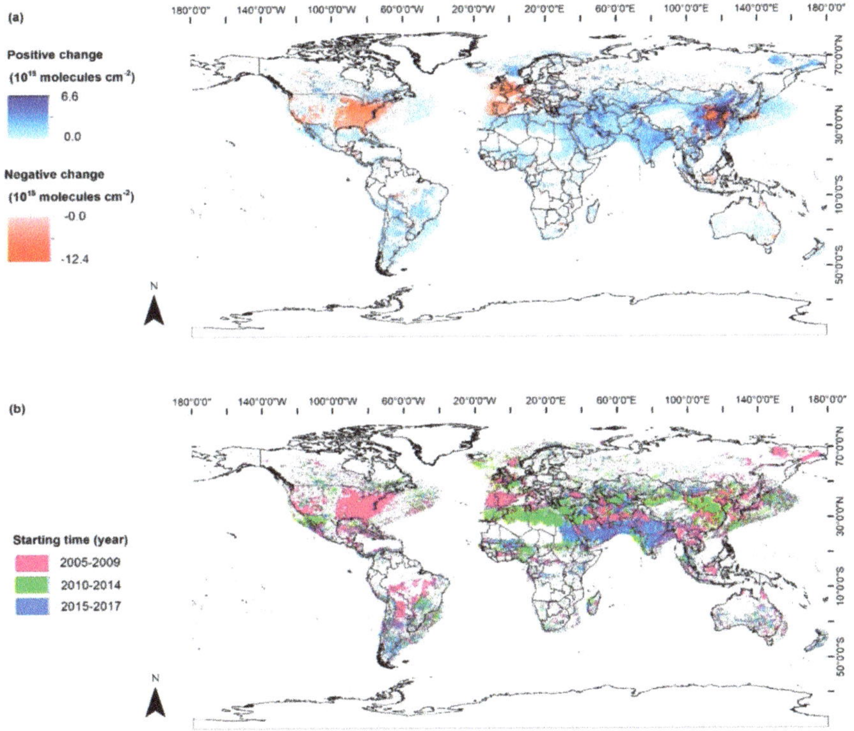

Figure 6. The breakpoint with greatest change in tropospheric NO_2 concentrations obtained by using the annual average tropospheric NO_2 concentration data series, 2005–2018, in Detecting Breakpoints and Estimating Segments in Trend (DBEST). (**a**) Magnitude of the change. (**b**) Starting time of the change.

Table 4. The values of the greatest breakpoint changes in tropospheric NO_2 concentrations (10^{15} molecules cm^{-2}), the within-country average and range of changes, as well as the distribution of the type of the changes detected by Detecting Breakpoints and Estimating Segments in Trend (DBEST).

	Major Change		Average Change	Range of Change Values	Change Type (%)	
	Positive	Negative			Abrupt	Non-Abrupt
USA	1.20	−5.60	−0.60	6.80	10.20	89.80
The Netherlands	-	−2.59	−1.54	1.59	35.29	64.71
Belgium	-	−2.50	−1.66	1.75	56.86	43.14
Germany	1.44	−3.28	−1.37	4.72	22.54	77.46
UK	0.98	−2.57	−0.98	3.56	14.77	85.23
Spain	0.54	−2.50	−0.54	3.04	9.10	90.90
Italy	1.23	−3.81	−0.91	5.04	17.70	82.30
France	0.53	−3.11	−0.83	3.64	9.25	90.75
India	2.13	−1.01	0.41	3.14	2.23	97.77
China	6.65	−12.41	0.28	19.06	22.13	77.87
Japan	0.76	−3.78	−0.73	4.54	16.02	83.98
Global	6.68	−12.41	0.09	19.06	4.15	95.85

The starting time of the major drops in tropospheric NO_2 concentrations is most often detected during the period of 2005–2009 for USA (89.6% of cells), Japan (78.8%) and Europe (57.8%) (Figure 6b). For India, the greatest positive change started most often during 2015–2017 (41.1% of cells). For China, the biggest positive change started mostly during 2008–2010 (54.3%) and then the greatest drop happened during 2011–2014 (88.7%).

In a global context, the years 2005 and 2007 were by far the years with the highest occurrence of negative breakpoints (27.7% and 17.4% respectively), indicating a major event during this period that had global effects and particularly in the focus areas (Figure 6a; Figure 7a). The time period with high occurrence of global positive breakpoints was 2008 to 2015, and the years 2008 and 2015 had the highest rates (12.4% and 12.2% respectively) (Figure 7b).

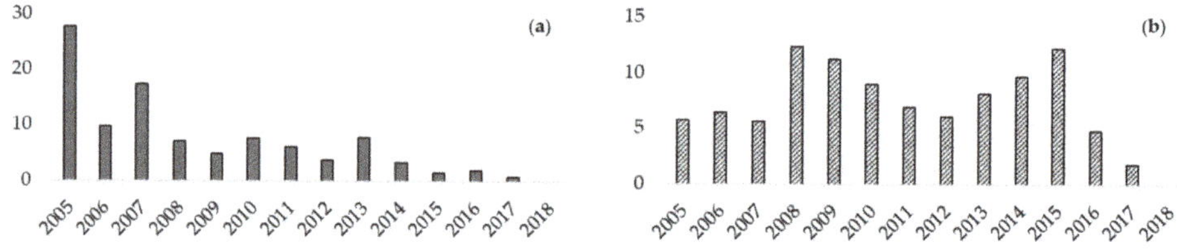

Figure 7. The temporal distribution of the global breakpoints with the greatest change in tropospheric NO_2 concentrations detected over the years 2005–2018. The values on the *y*-axis are in percentage (%). (**a**) The greatest negative changes. (**b**) The greatest positive changes.

4. Discussion

The relationship between the satellite-based and the ground-based datasets supports previous OMI validation studies. For instance, the Pearson's correlation coefficient R was 0.65, which is within the middle of the range (0.40–0.80) of several other studies [1,2,15,22]. The statistical comparison further indicated that OMI was more successful at estimating the temporal component than the spatial component (Figure 1b,c). This can partially be explained since the ground-based monitoring stations are focused on a certain emission source (e.g., traffic locations), whereas an OMI pixel (13×24 km^2) covers a larger area with potential emission sources both within and downwind from the pixel [1]. The strong relationships with the ground-based observations still indicate that OMI data are useful giving spatially explicit time-series of tropospheric NO_2 concentrations to study global patterns and trends.

The spatial distribution of average NO_2 concentrations found in this study (Figure 2) resembles those in other studies [5,19,25,26], confirming that the focus areas are indeed the main hotspots of tropospheric NO_2 concentrations globally. According to Krotkov et al. [25], the highest NO_2 concentrations coincide with urban areas with high populations and industrialized regions. NO_2 concentrations are generally much lower over oceans than that over land since there are no sources of NO_2 emissions except for passing ships [44]. This indicates that the trends observed along offshore boundaries are possibly caused by atmospheric deposition of NO_2 transported from their source by large-scale circulation [45]. According to Peters et al. [44], satellite instruments have issues with detecting trace gases over oceans because of the low NO_2 concentrations often being below the detection limit of the instruments (0.5×10^{15} molecules cm^{-2}).

The global and regional trends seen (Figure 3) generally agree with the results from previous studies. Previous studies have shown increasing trends over both India and China [5,19,25,26], where our results show increasing trends over both countries too (Figure 3). The decreasing trend with major drop in NO_2 that we observed over Eastern USA confirms the previous study by Krotkov et al. [46] reporting a dramatic decrease in OMI NO_2 from 2005 to 2015, as a result of both technological improvements and stricter regulation of emissions. In agreement with our trend results derived for Western Europe, recently Wang et al. [47] observed decreasing trends over Netherlands, Belgium, Germany and Italy, as detected in OMI NO_2 concentrations for 2012–2018. The trend results seem to be consistent among studies with data used from different satellite instruments and/or study periods [5,19,25,26].

Decreases of NO_2 concentrations can primarily be attributed to either local-, regional- or country-level environmental regulations, improvements in emission control technology (e.g., power plants and vehicles), or economic changes and the associated effects in energy usage [24,25]. Since the spatial distribution of average concentrations and significant decreasing trends correlate well, this indicates that environmental regulations and technological improvements in the countries with the most severe pollution have had a positive effect on concentrations of NO_2. However, it should also be noted that the two final years of this study period (2017–2018) were the years with the highest average global concentrations. This clearly shows the importance of continuous satellite-based monitoring of global patterns and trends in NO_2 concentrations, also for assessing the effects of regional environmental regulations and technological improvements to reduce emissions [48].

Linear regression models assume that changes occur linearly and gradually, which is not always the case [30,49]. Here, a polynomial fitting-based scheme (PolyTrend) was used to account for nonlinear trends. This polynomial approach thus helps to detect nonlinear trends in time-series that would not be identified by an ordinary least-squares (i.e., linear model) approach. The linear trend type was the dominant trend type globally (Figure 4; Table 3) as well as for Europe (except Spain), India and Japan, indicating monotonic (non-decreasing or non-increasing) trends over these areas. The nonlinear trends with a significant slope (quadratic and cubic) were mainly found over eastern parts of USA and Spain. Since the curve of these trends has one (quadratic) or two (cubic) bends, this indicates that the NO_2 concentration trends in these areas either started with an increase and then decreased or the opposite started with a decrease and then increased (quadratic), or with even more short-term changes in the direction of the trend (cubic). The latter case is in agreement with the regional trend curve for USA: a cubic trend starting with a short-term downward trend, then an upward trend, and then again another downward trend (Figure 5b). The identified areas with the concealed trends, mainly in the eastern parts of China and south of Spain, are new findings that, up to the best of our knowledge, have not been reported yet. The reason is that the OLS method is often used in trend studies, and such nonlinear trends are not detectable when OLS is applied for the entire studied period. If OLS applies here, no significant trend in 2005–2018 is detected. However, the concealed changes are credible patterns of nonlinear changes such as the greatest breakpoint changes we detected in NO_2 concentrations over China.

The majority of the detected significant breakpoints were non-abrupt indicating that the concentrations of NO_2 changed gradually, possibly due to stricter environmental regulations or economic cycles, as opposed to abrupt changes (e.g., in Belgium and Netherlands), which could be due to power plants or industries that have been either opened or shut down suddenly. The years 2005 to 2009 were by far the years with the highest occurrence of negative breakpoints, and regional-scale reductions of tropospheric NO_2 concentrations were also observed for USA, Europe and Japan during these years

(Figures 5–7). It has also previously been pointed out that 2008 was a year of significant reductions in NO_2 emissions (e.g., [21,22,50,51]) due to the start of the great economic recession [50,51]. This was an event, which caused large-scale economic reductions and affected anthropogenic activity globally, which in turn reduced the associated emissions of air pollution from, for example, vehicles, power plants and industries. According to the results of this study, the largest change magnitudes in NO_2 concentrations during 2005–2008 were found in USA and Japan. The European countries appear to have suffered less, based on the changes in tropospheric NO_2 concentrations (Figure 6, Table 4). The negative breakpoint we found over Eastern China with a four-year duration (2011–2014) is in general agreement with Li et al.'s [52] study of analyzing global change of tropospheric NO_2 from 2012 to 2017 using data from the Ozone Mapping Profiler Suite (OMPS) Nadir Mapper (NM) onboard the Suomi National Polar Partnership (SNPP). They reported a large decline of NO_2 in Eastern China started in 2013 and was almost entirely driven by wintertime decreases, thus indicating a decrease in anthropogenic emissions over the area. Souri et al. [53] in'their study of analyzing long-term trends of OMI NO_2 concentration 2005–2014 over East Asia, also found downward trends in Japan and more developed Chinese cities such as Guangzhou and Beijing, and upward trends in the majority of northern regions of China in 2010–2013. This supports the concealed trend (upward–downward) we observed for China. Another study by Krotkov et al. [46] also showed similar severe declines of NO_2 in Eastern China in 2011–2014 due to an economic shutdown and government efforts to restrain emissions from the power and industrial sectors. Likewise, the steepest increasing trend we observed was over India, and they reported a fast-growing trend from 2005 to 2015 for India's NO_2 level from coal power plants and smelters.

The time-series analysis methods used in this study (PolyTrend and DBEST) benefit from recent developments, as mentioned earlier, but like many other methods they also have weaknesses. They work on a pixel-by-pixel basis, and they consider each pixel's time-series data as an isolated entity in their trend classification and change detection procedure; the spatial behavior of adjacent areas is not used to improve the robustness of trend/change detection [54]. Thus, the obtained trend and breakpoint results should be interpreted with caution.

Future research could include multiple breakpoint detection analyses using data for pre- and post-pandemic phases of COVID-19 to study impacts of possible changes in anthropogenic sources of NO_2 emissions (e.g., transport, industry processes and energy production) on air pollution and tropospheric NO_2 concentration trends.

5. Conclusions

This study contributes to the ongoing research regarding spatiotemporal patterns and trends in tropospheric NO_2 concentrations using data from the OMI instrument, and it investigates how the tropospheric concentrations have changed globally and regionally over the period of 2005 through 2018. By applying novel techniques for analysis of time-series and their breakpoints, we quantified long-term nonlinear trends and provided information about distribution patterns in the point in time with the greatest changes.

1. Globally, the tropospheric NO_2 concentration showed a slightly increasing long-term trend (0.004×10^{15} molecules $cm^{-2}\ y^{-1}$) for the time period 2005–2018. A significant, positive change (0.03×10^{15} molecules cm^{-2}) was observed during 2008–2011.
2. Over Eastern USA, we found a negative trend of NO_2 concentration (-0.033×10^{15} molecules $cm^{-2}\ y^{-1}$) with two major breakpoint changes of -0.50×10^{15} and -0.08×10^{15} molecules cm^{-2} during 2005–2009 and 2013–2016, respectively.
3. Over Western Europe, the annual average NO_2 concentration decreased slowly (-0.008×10^{15} molecules $cm^{-2}\ y^{-1}$) and in a nonlinear manner including two major drops of -0.08×10^{15} and -0.16×10^{15} molecules cm^{-2} during 2006–2008 and 2016–2018, respectively. Most of the breakpoints changes detected over Netherlands and Belgium were negative and of abrupt type.

4. Over India, the steepest rising long-term trend in NO_2 concentration (0.040×10^{15} molecules cm^{-2} y^{-1}), among the other hot spot areas, was observed, and toward the end of the study period (2015–2017) the NO_2 concentration raised even at a higher rate.
5. Over China, the linear long-term trend was positive with a slight slope (0.014×10^{15} molecules cm^{-2} y^{-1}). However, by using the polynomial trend method, we found a nonlinear concealed trend containing one major positive change (0.78×10^{15} molecules cm^{-2}) during 2008–2011 and one big negative change (-0.81×10^{15} molecules cm^{-2}) thereafter in 2011–2016.
6. Over Japan, a considerable drop in NO_2 concentration (-0.47×10^{15} molecules cm^{-2}) was observed in 2005–2009, and the long-term NO_2 trend became the strongest downward trend (-0.049×10^{15} molecules cm^{-2} y^{-1}) as compared to all other focus areas.

Despite the breakpoint changes detected for the focus areas, the linear trend was the dominant trend type at global scale with a spatial coverage of 61.98%, out of which 54.47% were positive and 7.51% negative. The concealed trends, mainly observed over Eastern China and South Spain, ranked second. The years 2005 and 2007 were the years with the highest occurrence of negative breakpoints (27.7% and 17.4% respectively), indicating a major event during these years that had global effects and in the focus areas in particular. However, the trend thereafter reversed, and throughout the study period, the years 2017–2018 had the highest tropospheric NO_2 concentrations. This indicates that the anthropogenic contribution to air pollution is still a major issue, and that further actions are necessary to reduce this contribution. These techniques for analysis of time-series and their breakpoints could be used for studying underlying causes to regional patterns in trends, possibly providing insights to impact of environmental regulations and other actions to prevent air pollution, having substantial impact on human and environmental health.

Author Contributions: Conceptualization, T.T. and D.K.; methodology, S.J., T.T. and D.K.; software, S.J.; validation, D.K. and T.T.; formal analysis, S.J. and T.T.; investigation, S.J. and T.T.; data curation, T.T. and D.K.; writing—original draft preparation, D.K.; writing—review and editing, S.J. and T.T.; visualization, S.J., D.K., and T.T.; supervision, S.J. and T.T.; funding acquisition, T.T. All authors have read and agreed to the published version of the manuscript.

Funding: This research was funded by the Swedish National Space Board (SNSB Dnr 95/16). T.T. was also funded by the Danish Council for Independent Research (DFF, Grant ID: DFF-6111-00258).

Acknowledgments: We acknowledge NASA Goddard Space Flight Center, Goddard Earth Sciences Data and Information Service Center (GES DISC) for providing OMI/Aura NO_2 Cloud-Screened Total and Tropospheric Column L3 (OMNO2d) through EARTHDATA GES DISC data portal (https://disc.gsfc.nasa.gov/). The authors thank the United States Environmental Protection Agency (US EPA) for providing ground-based atmospheric NO_2 concentrations data (https://aqs.epa.gov/aqsweb/airdata/download_files.html#Annual). The authors are very grateful for the constructive feedback from three anonymous reviewers that helped improve the quality of the article.

Conflicts of Interest: The authors declare no conflict of interest.

References

1. Paraschiv, S.; Constantin, D.-E.; Paraschiv, S.-L.; Voiculescu, M. OMI and ground-based in-situ tropospheric nitrogen dioxide observations over several important European cities during 2005–2014. *Int. J. Environ. Res. Public Health* **2017**, *14*, 1415.
2. Bechle, M.J.; Millet, D.B.; Marshall, J.D. Remote sensing of exposure to NO_2: Satellite versus ground-based measurement in a large urban area. *Atmos. Environ.* **2013**, *69*, 345–353.
3. Schneider, P.; Lahoz, W.; van der A, R. Recent satellite-based trends of tropospheric nitrogen dioxide over large urban agglomerations worldwide. *Atmos. Chem. Phys. Discuss.* **2014**, *14*. [CrossRef]
4. WHO: Ambient Air Pollution: A Global Assessment of Exposure and Burden of Disease. Available online: http://www.who.int/iris/handle/10665/250141 (accessed on 21 August 2020).
5. Georgoulias, A.K.; van der A, R.J.; Stammes, P.; Boersma, K.F.; Eskes, H.J. Long-term trends and trend reversal detection in two decades of tropospheric NO_2 satellite observations. In *Geophysical Research Abstracts*; EBSCO Industries, Inc.: Birmingham, AL, USA, 2019; Volume 21.

6. Streets, D.G.; Canty, T.; Carmichael, G.R.; de Foy, B.; Dickerson, R.R.; Duncan, B.N.; Edwards, D.P.; Haynes, J.A.; Henze, D.K.; Houyoux, M.R. Emissions estimation from satellite retrievals: A review of current capability. *Atmos. Environ.* **2013**, *77*, 1011–1042.
7. Zhang, L.; Lee, C.S.; Zhang, R.; Chen, L. Spatial and temporal evaluation of long term trend (2005–2014) of OMI retrieved NO_2 and SO_2 concentrations in Henan Province, China. *Atmos. Environ.* **2017**, *154*, 151–166.
8. Liu, F.; Beirle, S.; Zhang, Q.; Dörner, S.; He, K.; Wagner, T. NOx lifetimes and emissions of cities and power plants in polluted background estimated by satellite observations. *Atmos. Chem. Phys.* **2016**, *16*, 5283.
9. Zhang, R.; Wang, Y.; Smeltzer, C.; Qu, H.; Koshak, W.; Boersma, K.F. Comparing OMI-based and EPA AQS in situ NO_2 trends: Towards understanding surface NOx emission changes. *Atmos. Meas. Tech.* **2018**, *11*, 3955–3967.
10. Isaksen, I.S.; Berntsen, T.K.; Dalsøren, S.B.; Eleftheratos, K.; Orsolini, Y.; Rognerud, B.; Stordal, F.; Søvde, O.A.; Zerefos, C.; Holmes, C.D. Atmospheric ozone and methane in a changing climate. *Atmosphere* **2014**, *5*, 518–535.
11. Solomon, S.; Portmann, R.; Sanders, R.; Daniel, J.; Madsen, W.; Bartram, B.; Dutton, E. On the role of nitrogen dioxide in the absorption of solar radiation. *J. Geophys. Res. Atmos.* **1999**, *104*, 12047–12058.
12. Ordónez, C.; Richter, A.; Steinbacher, M.; Zellweger, C.; Nüß, H.; Burrows, J.; Prévôt, A. Comparison of 7 years of satellite-borne and ground-based tropospheric NO_2 measurements around Milan, Italy. *J. Geophys. Res. Atmos.* **2006**, *111*. [CrossRef]
13. Safieddine, S.; Clerbaux, C.; George, M.; Hadji-Lazaro, J.; Hurtmans, D.; Coheur, P.F.; Wespes, C.; Loyola, D.; Valks, P.; Hao, N. Tropospheric ozone and nitrogen dioxide measurements in urban and rural regions as seen by IASI and GOME-2. *J. Geophys. Res. Atmos.* **2013**, *118*, 10,555–10,566. [CrossRef]
14. Yu, T.; Wang, W.; Ciren, P.; Sun, R. An assessment of air-quality monitoring station locations based on satellite observations. *Int. J. Remote Sens.* **2018**, *39*, 6463–6478. [CrossRef]
15. Lamsal, L.N.; Duncan, B.N.; Yoshida, Y.; Krotkov, N.A.; Pickering, K.E.; Streets, D.G.; Lu, Z. US NO_2 trends (2005–2013): EPA Air Quality System (AQS) data versus improved observations from the Ozone Monitoring Instrument (OMI). *Atmos. Environ.* **2015**, *110*, 130–143.
16. Boersma, K.; Eskes, H.; Dirksen, R.; van der A, R.; Veefkind, J.; Stammes, P.; Huijnen, V.; Kleipool, Q.; Sneep, M.; Claas, J. An improved tropospheric NO_2 column retrieval algorithm for the Ozone Monitoring Instrument. *Atmos. Meas. Tech.* **2011**, *4*, 1905–1928.
17. McPeters, R.; Kroon, M.; Labow, G.; Brinksma, E.; Balis, D.; Petropavlovskikh, I.; Veefkind, J.; Bhartia, P.; Levelt, P. Validation of the Aura Ozone Monitoring Instrument Total Column Ozone Product. *J. Geophys. Res. Atmos.* **2008**, *113*. [CrossRef]
18. Xie, Y.; Wang, W.; Wang, Q. Spatial Distribution and Temporal Trend of Tropospheric NO_2 over the Wanjiang City Belt of China. *Adv. Meteorol.* **2018**, *2018*, 6597186. [CrossRef]
19. Geddes, J.A.; Martin, R.V.; Boys, B.L.; van Donkelaar, A. Long-term trends worldwide in ambient NO_2 concentrations inferred from satellite observations. *Environ. Health Perspect.* **2016**, *124*, 281–289.
20. Munir, S.; Habeebullah, T.M.; Seroji, A.R.; Gabr, S.S.; Mohammed, A.M.; Morsy, E.A. Quantifying temporal trends of atmospheric pollutants in Makkah (1997–2012). *Atmos. Environ.* **2013**, *77*, 647–655. [CrossRef]
21. Tong, D.Q.; Lamsal, L.; Pan, L.; Ding, C.; Kim, H.; Lee, P.; Chai, T.; Pickering, K.E.; Stajner, I. Long-term NOx trends over large cities in the United States during the great recession: Comparison of satellite retrievals, ground observations, and emission inventories. *Atmos. Environ.* **2015**, *107*, 70–84. [CrossRef]
22. De Foy, B.; Lu, Z.; Streets, D.G. Impacts of control strategies, the Great Recession and weekday variations on NO_2 columns above North American cities. *Atmos. Environ.* **2016**, *138*, 74–86.
23. Gruzdev, A.; Elokhov, A. Validation of Ozone Monitoring Instrument NO_2 measurements using ground based NO_2 measurements at Zvenigorod, Russia. *Int. J. Remote Sens.* **2010**, *31*, 497–511. [CrossRef]
24. Duncan, B.N.; Lamsal, L.N.; Thompson, A.M.; Yoshida, Y.; Lu, Z.; Streets, D.G.; Hurwitz, M.M.; Pickering, K.E. A space-based, high-resolution view of notable changes in urban NOx pollution around the world (2005–2014). *J. Geophys. Res. Atmos.* **2016**, *121*, 976–996. [CrossRef]
25. Krotkov, N.A.; Lamsal, L.N.; Celarier, E.A.; Swartz, W.H.; Marchenko, S.V.; Bucsela, E.J.; Chan, K.L.; Wenig, M.; Zara, M. The version 3 OMI NO_2 standard product. *Atmos. Meas. Tech.* **2017**, *10*, 3133–3149. [CrossRef]
26. Miyazaki, K.; Eskes, H.; Sudo, K.; Boersma, K.F.; Bowman, K.; Kanaya, Y. Decadal changes in global surface NOx emissions from multi-constituent satellite data assimilation. *Atmos. Chem. Phys.* **2017**, *17*, 807–837.
27. Verbesselt, J.; Hyndman, R.; Newnham, G.; Culvenor, D. Detecting trend and seasonal changes in satellite image time series. *Remote Sens. Environ.* **2010**, *114*, 106–115. [CrossRef]

28. De Jong, R.; Verbesselt, J.; Schaepman, M.E.; De Bruin, S. Trend changes in global greening and browning: Contribution of short-term trends to longer-term change. *Glob. Chang. Biol.* **2012**, *18*, 642–655. [CrossRef]
29. Jamali, S.; Jönsson, P.; Eklundh, L.; Ardö, J.; Seaquist, J. Detecting changes in vegetation trends using time series segmentation. *Remote Sens. Environ.* **2015**, *156*, 182–195. [CrossRef]
30. Jamali, S.; Seaquist, J.; Eklundh, L.; Ardö, J. Automated mapping of vegetation trends with polynomials using NDVI imagery over the Sahel. *Remote Sens. Environ.* **2014**, *141*, 79–89. [CrossRef]
31. De Jong, R.; Verbesselt, J.; Zeileis, A.; Schaepman, M.E. Shifts in global vegetation activity trends. *Remote Sens.* **2013**, *5*, 1117–1133. [CrossRef]
32. Burrell, A.L.; Evans, J.P.; Liu, Y. Detecting dryland degradation using time series segmentation and residual trend analysis (TSS-RESTREND). *Remote Sens. Environ.* **2017**, *197*, 43–57. [CrossRef]
33. Horion, S.; Prishchepov, A.V.; Verbesselt, J.; de Beurs, K.; Tagesson, T.; Fensholt, R. Revealing turning points in ecosystem functioning over the Northern Eurasian agricultural frontier. *Glob. Chang. Biol.* **2016**, *22*, 2801–2817. [CrossRef] [PubMed]
34. NASA's Aura: New Eye for Clean Air. Available online: https://www.nasa.gov/vision/earth/lookingatearth/aura_first.html (accessed on 21 August 2020).
35. Levelt, P.; Noordhoek, R. *OMI Algorithm Theoretical Basis Document Volume I: OMI Instrument, Level 0-1b Processor, Calibration & Operations*; Technology Report ATBD-OMI-01, Version 2002, 1; NASA: Washington, DC, USA, 2002.
36. Ziemke, J.R.; Strode, S.A.; Douglass, A.R.; Joiner, J.; Vasilkov, A.; Oman, L.D.; Liu, J.; Strahan, S.E.; Bhartia, P.K.; Haffner, D.P. A cloud-ozone data product from Aura OMI and MLS satellite measurements. *Atmos. Meas. Tech.* **2017**, *10*, 4067. [CrossRef] [PubMed]
37. Dobber, M.R.; Dirksen, R.J.; Levelt, P.F.; van den Oord, G.H.; Voors, R.H.; Kleipool, Q.; Jaross, G.; Kowalewski, M.; Hilsenrath, E.; Leppelmeier, G.W. Ozone monitoring instrument calibration. *IEEE Trans. Geosci. Remote Sens.* **2006**, *44*, 1209–1238. [CrossRef]
38. Krotkov, N.A.; Lamsal, L.N.; Marchenko, S.V.; Celarier, E.A.; Bucsela, E.J.; Swartz, W.H.; Joiner, J. OMNO2d: OMI/Aura NO2 Cloud-Screened Total and Tropospheric Column L3 Global Gridded 0.25 Degree × 0.25 Degree V3. Available online: https://doi.org/10.5067/Aura/OMI/DATA3007 (accessed on 22 February 2019).
39. US Environmental Protection Agency. Air Quality System Data Mart. Available online: https://www.epa.gov/airdata (accessed on 29 March 2019).
40. Gilliam, J.; Hall, E. *Reference and Equivalent Methods Used to Measure National Ambient Air Quality Standards (NAAQS) Criteria Air Pollutants—Volume IUS Environmental Protection Agency, Washington, DC*; EPA/600/R-16/139; US Environmental Protection Agency: Washington, DC, USA, 2016.
41. Quality Assurance Guidance Document 2.3—Reference Method for the Determination of Nitrogen Dioxide in the Atmosphere (Chemiluminescence). Available online: https://www3.epa.gov/ttn/amtic/files/ambient/pm25/qa/no2.pdf (accessed on 29 March 2019).
42. Ialongo, I.; Herman, J.; Krotkov, N.; Lamsal, L.; Boersma, K.F.; Hovila, J.; Tamminen, J. Comparison of OMI NO_2 observations and their seasonal and weekly cycles with ground-based measurements in Helsinki. *Atmos. Meas. Tech.* **2016**, *9*, 5203–5212. [CrossRef]
43. Gideon, S. Estimating the dimension of a model. *Ann. Stat.* **1978**, *6*, 461–464.
44. Peters, E.; Wittrock, F.; Großmann, K.; Frieß, U.; Richter, A.; Burrows, J. Formaldehyde and nitrogen dioxide over the remote western Pacific Ocean: SCIAMACHY and GOME-2 validation using ship-based MAX-DOAS observations. *Atmos. Chem. Phys.* **2012**, *12*, 11179. [CrossRef]
45. Yuchechen, A.; Lakkis, S.G.; Canziani, P. Linear and non-linear trends for seasonal NO_2 and SO_2 concentrations in the Southern Hemisphere (2004−2016). *Remote Sens.* **2017**, *9*, 891. [CrossRef]
46. Krotkov, N.A.; McLinden, C.A.; Li, C.; Lamsal, L.N.; Celarier, E.A.; Marchenko, S.V.; Swartz, W.H.; Bucsela, E.J.; Joiner, J.; Duncan, B.N. Aura OMI observations of regional SO_2 and NO_2 pollution changes from 2005 to 2015. *Atmos. Chem. Phys.* **2016**, *16*, 4605. [CrossRef]
47. Wang, Y.; Wang, J. Tropospheric SO_2 and NO_2 in 2012–2018: Contrasting views of two sensors (OMI and OMPS) from space. *Atmos. Environ.* **2020**, *223*, 117214. [CrossRef]
48. Martins, D.K.; Najjar, R.G.; Tzortziou, M.; Abuhassan, N.; Thompson, A.M.; Kollonige, D.E. Spatial and temporal variability of ground and satellite column measurements of NO_2 and O_3 over the Atlantic Ocean during the Deposition of Atmospheric Nitrogen to Coastal Ecosystems Experiment. *J. Geophys. Res. Atmos.* **2016**, *121*, 14175–14187. [CrossRef]

49. Jamali, S. Analyzing Vegetation Trends with Sensor Data from Earth Observation Satellites. Ph.D. Thesis, Lund University, Lund, Sweden, 2014.
50. Bishop, G.A.; Stedman, D.H. The recession of 2008 and its impact on light-duty vehicle emissions in three western United States cities. *Environ. Sci. Technol.* **2014**, *48*, 14822–14827. [CrossRef] [PubMed]
51. Castellanos, P.; Boersma, K.F. Reductions in nitrogen oxides over Europe driven by environmental policy and economic recession. *Sci. Rep.* **2012**, *2*, 265. [CrossRef]
52. Lin, N.; Wang, Y.; Zhang, Y.; Yang, K. A large decline of tropospheric NO_2 in China observed from space by SNPP OMPS. *Sci. Total Environ.* **2019**, *675*, 337–342. [CrossRef]
53. Souri, A.H.; Choi, Y.; Jeon, W.; Woo, J.H.; Zhang, Q.; Kurokawa, J.I. Remote sensing evidence of decadal changes in major tropospheric ozone precursors over East Asia. *J. Geophys. Res. Atmos.* **2017**, *122*, 2474–2492. [CrossRef]
54. Abel, C.; Horion, S.; Tagesson, T.; Brandt, M.; Fensholt, R. Towards improved remote sensing based monitoring of dryland ecosystem functioning using sequential linear regression slopes (SeRGS). *Remote Sens. Environ.* **2019**, *224*, 317–332. [CrossRef]

remote sensing

MDPI

Technical Note

The Cross-Border Transport of $PM_{2.5}$ from the Southeast Asian Biomass Burning Emissions and Its Impact on Air Pollution in Yunnan Plateau, Southwest China

Qingjian Yang [1], Tianliang Zhao [1,*], Zhijie Tian [2], Kanike Raghavendra Kumar [3], Jiacheng Chang [4], Weiyang Hu [5], Zhuozhi Shu [1] and Jun Hu [6]

1 Key Laboratory for Aerosol-Cloud-Precipitation of China Meteorological Administration, Nanjing University of Information Science and Technology, Nanjing 210044, China; 20211103009@nuist.edu.cn (Q.Y.); zzshu91@nuist.edu.cn (Z.S.)
2 China Institute for Radiation Protection, Taiyuan 030006, China; sylviayung@nuist.edu.cn
3 Department of Physics, Koneru Lakshmaiah Education Foundation (KLEF), Vaddeswaram, Guntur 522 502, Andhra Pradesh, India; rkkanike@kluniversity.in
4 School of Environment and Energy, South China University of Technology, Guangzhou 510006, China; 202110188646@mail.scut.edu.cn
5 State Key Laboratory of Pollution Control and Resource Reuse and School of the Environment, Nanjing University, 163 Xianlin Rd., Nanjing 210023, China; dg21250013@smail.nju.edu.cn
6 Fujian Academy of Environmental Sciences, Fuzhou 350011, China; jun_hu@ou.edu
* Correspondence: tlzhao@nuist.edu.cn

Abstract: Southeast Asia is one of the largest biomass burning (BB) regions in the world, and the air pollutants generated by this BB have an important impact on air pollution in southern China. However, the mechanism of the cross-border transport of BB pollutants to neighboring regions is yet to be understood. Based on the MODIS remote sensing products and conventional observation data of meteorology and the environment, the WRF-Chem and FLEXPART-WRF models were used to simulate a typical $PM_{2.5}$ pollution episode that occurred during 24–26 March 2017 to analyze the mechanism of cross-border transport of BB pollutants over Yunnan Plateau (YP) in southwest China. During this air pollution episode, in conjunction with the flourishing BB activities over the neighboring Indo-China Peninsula (ICP) regions in Southeast Asia, and driven by the southwesterly winds prevailing from the ICP to YP, the cross-border transport of pollutants was observed along the transport pathway with the lifting plateau topography in YP. Based on the proximity to the BB sources in ICP, YP was divided into a source region (SR) and a receptor region (RR) for the cross-border transport, and the negative and positive correlation coefficients (R) between $PM_{2.5}$ concentrations and wind speeds, respectively, were presented, indicating the different impacts of BB emissions on the two regions. XSBN and Kunming, the representative SR and RR sites in the border and hinterland of YP, respectively, have distinct mechanisms that enhance $PM_{2.5}$ concentrations of air pollution. The SR site is mainly affected by the ICP BB emissions with local accumulation in the stagnant meteorological conditions, whereas the RR site is dominated by the regional transport of $PM_{2.5}$ with strong winds and vertical mixing. It was revealed that the large $PM_{2.5}$ contributions of ICP BB emissions lift from the lower altitudes in SR to the higher altitudes in RR for the regional transport of $PM_{2.5}$. Moreover, the contributions of regional transport of $PM_{2.5}$ decrease with the increase in transport distance, reflecting an important role of transport distance between the source–receptor areas in air pollution change.

Keywords: Yunnan Plateau; biomass burning; cross-border transport; $PM_{2.5}$; WRF-Chem

Citation: Yang, Q.; Zhao, T.; Tian, Z.; Kumar, K.R.; Chang, J.; Hu, W.; Shu, Z.; Hu, J. The Cross-Border Transport of $PM_{2.5}$ from the Southeast Asian Biomass Burning Emissions and Its Impact on Air Pollution in Yunnan Plateau, Southwest China. *Remote Sens.* **2022**, *14*, 1886. https://doi.org/10.3390/rs14081886

Academic Editors: Maria João Costa and Daniele Bortoli

Received: 10 March 2022
Accepted: 11 April 2022
Published: 14 April 2022

1. Introduction

Haze pollution caused by aerosol especially $PM_{2.5}$ has significant adverse effects on environmental change and human health [1,2]. In this regard, research has paid a large amount of attention to the air pollution in emission source areas such as the Yangtze River Delta, the Sichuan Basin, and the North China plain [3–5]. However, the Yunnan Plateau (YP), which is a relatively clean region in Southwest China having an aerosol optical depth (AOD) of ~0.1–0.2 [6,7], has lacked attention and studies on the mechanisms responsible for air pollution change.

The regional transport of atmospheric pollutants is one of the major elements affecting the air quality in China [8,9], and has become a critical part of the field of the atmospheric environment [10–12]. Long-range transport, including cross-border transport of atmospheric pollutants, can influence air quality in a large region [13–15]. As a result, strong winds can easily transport $PM_{2.5}$ from upstream source regions to downwind areas, resulting in a rise in $PM_{2.5}$ concentrations [16,17]. Meanwhile, strong winds can also play a role in sweeping local pollutants [18,19]. When the wind speed at source regions increases, air pollutants can be carried to the downstream regions, thus causing a reduction in pollutants in source regions. Therefore, it is an interesting topic of study to understand the role played by winds in the regional atmospheric environment.

Southeast Asia is one of the largest biomass burning (BB) regions in the world, with active fire activities in the spring [20]. As one of the major sources of $PM_{2.5}$ emissions [21], BB emissions can contribute 70–80% to the total $PM_{2.5}$ in source regions [22,23]. Meanwhile, $PM_{2.5}$ generated by a large amount of BB emissions also has an impact on the downwind regions, due to the regional transport driven by atmospheric circulation [24]. Air pollutants produced from Southeast Asian BB emissions can be transported to Southwest China and the Yangtze River Delta over long distances, and even to Taiwan Province, Japan, and the entire East Asia. Although some studies found that air pollutants from BB in Southeast Asia have a significant impact on the atmospheric environment of Southeast China and the northwestern Pacific, few studies have investigated the influence of Southeast Asian BB on air quality in southwest China, and especially the YP region [25–28].

The YP region is located on the southwest border of China, and has a complex topography that gradually decreases from north to south, and southwesterly winds prevail in the YP region throughout the year [29]. The Indo-China Peninsula (ICP) in Southeast Asia, which adjoins the YP region, has shown high AOD values in the spring due to BB emissions [30]. Therefore, the YP region is inevitably influenced by the regional transport of air pollutants from Southeast Asia, especially ICP governed by southwesterly winds. However, the current studies on BB in Southeast Asia have mainly focused on the long-range transport of air pollutants with the effect on atmospheric chemical compositions [31,32], aerosols radiation [33–35], and climatic forcing [36,37]. Due to the lack of studies and analyses on the mechanism of cross-border transport of BB emissions from ICP to the bordering YP, the extent of the influence of BB on air quality in Southwest China is still not well understood. Therefore, in a region such as that of YP, which has the complex terrain of a plateau in a relatively clean environment, the underlying mechanism of air pollution is worthy of in-depth investigation associated with the cross-border or transboundary transport of $PM_{2.5}$ from the Southeast Asian BB emissions.

In this study, we utilized the satellite-based MODIS remote sensing products and the observational data of meteorology and air pollutants to investigate an air pollution event that occurred during 24–26 March 2017 in the YP region associated with the cross-border transport of $PM_{2.5}$ from the BB emission sources in ICP. By utilizing the Weather Research and Forecasting model coupled with Chemistry (WRF-Chem) and the Flexible Particle dispersion model (FLEXPART) driven by WRF (FLEXPART-WRF), we simulated the spatial–temporal variations in $PM_{2.5}$ over Southwest China and Southeast Asia. The present study explored how the regional $PM_{2.5}$ transported from BB emission sources in ICP affects the air quality in the downwind YP region, and the extent to which the regional transport of $PM_{2.5}$ from BB emissions affected $PM_{2.5}$ concentrations in the air pollution episode in YP.

2. Materials and Methods

2.1. Ground-Based Observation Data

To investigate the distribution of meteorological factors and $PM_{2.5}$, and to validate the performance of the WRF-Chem simulation, hourly data of surface $PM_{2.5}$ concentrations were obtained from the China National Environmental Monitoring Centre (CNEMC). Hourly near-surface wind speed, relative humidity, and air temperature data were derived from the China Meteorological Administration (CMA). The time we mention henceforth is YP's local time (UTC + 08:00 h).

2.2. MODIS Remote Sensing Products

The MODIS instrument is a multispectral sensor aboard the Aqua and TERRA satellites. It contains 36 wavelength bands from 400 to 1440 nm and allows for the retrieval of aerosol items to cover the entire globe in 1–2 days. The AOD products derived from MODIS have been widely used at global or regional scales [38,39]. In this study, the MODIS Dark-Target/Deep-Blue combined data of Collection 6.1 averaged from Terra and Aqua were utilized to identify AOD's geographic distribution and to validate the performance of the WRF-Chem Model.

2.3. Model Configuration

2.3.1. WRF-Chem Model

Here, the WRF-Chem online coupling model version 3.9.1 [40] was utilized to simulate an air pollution event that occurred over the YP. Two nested domains were used in the configuration. The coarse domain with a horizontal resolution of 27 × 27 km covered Southwest China and Southeast Asia, wherein the nested fine domain with a 9 × 9 km horizontal resolution included most of the YP and its surrounding ICP regions (Figure 1a). Thirty-two vertical hybrid layers were set from the surface to 50 hPa. The initial and boundary conditions of the WRF-Chem simulation were obtained from the ERA-Interim with a horizontal resolution of 0.75° × 0.75°.

Multiple physical schemes are utilized in the WRF-Chem simulation, such as the YSU boundary layer scheme [41], the Morrison 2 microphysics [42], the RRTMG radiation scheme [43], and the unified Noah land surface model [44]. The RADM2 chemical scheme [45] was selected for the atmospheric gas-phase chemistry mechanism. Table 1 lists the primary parameterization schemes utilized in the modeling configuration.

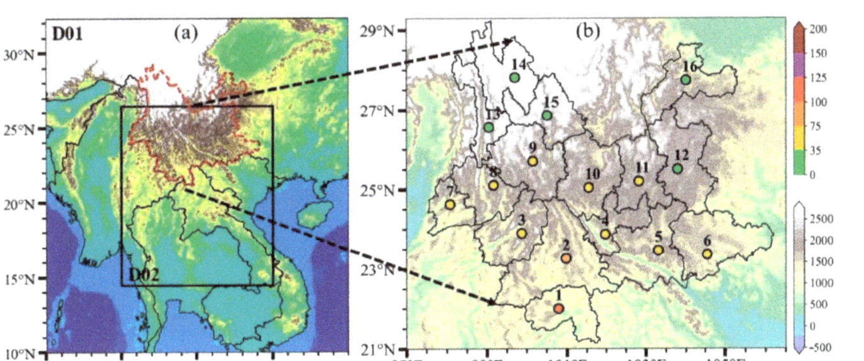

Figure 1. (**a**) Two nesting domains of WRF-Chem modeling with the terrain heights (m in a.s.l.) over the YP (outlined with dash red line) and surrounding regions in southwest China and Southeast Asia. (**b**) Spatial distribution of averaged surface $PM_{2.5}$ concentrations (color shaded circles with black edges, μg m^{-3} in the upper color bar) observed at 16 urban sites (Table 2) from 24 to 26 March 2017 over YP, and spatial distribution of topographic height over YP and surrounding areas (color contours, m in a.s.l. in the lower color bar).

Table 1. Parameterization schemes for the WRF-Chem simulation.

Options	Schemes
Microphysics	Morrison 2-moment scheme (Morrison)
Longwave radiation	Rapid Radiative Transfer Model for GCMs (RRTMG)
Shortwave radiation	Rapid Radiative Transfer Model for GCMs (RRTMG)
Land-surface	Noah Land Surface Model (Noah)
Boundary layer	Yonsei University scheme (YSU)
Cumulus	Improved version of the Grell–Devenyi ensemble scheme (Grell 3-D)
Photolysis	Madronich photolysis scheme (Madronich)
Chemistry	The regional acid deposition model, version 2 (RADM2)
Aerosol particles	The Modal Aerosol Dynamics Model for Europe (MADE/SORGAM)

Table 2. Names of all sites in YP and their corresponding site numbers.

Number	1	2	3	4	5	6
Name	Xishuangbanna	Puer	Lincang	Yuxi	Honghe	Wenshan
Number	7	8	9	10	11	12
Name	Dehong	Baoshan	Dali	Chuxiong	Kunming	Qujing
Name	13	14	15	16		
Number	Nujiang	Diqing	Lijiang	Zhaotong		

2.3.2. Air Pollutant Emission Inventories

Three emissions were used to drive the WRF-Chem modeling. The anthropogenic emission data were taken from MIX [46], which covers more than 30 different countries and regions in Asia, based on a multi-scale data coupling method to include local source inventories, such as ANL-India (India), CAPSS (Korea), REAS2 (Japan, Taiwan, China), MEIC (anthropogenic sources in mainland China), and PKU-NH3 (ammonia emission inventory in China). MIX consists of emissions from on-road mobile sources, agricultural activities, power plants, industrial processes, and residential combustion.

Using the Model of Emissions of Gases and Aerosols from Nature (MEGAN), which includes more than 20 biogenic species [47], the online biogenic emissions were calculated. The hourly BB emissions were obtained from the Fire Inventory from NCAR (FINN) [48]. The FINN was produced using land cover types and fire point emissions monitored by the MODIS satellites (Terra and Aqua), combined with emission factors and combustible loads, and includes particulate matter and gas emissions from biomass burning in agriculture, forests, etc. The horizontal resolution of FINN is 1 km, and the vertical distribution of the fire emission pollutants is calculated by the online plume-rise parameterization [49,50].

2.3.3. Numerical Experiments

Based on the modeling configurations, two numerical experiments were conducted during 21–27 March 2017, of which the first two days were used as the spin-up time of modeling a $PM_{2.5}$ pollution episode. The experiments were: (1) a control experiment (CE), with the MIX anthropogenic emission inventory, the MEGAN biogenic emission inventory, and the FINN BB emission inventory in the modeling configuration; (2) a sensitivity experiment (SE), which was the same as CE but with the BB emissions turned off over two domains.

Through the comparison of the $PM_{2.5}$ concentrations between CE and SE, the contribution rates of BB emissions to $PM_{2.5}$ concentrations for the air pollution episode were evaluated by Equation (1):

$$\text{Contribution rates} = \frac{PM_{2.5\_CE} - PM_{2.5\_SE}}{PM_{2.5\_CE}} \times 100\% \quad (1)$$

where $PM_{2.5\_CE}$ and $PM_{2.5\_SE}$ represent the results from CE and SE, respectively.

2.3.4. FLEXPART-WRF Models

The Flexible Particle dispersion model (FLEXPART) [51,52] is a Lagrangian particle diffusion model considering the processes of wet and dry depositions, turbulent diffusion, and tracer transport [53]. FLEXPART driven by WRF (FLEXPART-WRF) has been widely utilized to examine the potential sources and the long-distance transport of air pollutants [54–56]. Based on this backward trajectory model, we followed the method proposed by Chen et al. [57] and Yu et al. [55] to identify the upstream emission sources of air pollution in the YP region.

2.4. *WRF-Chem Modeling Validation*

A credible simulation of meteorology is essential for the simulation of air pollutants with WRF-Chem [58]. Therefore, the meteorological simulation in typical sites (sites 1–11 in Figure 1b and Table 2 with average surface $PM_{2.5}$ concentrations over 35 μg m^{-3} during the pollution episode) was validated by comparing the modeling results with meteorological observations of 10 m wind speed (WS10), 2 m relative humidity (RH2), and 2 m air temperature (T2). Table 3 and Figure 2a–c list the statistical measures used to compare the observed and simulated meteorological variables. R, RMSE, MB, and ME denote the correlation coefficient, the root mean square, the mean bias, and the mean error, respectively. The CE results of surface $PM_{2.5}$ concentrations were validated with the observational data in typical sites, and the statistical verification is shown in Table 4 and Figure 2d with the R, RMSE, mean fractional bias (MFB), and mean fractional error (MFE).

Table 3. Statistical metrics between observed and simulated meteorological parameters averaged over 11 typical sites during 24–26 March 2017. The "*" indicates R passed the 99.9% confidence level.

	R	RMSE	MB	ME
T2 (°C)	0.92 *	2.71	−0.25	2.16
WS10 (m s^{-1})	0.65 *	2.03	0.10	1.51
RH2 (%)	0.80 *	14.11	−1.31	11.10

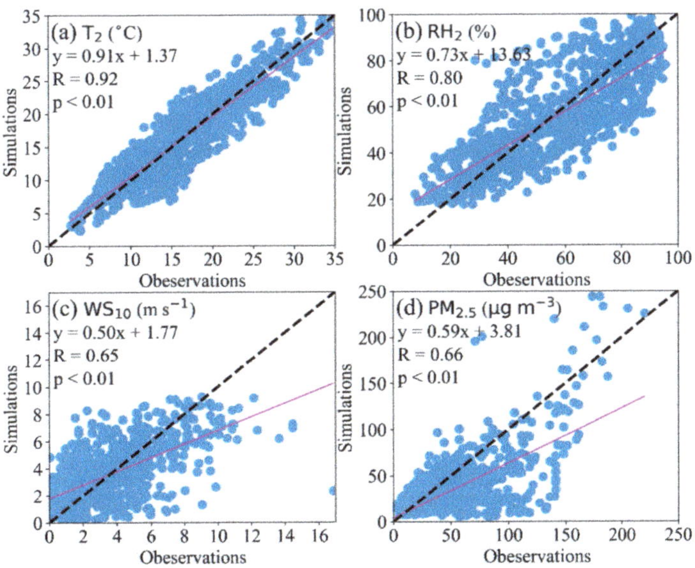

Figure 2. Scatter plots between observations and simulations of hourly meteorological parameters and $PM_{2.5}$ at 11 typical sites during 24–26 March 2017 (blue dots) with liner fitting lines (purple solid lines) passing 99% significance level and 1:1 line (black dash lines) between observations and simulations.

Table 4. Statistical metrics of the comparisons from hourly observed and simulated surface $PM_{2.5}$ concentrations at 11 typical sites during 24–26 March 2017. The "*" indicates R passed the 99.9% confidence level.

	R	RMSE ($\mu g\ m^{-3}$)	MFB (%)	MFE (%)
$PM_{2.5}$	0.66 *	31.09	−21.30	33.26

Due to the different physical properties of AOD and $PM_{2.5}$, there is no good linear relationship between them [59]. Consequently, the vertically integrated concentrations of $PM_{2.5}$ above 700 hPa averaged from 24 to 26 March 2017 were compared with the MODIS AOD shown in Figure 3b,c. The WRF-Chem simulation was evaluated to reasonably capture the spatial distribution of AOD. Both AOD and $PM_{2.5}$ concentrations increase gradually from northern to southern YP, reaching maximum values in Laos and Vietnam near the southeastern border of YP, and relatively large values in Beibu Gulf and Guangxi province.

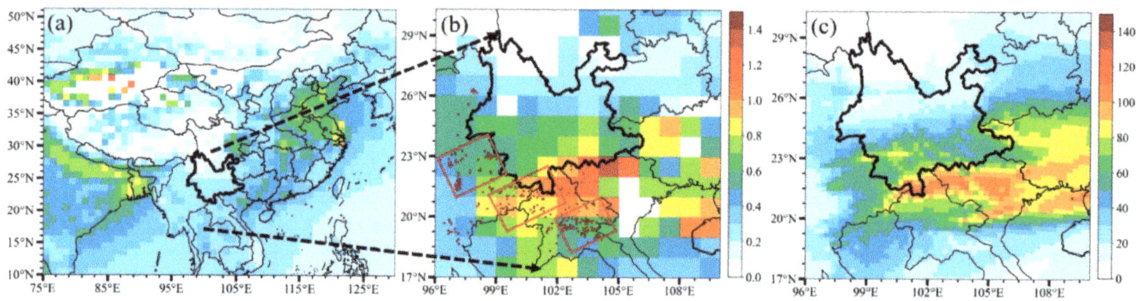

Figure 3. (**a**) Spatial distribution of the monthly mean of MODIS AOD in 2017 over East and Southeast Asia. (**b**) Spatial distribution of the daily mean of MODIS AOD and BB $PM_{2.5}$ emissions from FINN (red dots, emission rate > 3.5 $\mu g\ m^{-3}\ s^{-1}$) during 24–26 March 2017 over Southeast Asia, with the main BB emission sources marked in red rectangles and YP outlined with a bold black line. (**c**) Spatial distribution of the hourly mean of $PM_{2.5}$ concentrations ($\mu g\ m^{-3}$) modeled from 24 to 26 March 2017 over YP and its surrounding areas; the $PM_{2.5}$ concentrations were obtained by vertical integration from 700 hPa upwards.

On the whole, the validation indicates that the meteorological variations and development of $PM_{2.5}$ concentrations were reasonably reproduced by the simulation results of WRF-Chem during the air pollution episode, satisfying Boylan's recommendation for good modeling performance [60]. Thus, the WRF-Chem results could be utilized in the investigation of the cross-border transport of $PM_{2.5}$ from BB sources in Southeast Asia and the influence on air pollution in YP, a clean region in Southwest China.

3. Results and Discussion

3.1. A Springtime Air Pollution Event Observed in YP

As shown in Figure 3a, in 2017, the AOD values were high in central-east China compared to the low AOD values in YP and Southeast Asia. Previous studies on the distributions of $PM_{2.5}$ concentrations and AOD over China showed that the YP, which has low $PM_{2.5}$ pollution and low AOD, presents a clean atmospheric environment compared to other regions of China [61,62].

However, against the background of such a clean atmospheric environment, an air pollution event occurred over YP during 24–26 March 2017. Based on the observation data obtained from CNEMC (Figure 1b) and MODIS (Figure 3b), the distribution of daily average $PM_{2.5}$ concentrations and AOD during the pollution period over YP showed decreasing values with increasing distances to the Southeast Asian BB sources and the uplifting topographic heights from the southern to northern YP. The red dots in Figure 3b show the spatial distribution of average BB $PM_{2.5}$ emissions obtained from FINN during the pollution period. There are few BB emissions in YP, whereas the three major emission-intensive areas are in the neighboring ICP regions, which are in northern Myanmar, eastern Myanmar, and northern Laos. We further summed the intensity of BB emissions in Domain 2 (Figure 1a) to obtain the hourly variation curve of BB emissions (Figure 4a). From 21 to 27 March 2017, the BB emission intensity and the $PM_{2.5}$ concentrations in Xishuangbanna (XSBN, site 1) showed consistent daily variations. The $PM_{2.5}$ concentrations in XSBN increased with increasing BB emission intensity (black arrows in Figure 4a). When the intensity of BB emissions decreased and remained at a low level, the $PM_{2.5}$ concentrations also decreased rapidly. Moreover, the daily maximum $PM_{2.5}$ concentrations and daily maximum BB emission intensity also had a good one-to-one correspondence. The pollution episode that occurred during 24–26 March over YP is in good agreement with the most active BB activities compared to other days. The lagged correlation between BB emission intensity and $PM_{2.5}$ concentrations was further calculated with a lag time of 10 h. The changes in the two values were estimated to have a good positive correlation (R = 0.45), passing the significant level of 0.01, indicating the mechanism behind this pollution event in YP has a strong link to BB emissions over ICP.

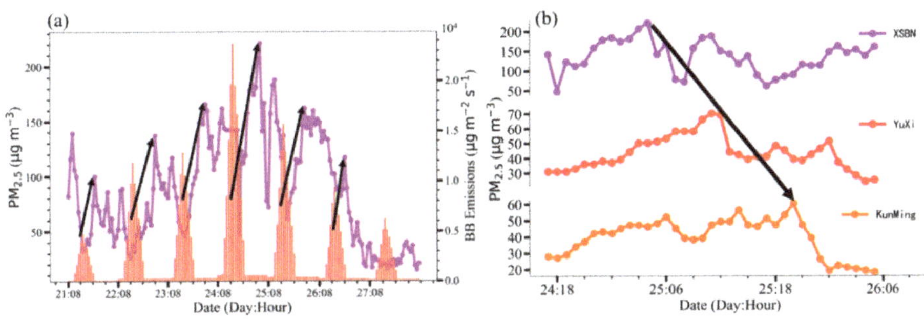

Figure 4. (**a**) Hourly changes in $PM_{2.5}$ concentrations observed in XSBN (purple line) and BB $PM_{2.5}$ emission rate averaged in Domain 2 from FINN (red bar) The black arrows indicate the $PM_{2.5}$ concentrations increase with increasing BB emission intensity. (**b**) Hourly changes in $PM_{2.5}$ concentrations in three downstream sites, XSBN, Yuxi, and Kunming, from 18:00 of 24 March to 06:00 of 26 March. The black arrow indicates the intervals of the lag time along with the regional $PM_{2.5}$ transport from XSBN to Kunming during the air pollution episode.

It is noteworthy that the AOD reached its maximum in northern Laos and northwestern Vietnam near the southeastern border of YP (Figure 3b), which is attributed to the convergence of the southeastern and southwestern winds (Figure 5c) and the obstruction effect of the large topographic height in both northern and eastern parts of this area (Figure 1b). The dual effect of these two factors led to the accumulation of pollutants because southwesterly winds prevailed in ICP and YP from the low to high altitude, and the high mountains largely blocked the further transport of pollutants from this area to the YP region. The present work mainly focuses on the effect of BB emissions on the air quality of YP under the prevailing southwesterly winds. The pollution mechanism of this high AOD area under the influence of BB emissions could be the material or objective for further study.

To further understand the influence of prevailing southwesterly winds on the pollution event on YP, we explored the hourly variations in $PM_{2.5}$ concentrations from 18:00 h local time on 24 March to 06:00 h on 26 March at three observational sites, XSBN, Yuxi, and Kunming, which are in the major pollutants' transport pathway (sites 1,4,11 in Figure 1b). Driven by southwesterly winds, the surface $PM_{2.5}$ peaks advanced northwards at 02:00 h on 25 March, from XSBN, at 10:00 on 25 March to Yuxi, and at 20:00 on 25 March to Kunming, with a quasi-9 h time lag (Figure 4b). At 02:00 on 25 March (Figure 5a), the $PM_{2.5}$

concentrations reached more than 150 μg m^{-3} in many areas of ICP, and the $PM_{2.5}$ in XSBN, the southernmost border city in YP, peaked first. At this time, Yuxi and Kunming were relatively clean, and the southwesterly winds prevailing in ICP and YP were conducive to the cross-border transport of pollutants from the upward ICP to the downwind YP region. About 8 h later, the pollutants were transported to Yuxi at 10:00 h on 25 March (Figure 5b). With the strengthening of southwesterly winds, the pollutants were finally transported to Kunming at 20:00 h on 25 March (Figure 5c), and the $PM_{2.5}$ concentrations in most cities of the central and southern YP reported an increase in $PM_{2.5}$. This phenomenon reflects an obvious characteristic of transport-type pollution events, and previous studies on transport-type pollution events showed that pollutant concentrations have a good correlation with wind speed [17,55]. This provided a motivation to further understand the relationship between $PM_{2.5}$ concentrations and wind speed during the pollution event that occurred over YP, which is explained in the next sections.

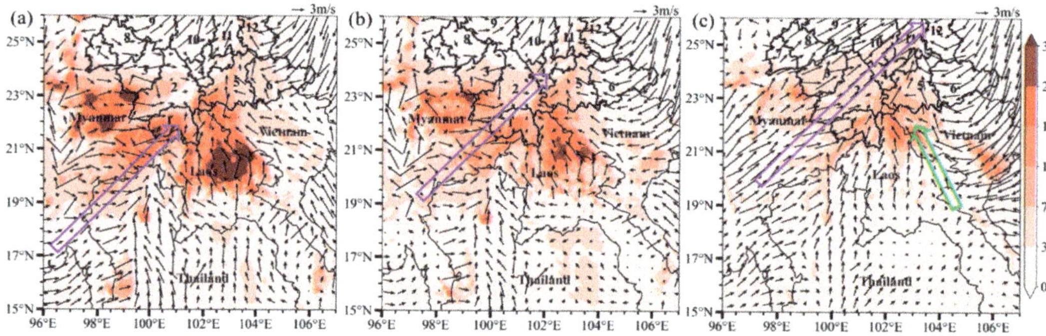

Figure 5. Spatial distribution of surface $PM_{2.5}$ concentrations and 10 m wind vectors simulated at (**a**) 02:00 of 25 March, (**b**) 10:00 of 25 March, and (**c**) 20:00 of 25 March. The purple arrows highlight the major southwesterly winds, and the green arrow in (**c**) highlights the major southeasterly winds.

3.2. Correlation between Wind Speeds and $PM_{2.5}$ Concentrations

The near-surface wind speeds at 10 m are correlated with the $PM_{2.5}$ concentrations over YP, and the spatial distribution of the correlation coefficients (R) is shown in Figure 6a. The $PM_{2.5}$ concentrations in XSBN, the closest site to the BB emissions in ICP, have a significant negative correlation with wind speed, with an R-value of −0.71, passing the significance level at 99%. This indicates that, when the near-surface wind speed increases, more $PM_{2.5}$ concentrations are exported from XSBN, presenting a similar effect over the "source" region (SR) of $PM_{2.5}$ emissions. The border sites on the southwestern part of YP, which are close to the fire activities, all showed the similar effect. In the hinterland of YP, which is further from the fire activities, positive correlations were observed between the near-surface wind speeds and the $PM_{2.5}$ concentrations.

Previous studies showed that a major transport channel of BB pollutants from the ICP to southern China exists around 700 hPa [25], and the transport height is elevated with the increase in distance between YP and ICP. Therefore, we further calculated the R between surface $PM_{2.5}$ concentrations and 700 hPa wind speed (Figure 6b), and the significant positive correlations with R > 0.5 over the central and eastern regions of YP, where the strong winds in the free troposphere play a crucial role in transporting air pollutants to the surface. As a result, these areas are depicted as a "receptor" region (RR) in regional $PM_{2.5}$ transport, which is in good agreement with the study of Yu et al. [55]. The specific mechanisms of air pollution that occurred in the "source" and "receptor" regions in the regional $PM_{2.5}$ transport are described in the following sections.

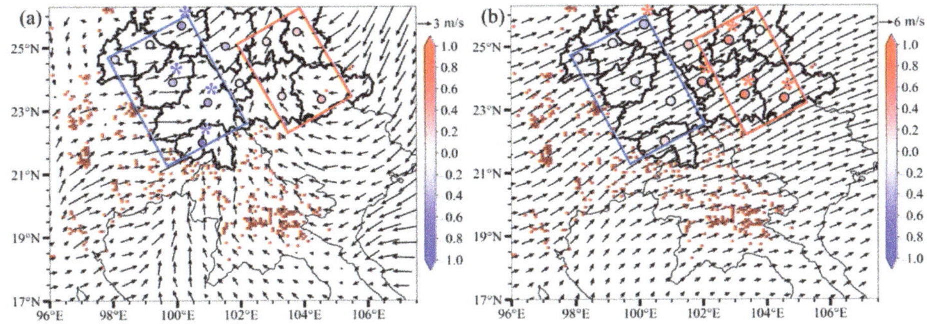

Figure 6. Spatial distribution of correlation coefficients (R) between surface $PM_{2.5}$ concentrations and (**a**) 10 m wind speed and (**b**) 700 hPa wind speed in sites over YP (scatters), and distribution of averaged (**a**) 10 m wind fields and (**b**) 700 hPa wind fields during 24–26 March 2017. The blue and red rectangles in both (**a**,**b**) represent SR and RR, respectively. The red dots are the same as in Figure 2b. The "*" indicates R passed the 99% confidence level.

3.3. The Different Mechanisms of $PM_{2.5}$ Pollution in SR and RR

For XSBN, the representative site in SR, the pollution episode is divided into three periods (Figure 7a), namely the formation period (P1), the maintenance period (P2), and the dissipation period (P3). In P1, the weak wind speed decreased, the boundary layer height was mostly below 1000 m, and the $PM_{2.5}$ concentrations increased rapidly. Previous studies showed that the transport distance plays a significant role in the regional transport of $PM_{2.5}$ from BB emissions. On strong BB days, the mean $PM_{2.5}$ concentrations increase sharply when the distance between the source region and the downwind region is less than 100 km [63]. Meanwhile, Figure 4a shows a synergistic daily variation in $PM_{2.5}$ in XSBN and BB emission intensity in ICP bordering YP. As a result, the BB emissions can affect the $PM_{2.5}$ concentrations in XSBN under the weak wind speed and low boundary layer height. The $PM_{2.5}$ emitted from fire activities is also transported over XSBN by channels at high altitudes. The boundary layer height begins to increase in the second half of P1, once above 3000 m, which is conducive to the development of turbulence. The turbulence further promotes the vertical mixing of $PM_{2.5}$ and facilitates the diffusion of $PM_{2.5}$ from high altitude to the ground, further aggravating the pollution.

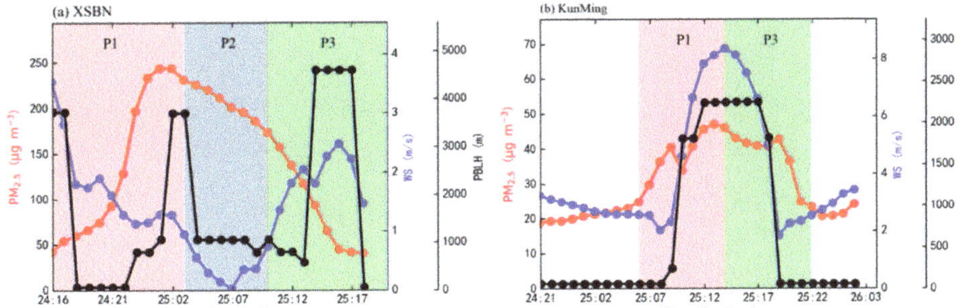

Figure 7. Hourly variations in $PM_{2.5}$ concentrations (red lines), wind speed (WS, blue lines), and planetary boundary layer height (PBLH, black lines) at two representative sites: (**a**) XSBN in SR and (**b**) Kunming in RR, from 16:00 of 24 March to 03:00 of 26 March. P1, P2, and P3 indicate the formation, maintenance, and dissipation periods of the air pollution episode, respectively.

In P2, the fire activities ended in ICP (Figure 4a), and the vertical mixing process was weakened simultaneously with decreasing boundary layer height. However, the $PM_{2.5}$

concentrations decreased slowly with the decreasing wind speed and boundary layer height. Hence, XSBN shows a stable meteorological condition that is conducive to the maintenance of air pollution [64,65]. Under this condition, the $PM_{2.5}$ concentrations were continuously over 150 μg m^{-3} and the $PM_{2.5}$ residue lies on the surface. In P3, the wind speed and boundary layer height increased simultaneously; as a result, the $PM_{2.5}$ concentrations decreased rapidly under the dual effect of horizontal and vertical diffusion. This resulted in the end of P2. To summarize, the changes in $PM_{2.5}$ in XSBN are mainly affected by BB emissions, stagnant meteorological conditions, vertical mixing, and strong winds.

In Kunming, the representative site in RR (Figure 7b), although the $PM_{2.5}$ concentrations increased during the event, the air quality still maintained a good level, which is closely related to the fact that the site is far away from fire activities. In this area, meteorological conditions play a major role in controlling the $PM_{2.5}$ changes under specific processes. Kunming only experienced two periods i.e., the P1 and P3. With increasing wind speed and boundary layer height, the prevailing southwesterly winds transported $PM_{2.5}$ over Kunming and then increased the ground $PM_{2.5}$ concentrations through vertical mixing of turbulence. After that, with decreasing boundary layer height and wind speed, the vertical mixing effect diminished, and the $PM_{2.5}$ concentrations decreased. As a result, the changes in $PM_{2.5}$ in Kunming were mainly affected by the regional transport of $PM_{2.5}$ due to strong winds and vertical mixing.

3.4. Patterns of Regional $PM_{2.5}$ Transport to Different YP Sites

The representative SR and RR sites in the border and hinterland of YP, XSBN, and Kunming, respectively, were selected to estimate the contributions of regional $PM_{2.5}$ transport to air pollution in YP. The estimation was based on the air particle residence time during the pollution period simulated by the FLEXPART-WRF model and three air pollutant emission inventories described in Section 2.3.2. Each simulation was run for a 48-h rearward trajectory of 50,000 air particles being released from two sites, and the air particle residence time was calculated in a $0.1^{\circ} \times 0.1^{\circ}$ horizontal spatial resolution. The air particle residence time was further multiplied with the $PM_{2.5}$ emission fluxes from three air pollutant emission inventories to quantify the contribution of regional $PM_{2.5}$ transport to $PM_{2.5}$ concentrations over the YP. Detailed methods can be found in Chen et al. [57] and Yu et al. [55].

Governed by the prevailing southwesterly winds, the regional transport of $PM_{2.5}$ from the BB emission source regions in ICP provided a significant contribution to the elevated $PM_{2.5}$ concentrations over XSBN and Kunming during 24–26 March 2017 (Figure 8). For XSBN, the major pathway is the southwesterly route from southern and eastern Myanmar, wherein the eastern regions of Myanmar bordering XSBN contribute the most. For Kunming, the $PM_{2.5}$ concentrations are dominated by multiple sources, and the major pathway is the southwesterly route from eastern Myanmar. Moreover, there are two additional minor sources from the northern regions of ICP bordering YP and the domestic area in the east of Kunming. The short-range transport of $PM_{2.5}$ has a major impact on $PM_{2.5}$ concentrations over XSBN, whereas the $PM_{2.5}$ concentrations over Kunming are dominated by the long-range transport of $PM_{2.5}$.

3.5. Contribution of BB Emissions to $PM_{2.5}$ Concentrations over YP

The contribution rates of BB emissions to $PM_{2.5}$ concentrations in YP were quantitatively estimated by Equation (1). The spatial distribution of the contribution rates to surface $PM_{2.5}$ concentrations in YP sites (Figure 9a) is highly similar to the distribution of $PM_{2.5}$ concentrations in Figure 1b. The contribution rates gradually decrease along the transport pathway following the lifting plateau topography in YP. The regional average contribution rate over SR is larger than that in RR, with a difference of 23% (Table 5), and the regional average contribution rate over the whole YP is up to 69%. Three sites with low contribution rates (below 50%) are identified as Yuxi (site 4), Kunming (site 11), and Qujing (site 12). In Yuxi and Kunming, which are the industrial cities of YP, the local anthropogenic emissions are relatively higher, causing the reduction in BB contributions to $PM_{2.5}$ concentrations. For

Qujing, the farthest city from BB emission sources, the long transport distance and greater topography height reduce the impact of BB contributions to local $PM_{2.5}$ concentrations. However, the contribution rate is still up to 41% against the clean background, which further confirms the important impact of BB emissions on the air quality over YP.

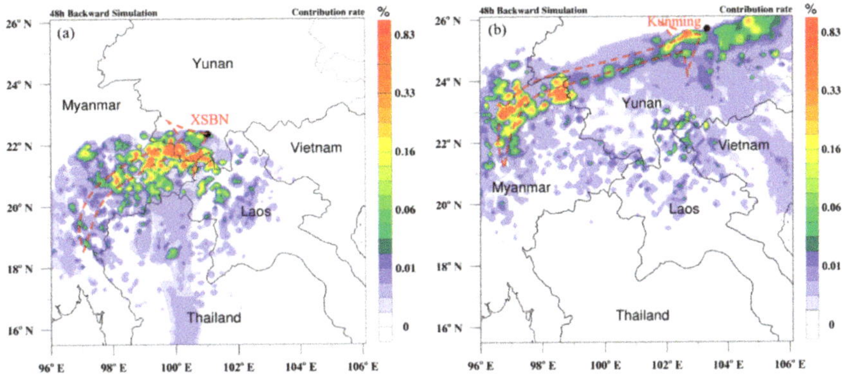

Figure 8. Spatial distribution of contribution rates (color contours) to $PM_{2.5}$ concentrations in (**a**) XSBN and (**b**) Kunming with the major pathways of regional transport (red dash arrows) simulated by the FLEXPART-WRF model from 08:00 of 24 March to 08:00 of 26 March.

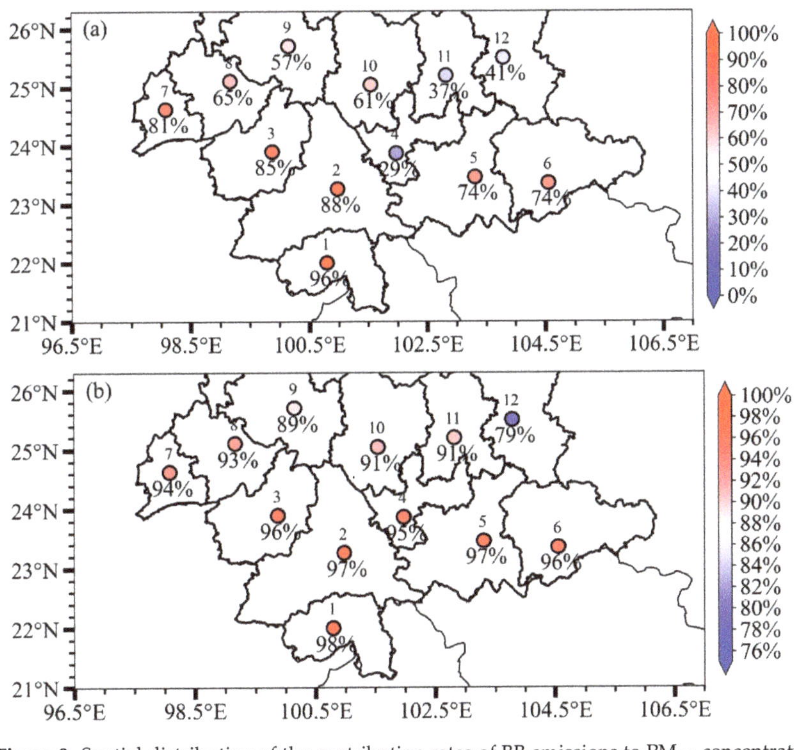

Figure 9. Spatial distribution of the contribution rates of BB emissions to $PM_{2.5}$ concentrations at 12 sites (the numbers 1–12) over YP: (**a**) at the surface and (**b**) at 700 hPa.

Table 5. Regional average contribution rates of BB emissions over the Southeast Asian region to $PM_{2.5}$ concentrations at the surface and 700 hPa over SR and RR, and the regional average increments from the surface to 700 hPa in the two regions.

	SR	RR
Surface	79%	56%
700 hPa	94%	90%
Increments	16%	34%

The pollutants from BB emissions have a characteristic of vertical distribution. As a result, the contribution rates of BB emissions to $PM_{2.5}$ concentrations at 700 hPa were further analyzed (Figure 9b). Two distinctive features can be noticed: (1) The contribution rates in most sites and the regional average contribution rate of YP increase with increasing altitude. However, the increments in the SR sites are much smaller than those in RR, wherein the regional average increment in SR is 16% from surface to 700 hPa, whereas that in RR is 34% (Table 5). The regional $PM_{2.5}$ transport at high altitudes has a larger impact on RR sites, which is consistent with the pollution mechanism discussed in Section 3.3. (2) In Yuxi and Kunming, where the contribution rates of BB emissions to $PM_{2.5}$ concentrations at the surface are relatively small, the contribution rates increase more than 50% from the surface to 700 hPa, indicating that BB emissions have a much greater impact on high-altitude $PM_{2.5}$ concentrations than anthropogenic emissions.

Moreover, the regional average contribution rate over SR is larger than that in RR at both the surface and 700 hPa (Table 5), but the difference between SR and RR (SR minus RR) at the surface (23%) is much greater than that at 700 hPa (4%) due to the obstruction effect of topographic height along the transport pathway. The contributions of regional transport of $PM_{2.5}$ from BB activities decrease with increasing transport distance, reflecting an important role of transport distance between the source–receptor areas in changing the air pollution.

4. Conclusions

Using MODIS remote sensing products and ground-based observations, and conducting model simulations with WRF-Chem and WRF-FLEXPART, the present study examined an air pollution event that occurred over YP, resulting from the cross-border transport of $PM_{2.5}$ due to BB activities from ICP to YP. The aim was to explore how BB emissions in ICP affect the air quality in the neighboring YP.

Under the prevailing southwesterly winds, the BB sources in ICP have different impacts on the $PM_{2.5}$ concentrations over SR and RR. XSBN and Kunming, the representative sites in SR and RR, respectively, have distinct mechanisms enhancing $PM_{2.5}$ concentrations of air pollution. The SR site is mainly affected by Southeast Asian BB emissions with local accumulation in the stagnant meteorological conditions, whereas the RR site is dominated by the regional $PM_{2.5}$ transport with strong winds and vertical mixing. XSBN and Kunming also have different major pathways of regional $PM_{2.5}$ transport. The $PM_{2.5}$ concentrations in XSBN are mainly affected by short-range transport of $PM_{2.5}$, whereas long-range transport of $PM_{2.5}$ plays a dominating role in Kunming. The regional average $PM_{2.5}$ contributions of ICP BB emissions to surface $PM_{2.5}$ over SR is larger than that in RR, with a difference of 23%; in addition, the regional average increments in the contribution from the surface to 700 hPa are 16% in SR and 34% in RR. It is revealed that the large $PM_{2.5}$ contributions of ICP BB emissions lift from the lower altitudes in SR to the higher altitudes in RR in regional $PM_{2.5}$ transport. Moreover, the contributions of regional transport of $PM_{2.5}$ decrease with an increase in transport distance, reflecting an important role of transport distance between the source–receptor areas in changing the scenario of air pollution.

Based on the investigation of a springtime air pollution event in YP, which differs from other regions such as Eastern China where pollution events happen frequently, the study revealed the underlying mechanism of the pollution episode in YP and the extent to which the regional transport of $PM_{2.5}$ from BB emissions affects $PM_{2.5}$ concentrations in YP. However, the MIX anthropogenic emissions in YP and ICP were produced based on data from 2010, which contain more uncertainties compared to those of Eastern China. As a result, future studies involving air pollution simulations can be greatly enhanced by a more accurate emission inventory. To further understand the mechanisms in the regional transport of $PM_{2.5}$ from BB activities, future exploration can be conducted with the support of multi-source satellite data, long-term ground observations, and a modeling study with refined model schemes and data assimilation.

Author Contributions: Data curation, Q.Y. and Z.T.; Funding acquisition, T.Z.; Methodology, Q.Y., T.Z., J.C. and W.H.; Visualization, Z.S. and J.H.; Writing—original draft, Q.Y.; Writing—review and editing, T.Z. and K.R.K. All authors have read and agreed to the published version of the manuscript.

Funding: This research was financially funded by grants received from the National Key Research and Development Program of China (2019YFC0214604), the National Natural Science Foundation of China (41830965; 91744209). One of the authors KRK is grateful to the Science and Engineering Research Board (S ERB), a statutory body under the Department of Science and Technology (DST), India for providing financial grants through the Start-Up Research Grant (SRG; File No. SRG/2020/001445) scheme and the DST, Govt. of India, for the award of the DST-FIST Level-1 (Grant No.SR/FST/PS-1/2018/35) scheme to the Department of Physics, KLEF.

Data Availability Statement: MODIS L3 Atmosphere products (AOD) are available at https://ladsweb.modaps.eosdis.nasa.gov/search/ (accessed on 1 March 2022). ERA-Interim reanalysis data are available at https://www.ecmwf.int/en/forecasts/datasets/reanalysis-datasets/era-interim (accessed on 1 March 2022). The $PM_{2.5}$ datasets and near-surface meteorological data are available at http://www.cnemc.cn and http://data.cma.cn/ (accessed on 1 March 2022).

Conflicts of Interest: The authors declare no conflict of interest.

References

1. Guo, Y.; Zeng, H.; Zheng, R.; Li, S.; Barnett, A.G.; Zhang, S.; Zou, X.; Huxley, R.R.; Chen, W.; Williams, G. Lung cancer incidence and ambient air pollution in China: A spatial age–period cohort study 1990–2009. *Lancet* **2015**, *386*, S5. [CrossRef]
2. Dawson, J.P.; Adams, P.J.; Pandis, S.N. Sensitivity of PM 2.5 to climate in the Eastern US: A modeling case study. *Atmos. Chem. Phys.* **2007**, *7*, 4295–4309. [CrossRef]
3. Fu, Q.; Zhuang, G.; Wang, J.; Xu, C.; Huang, K.; Li, J.; Hou, B.; Lu, T.; Streets, D.G. Mechanism of formation of the heaviest pollution episode ever recorded in the Yangtze River Delta, China. *Atmos. Environ.* **2008**, *42*, 2023–2036. [CrossRef]
4. Shu, Z.; Liu, Y.; Zhao, T.; Xia, J.; Wang, C.; Cao, L.; Wang, H.; Zhang, L.; Zheng, Y.; Shen, L. Elevated 3D structures of PM 2.5 and impact of complex terrain-forcing circulations on heavy haze pollution over Sichuan Basin, China. *Atmos. Chem. Phys.* **2021**, *21*, 9253–9268. [CrossRef]
5. Zhang, J.P.; Zhu, T.; Zhang, Q.H.; Li, C.C.; Shu, H.L.; Ying, Y.; Dai, Z.P.; Wang, X.; Liu, X.Y.; Liang, A.M. The impact of circulation patterns on regional transport pathways and air quality over Beijing and its surroundings. *Atmos. Chem. Phys.* **2012**, *12*, 5031–5053. [CrossRef]
6. Li, C.; Mao, J.; Lau, K.-H.A.; Chen, J.-C.; Yuan, Z.; Liu, X.; Zhu, A.; Liu, G. Characteristics of distribution and seasonal variation of aerosol optical depth in eastern China with MODIS products. *Chinese Sci. Bull.* **2003**, *48*, 2488–2495.
7. Luo, Y.; Zheng, X.; Zhao, T.; Chen, J. A climatology of aerosol optical depth over China from recent 10 years of MODIS remote sensing data. *Int. J. Climatol.* **2014**, *34*, 863–870. [CrossRef]
8. Huang, X.; Ding, A.; Wang, Z.; Ding, K.; Gao, J.; Chai, F.; Fu, C. Amplified transboundary transport of haze by aerosol–boundary layer interaction in China. *Nat. Geosci.* **2020**, *13*, 428–434. [CrossRef]
9. Miao, Y.; Guo, J.; Liu, S.; Liu, H.; Zhang, G.; Yan, Y.; He, J. Relay transport of aerosols to Beijing-Tianjin-Hebei region by multi-scale atmospheric circulations. *Atmos. Environ.* **2017**, *165*, 35–45. [CrossRef]
10. Kang, H.; Zhu, B.; Gao, J.; He, Y.; Wang, H.; Su, J.; Pan, C.; Zhu, T.; Yu, B. Potential impacts of cold frontal passage on air quality over the Yangtze River Delta, China. *Atmos. Chem. Phys.* **2019**, *19*, 3673–3685. [CrossRef]
11. Lu, M.; Tang, X.; Wang, Z.; Gbaguidi, A.; Liang, S.; Hu, K.; Wu, L.; Wu, H.; Huang, Z.; Shen, L. Source tagging modeling study of heavy haze episodes under complex regional transport processes over Wuhan megacity, Central China. *Environ. Pollut.* **2017**, *231*, 612–621. [CrossRef] [PubMed]
12. Zhang, H.; Cheng, S.; Yao, S.; Wang, X.; Zhang, J. Multiple perspectives for modeling regional PM2. 5 transport across cities in the Beijing–Tianjin–Hebei region during haze episodes. *Atmos. Environ.* **2019**, *212*, 22–35. [CrossRef]
13. Squizzato, S.; Masiol, M.; Innocente, E.; Pecorari, E.; Rampazzo, G.; Pavoni, B. A procedure to assess local and long-range transport contributions to PM2. 5 and secondary inorganic aerosol. *J. Aerosol Sci.* **2012**, *46*, 64–76. [CrossRef]

14. Lang, J.; Cheng, S.; Li, J.; Chen, D.; Zhou, Y.; Wei, X.; Han, L.; Wang, H. A monitoring and modeling study to investigate regional transport and characteristics of PM2. 5 pollution. *Aerosol Air Qual. Res.* **2013**, *13*, 943–956. [CrossRef]
15. Wu, J.; Li, G.; Cao, J.; Bei, N.; Wang, Y.; Feng, T.; Huang, R.; Liu, S.; Zhang, Q.; Tie, X. Contributions of trans-boundary transport to summertime air quality in Beijing, China. *Atmos. Chem. Phys.* **2017**, *17*, 2035–2051. [CrossRef]
16. Wang, Y.; Bao, S.; Wang, S.; Hu, Y.; Shi, X.; Wang, J.; Zhao, B.; Jiang, J.; Zheng, M.; Wu, M. Local and regional contributions to fine particulate matter in Beijing during heavy haze episodes. *Sci. Total Environ.* **2017**, *580*, 283–296. [CrossRef]
17. Hu, W.; Zhao, T.; Bai, Y.; Kong, S.; Xiong, J.; Sun, X.; Yang, Q.; Gu, Y.; Lu, H. Importance of regional PM2. 5 transport and precipitation washout in heavy air pollution in the Twain-Hu Basin over Central China: Observational analysis and WRF-Chem simulation. *Sci. Total Environ.* **2021**, *758*, 143710. [CrossRef]
18. Lin, C.-Y.; Liu, S.C.; Chou, C.C.K.; Liu, T.H.; Lee, C.-T.; Yuan, C.-S.; Shiu, C.-J.; Young, C.-Y. Long-range transport of Asian dust and air pollutants to Taiwan. *Terr. Atmos. Ocean. Sci.* **2004**, *15*, 759–784. [CrossRef]
19. Sánchez-Ccoyllo, O.R.; de Fatima Andrade, M. The influence of meteorological conditions on the behavior of pollutants concentrations in São Paulo, Brazil. *Environ. Pollut.* **2002**, *116*, 257–263. [CrossRef]
20. Gustafsson, O.; Krusa, M.; Zencak, Z.; Sheesley, R.J.; Granat, L.; Engstrom, E.; Praveen, P.S.; Rao, P.; Leck, C.; Rodhe, H. Brown clouds over South Asia: Biomass or fossil fuel combustion? *Science* **2009**, *323*, 495–498. [CrossRef]
21. Reisen, F.; Meyer, C.P.M.; Keywood, M.D. Impact of biomass burning sources on seasonal aerosol air quality. *Atmos. Environ.* **2013**, *67*, 437–447. [CrossRef]
22. Fu, J.S.; Hsu, N.C.; Gao, Y.; Huang, K.; Li, C.; Lin, N.-H.; Tsay, S.-C. Evaluating the influences of biomass burning during 2006 BASE-ASIA: A regional chemical transport modeling. *Atmos. Chem. Phys.* **2012**, *12*, 3837–3855. [CrossRef]
23. Zhang, Y.; Yang, Y.; Li, J.; Wang, Y.; Wang, Z. Modeling the Impacts of Biomass Burning in Southeast Asia on PMover China in Spring. *Res. Environ. Sci.* **2016**, *29*, 952–962.
24. Tian, D.; Hu, Y.; Wang, Y.; Boylan, J.W.; Zheng, M.; Russell, A.G. Assessment of biomass burning emissions and their impacts on urban and regional PM2. 5: A Georgia case study. *Environ. Sci. Technol.* **2009**, *43*, 299–305. [CrossRef]
25. Lin, C.-Y.; Hsu, H.-M.; Lee, Y.H.; Kuo, C.H.; Sheng, Y.-F.; Chu, D.A. A new transport mechanism of biomass burning from Indochina as identified by modeling studies. *Atmos. Chem. Phys.* **2009**, *9*, 7901–7911. [CrossRef]
26. Tsay, S.C.; Hsu, N.C.; Lau, W.K.; Li, C.; Gabriel, P.M.; Ji, Q.; Holben, B.N.; Judd Welton, E.; Nguyen, A.X.; Janjai, S. From BASE-ASIA toward 7-SEAS: A satellite-surface perspective of boreal spring biomass-burning aerosols and clouds in Southeast Asia. *Atmos. Environ.* **2013**, *78*, 20–34. [CrossRef]
27. Deng, X.; Tie, X.; Zhou, X.; Wu, D.; Zhong, L.; Tan, H.; Fei, L.; Huang, X.; Bi, X.; Tao, D. Effects of Southeast Asia biomass burning on aerosols and ozone concentrations over the Pearl River Delta (PRD) region. *Atmos. Environ.* **2008**, *42*, 8493–8501. [CrossRef]
28. Chan, C.Y.; Chan, L.Y.; Harris, J.M.; Oltmans, S.J.; Zheng, X.D. Characteristics of biomass burning emission sources, transport, and chemical speciation in enhanced springtime tropospheric ozone profile over Hong Kong. *J. Geophys. Res. Atmos.* **2003**, *108*, ACH-3. [CrossRef]
29. Wang, Y. *Yunnan Mountain Climate*; Yunnan Science and Technology Press: Kunming, China, 2006. (In Chinese)
30. Zhu, J.; Xia, X.; Che, H.; Wang, J.; Zhang, J.; Duan, Y. Study of aerosol optical properties at Kunming in southwest China and long-range transport of biomass burning aerosols from North Burma. *Atmos. Res.* **2016**, *169*, 237–247. [CrossRef]
31. Nguyen, L.S.P.; Huang, H.Y.; Lei, T.L.; Bui, T.T.; Wang, S.H.; Chi, K.H.; Sheu, G.R.; Lee, C.T.; Ou-Yang, C.F.; Lin, N.H. Characterizing a landmark biomass-burning event and its implication for aging processes during long-range transport. *Atmos. Environ.* **2020**, *241*, 117766. [CrossRef]
32. Huang, K.; Fu, J.S.; Hsu, N.C.; Gao, Y.; Dong, X.; Tsay, S.C.; Lam, Y.F. Impact assessment of biomass burning on air quality in Southeast and East Asia during BASE-ASIA. *Atmos. Environ.* **2013**, *78*, 291–302. [CrossRef]
33. Zhang, Y.; Peng, Y.; Song, W.; Zhang, Y.L.; Ponsawansong, P.; Prapamontol, T.; Wang, Y. Contribution of brown carbon to the light absorption and radiative effect of carbonaceous aerosols from biomass burning emissions in Chiang Mai, Thailand. *Atmos. Environ.* **2021**, *260*, 118544. [CrossRef]
34. Ding, K.; Huang, X.; Ding, A.; Wang, M.; Su, H.; Kerminen, V.M.; Petäjä, T.; Tan, Z.; Wang, Z.; Zhou, D.; et al. Aerosol-boundary-layer-monsoon interactions amplify semi-direct effect of biomass smoke on low cloud formation in Southeast Asia. *Nat. Commun.* **2021**, *12*, 6416. [CrossRef] [PubMed]
35. Ge, C.; Wang, J.; Reid, J.S. Mesoscale modeling of smoke transport over the Southeast Asian Maritime Continent: Coupling of smoke direct radiative effect below and above the low-level clouds. *Atmos. Chem. Phys.* **2014**, *14*, 159–174. [CrossRef]
36. Zhu, A.; Xu, H.; Deng, J.; Ma, J.; Li, S. El Niño–Southern Oscillation (ENSO) effect on interannual variability in spring aerosols over East Asia. *Atmos. Chem. Phys.* **2021**, *21*, 5919–5933. [CrossRef]
37. Dong, X.; Fu, J.S.; Huang, K.; Zhu, Q.; Tipton, M. Regional climate effects of biomass burning and dust in East Asia: Evidence from modeling and observation. *Geophys. Res. Lett.* **2019**, *46*, 11490–11499. [CrossRef]
38. Bilal, M.; Nazeer, M.; Qiu, Z.; Ding, X.; Wei, J. Global validation of MODIS C6 and C6. 1 merged aerosol products over diverse vegetated surfaces. *Remote Sens.* **2018**, *10*, 475. [CrossRef]
39. Ma, Y.; Li, Z.; Li, Z.; Xie, Y.; Fu, Q.; Li, D.; Zhang, Y.; Xu, H.; Li, K. Validation of MODIS aerosol optical depth retrieval over mountains in central China based on a sun-sky radiometer site of SONET. *Remote Sens.* **2016**, *8*, 111. [CrossRef]
40. Grell, G.A.; Peckham, S.E.; Schmitz, R.; McKeen, S.A.; Frost, G.; Skamarock, W.C.; Eder, B. Fully coupled "online" chemistry within the WRF model. *Atmos. Environ.* **2005**, *39*, 6957–6975. [CrossRef]

41. Hong, S.Y.; Noh, Y.; Dudhia, J. A New Vertical Diffusion Package with an Explicit Treatment of Entrainment Processes. *Mon. Weather Rev.* **2005**, *134*, 2318. [CrossRef]
42. Morrison, H.; Thompson, G.; Tatarskii, V. Impact of Cloud Microphysics on the Development of Trailing Stratiform Precipitation in a Simulated Squall Line: Comparison of One- and Two-Moment Schemes. *Mon. Weather Rev.* **2009**, *137*, 991–1007. [CrossRef]
43. Iacono, M.J.; Delamere, J.S.; Mlawer, E.J.; Shephard, M.W.; Collins, W.D. Radiative Forcing by Long-Lived Greenhouse Gases: Calculations with the AER Radiative Transfer Models. *J. Geophys. Res. Atmos.* **2008**, *113*, D13103. [CrossRef]
44. Chen, F.; Dudhia, J. Coupling an Advanced Land Surface–Hydrology Model with the Penn State–NCAR MM5 Modeling System. Part I: Model Implementation and Sensitivity. *Mon. Weather Rev.* **2001**, *129*, 569–585. [CrossRef]
45. Stockwell, W.R.; Middleton, P.; Chang, J.S.; Tang, X. The second generation regional acid deposition model chemical mechanism for regional air quality modeling. *J. Geophys. Res. Atmos.* **1990**, *95*, 16343. [CrossRef]
46. Li, M.; Zhang, Q.; Kurokawa, J.I.; Woo, J.H.; He, K.; Lu, Z.; Ohara, T.; Song, Y.; Streets, D.G.; Carmichael, G.R. MIX: A mosaic Asian anthropogenic emission inventory under the international collaboration framework of the MICS-Asia and HTAP. *Atmos. Chem. Phys.* **2017**, *17*, 935–963. [CrossRef]
47. Guenther, A.; Karl, T.; Harley, P.; Wiedinmyer, C.; Palmer, P.I.; Geron, C. Estimates of global terrestrial isoprene emissions using MEGAN (Model of Emissions of Gases and Aerosols from Nature). *Atmos. Chem. Phys.* **2006**, *6*, 3181–3210. [CrossRef]
48. Wiedinmyer, C.; Akagi, S.K.; Yokelson, R.J.; Emmons, L. The Fire INventory from NCAR (FINN): A high resolution global model to estimate the emissions from open burning. *Geosci. Model Dev.* **2011**, *4*, 625–641. [CrossRef]
49. Freitas, S.R.; Longo, K.M.; Chatfield, R.; Latham, D.; Dias, M.S.; Andreae, M.O.; Prins, E.; Santos, J.C.; Gielow, R.; Carvalho, J.A. Including the sub-grid scale plume rise of vegetation fires in low resolution atmospheric transport models. *Atmos. Chem. Phys.* **2007**, *7*, 447–450. [CrossRef]
50. Freitas, S.R.; Longo, K.M.; Trentmann, J.; Latham, D. Technical Note: Sensitivity of 1-D smoke plume rise models to the inclusion of environmental wind drag. *Atmos. Chem. Phys.* **2010**, *10*, 585–594. [CrossRef]
51. Stohl, A. A backward modeling study of intercontinental pollution transport using aircraft measurements. *J. Geophys. Res. Atmos.* **2003**, *108*, 108. [CrossRef]
52. Stohl, A.; Forster, C.; Frank, A.; Seibert, P.; Wotawa, G. Technical note: The Lagrangian particle dispersion model FLEXPART version 6.2. *Atmos. Chem. Phys.* **2005**, *5*, 2461–2474. [CrossRef]
53. Brioude, J.; Arnold, D.; Stoh, A.; Cassiani, M.; Morton, D.; Seibert, P.; Angevine, W.; Evan, S.; Dingwell, A.; Fast, J.D. The Lagrangian particle dispersion model FLEXPART-WRF version 3.1. *Geosci. Model Dev.* **2013**, *6*, 1889–1904. [CrossRef]
54. Sauvage, B.; Fontaine, A.; Eckhardt, S.; Auby, A.; Boulanger, D.; Petetin, H.; Paugam, R.; Athier, G.; Cousin, J.M.; Darras, S. Source attribution using FLEXPART and carbon monoxide emission inventories: SOFT-IO version 1.0. *Atmos. Chem. Phys.* **2017**, *17*, 15271–15292. [CrossRef]
55. Yu, C.; Zhao, T.; Bai, Y.; Zhang, L.; Chang, J. Heavy air pollution with a unique "non-stagnant" atmospheric boundary layer in the Yangtze River middle basin aggravated by regional transport of PM2.5 over China. *Atmos. Chem. Phys.* **2020**, *20*, 7217–7230. [CrossRef]
56. Shen, L.; Hu, W.; Zhao, T.; Bai, Y.; Wang, H.; Kong, S.; Zhu, Y. Changes in the Distribution Pattern of PM2.5 Pollution over Central China. *Remote Sens.* **2021**, *13*, 4855. [CrossRef]
57. Chen, S.; Zhou, G.; Zhu, B.; Geng, F.; Chang, L. A method for fast quantification of air pollutant sources. *Acta Sci. Circumstantiae* **2017**, *37*, 2474–2481.
58. Hanna, S.R.; Yang, R. Evaluations of Mesoscale Models' Simulations of Near-Surface Winds, Temperature Gradients, and Mixing Depths. *J. Appl. Meteorol.* **2001**, *40*, 1095–1104. [CrossRef]
59. Zhai, S.; Jacob, D.J.; Brewer, J.F.; Li, K.; Moch, J.M.; Kim, J.; Lee, S.; Lim, H.; Lee, H.C.; Kuk, S.K. Relating geostationary satellite measurements of aerosol optical depth (AOD) over East Asia to fine particulate matter (PM 2.5): Insights from the KORUS-AQ aircraft campaign and GEOS-Chem model simulations. *Atmos. Chem. Phys.* **2021**, *21*, 16775–16791. [CrossRef]
60. Boylan, J.W.; Russell, A.G. PM and light extinction model performance metrics, goals, and criteria for three-dimensional air quality models. *Atmos. Environ.* **2006**, *40*, 4946–4959. [CrossRef]
61. Ye, W.-F.; Ma, Z.-Y.; Ha, X.-Z. Spatial-temporal patterns of PM2. 5 concentrations for 338 Chinese cities. *Sci. Total Environ.* **2018**, *631*, 524–533. [CrossRef]
62. de Leeuw, G.; Sogacheva, L.; Rodriguez, E.; Kourtidis, K.; Georgoulias, A.K.; Alexandri, G.; Amiridis, V.; Proestakis, E.; Marinou, E.; Xue, Y. Two decades of satellite observations of AOD over mainland China using ATSR-2, AATSR and MODIS/Terra: Data set evaluation and large-scale patterns. *Atmos. Chem. Phys.* **2018**, *18*, 1573–1592. [CrossRef]
63. Li, X.; Cheng, T.; Shi, S.; Guo, H.; Han, Z. Evaluating the impacts of burning biomass on PM2.5 regional transport under various emission conditions. *Sci. Total Environ.* **2021**, *793*, 148481. [CrossRef] [PubMed]
64. Huang, Q.; Cai, X.; Wang, J.; Song, Y.; Zhu, T. Climatological study of the Boundary-layer air Stagnation Index for China and its relationship with air pollution. *Atmos. Chem. Phys.* **2018**, *18*, 7573–7593. [CrossRef]
65. Xu, X.; Zhao, T.; Liu, F.; Gong, S.L.; Kristovich, D.; Lu, C.; Guo, Y.; Cheng, X.; Wang, Y.; Ding, G. Climate modulation of the Tibetan Plateau on haze in China. *Atmos. Chem. Phys.* **2016**, *16*, 1365–1375. [CrossRef]

remote sensing

Technical Note

Aerial Mapping of Odorous Gases in a Wastewater Treatment Plant Using a Small Drone

Javier Burgués [1,2,3,*], María Deseada Esclapez [4], Silvia Doñate [4], Laura Pastor [4] and Santiago Marco [1,2,3]

1 Institute for Bioengineering of Catalonia (IBEC), Baldiri Reixac 10-12, 08028 Barcelona, Spain; smarco@ibecbarcelona.eu
2 The Barcelona Institute of Science and Technology, Carrer del Comte d'Urgell 187, 08036 Barcelona, Spain
3 Department of Electronics and Biomedical Engineering, Universitat de Barcelona, Marti i Franqués 1, 08028 Barcelona, Spain
4 Depuración de Aguas del Mediterráneo (DAM), Avenida Benjamín Franklin 21, Parque Tecnológico, 46980 Paterna, Spain; desi.esclapez@dam-aguas.es (M.D.E.); silvia.donate@dam-aguas.es (S.D.); laura.pastor@dam-aguas.es (L.P.)
* Correspondence: jburgues@ibecbarcelona.eu; Tel.: +34-934-029-070

Abstract: Wastewater treatment plants (WWTPs) are sources of greenhouse gases, hazardous air pollutants and offensive odors. These emissions can have negative repercussions in and around the plant, degrading the quality of life of surrounding neighborhoods, damaging the environment, and reducing employee's overall job satisfaction. Current monitoring methodologies based on fixed gas detectors and sporadic olfactometric measurements (human panels) do not allow for an accurate spatial representation of such emissions. In this paper we use a small drone equipped with an array of electrochemical and metal oxide (MOX) sensors for mapping odorous gases in a mid-sized WWTP. An innovative sampling system based on two (10 m long) flexible tubes hanging from the drone allowed near-source sampling from a safe distance with negligible influence from the downwash of the drone's propellers. The proposed platform is very convenient for monitoring hard-to-reach emission sources, such as the plant's deodorization chimney, which turned out to be responsible for the strongest odor emissions. The geo-localized measurements visualized in the form of a two-dimensional (2D) gas concentration map revealed the main emission hotspots where abatement solutions were needed. A principal component analysis (PCA) of the multivariate sensor signals suggests that the proposed system can also be used to trace which emission source is responsible for a certain measurement.

Keywords: drone; UAV; gas sensors; odour; air pollution; industrial emissions; mapping; environmental monitoring

Citation: Burgués, J.; Esclapez, M.D.; Doñate, S.; Pastor, L.; Marco, S. Aerial Mapping of Odorous Gases in a Wastewater Treatment Plant Using a Small Drone. *Remote Sens.* **2021**, *13*, 1757. https://doi.org/10.3390/rs13091757

Academic Editor: Maria João Costa

Received: 2 April 2021
Accepted: 29 April 2021
Published: 30 April 2021

1. Introduction

The monitoring of emissions to air is a key element in preventing and reducing pollution from industrial installations, in ensuring a high level of protection of the environment, and in minimizing odor impact to the surrounding population. Industrial activities such as production of energy, intensive rearing of poultry and pigs or waste management are sources of greenhouse gases (GHGs), hazardous air pollutants (HAPs) and offensive odors. In 2017, emissions from waste management sites made up 3% of total GHG emissions and 5% of particulate matter (PM) emissions in Spain [1]. These facilities are also responsible for many citizen complaints to the local authorities regarding odor annoyance episodes [2]. The objectives of monitoring are many and diverse. For example, monitoring can be applied to assess compliance with environmental permit requirements; check the performance of odor abatement systems; determine the relative contribution of different sources to the overall emissions; report emissions for national and international inventories, e.g., the Pollutant Release and Transfer Registers (PRTRs); and many others [3].

In Europe, industrial air emissions are regulated by the Industrial Emissions Directive 2010/75/EU (IED) [4]. The IED and national regulations impose requirements on the monitoring approach to be used for a particular installation, for example the requirement for continuous monitoring of certain pollutants with specific instruments. The accepted monitoring methodologies and reference instruments for each type of gas are described in the Best Available Technique (BAT) document [5]. The quantification of the total emissions of an installation often requires the assessment of channeled (point-like) emissions and diffuse emissions including fugitive emissions. Channeled emissions are relatively easy to monitor with automated measuring systems (AMS) permanently installed on-site. However, the quantification of diffuse emissions might not be easy with AMS and is, in general, labor- and cost-intensive due to the number of potential sources.

To simplify the measurement of diffuse emissions, the European IED specifies that *"measurements techniques based on the use of a transportable measurement platform, despite being less accurate than reference methods, may be used to supplement the information supplied by fixed measurements for the determination of the spatial concentration distribution or for the assessment of diffusive gas emissions"*. The advantage of a portable instrument over a set of fixed analyzers installed on different locations of the plant is the lower investment and operational costs, as well as higher spatial resolution of the measurements. However, manually scanning an entire plant with a portable instrument is a tedious and risky task. The use of terrestrial robots may seem the most obvious solution to this problem, however their limited maneuverability hinders their practical application in realistic scenarios which often include obstacles (e.g., buildings, stairs, trees, etc.) and elevated emission sources (e.g., chimneys, flares).

Aerial surveys with small drones (<10 kg) equipped with gas detectors are a promising cost-effective and safe alternative for emission monitoring in industrial plants [6]. Both fixed- and rotary-wing drones can be used, however rotorcrafts are preferred for this application due to key practical advantages such as vertical take-off and landing (VTOL), autonomous hovering, high maneuverability, and low cruise speed. Drones equipped with laser-based methane detectors have been demonstrated with great success in the oil and gas (O&G) industry, e.g., for quantifying whole-site methane emissions [7] and detecting fugitive methane leaks [8–10]. The main O&G companies are already testing this technology in their plants [11–13]. Similar platforms have been recently used in solid waste landfills (SWLs) for identifying surface methane hotspots [14].

Wastewater treatment plants (WWTPs) are another scenario where small drones could improve the monitoring of plant emissions/odors. To the best of our knowledge, there are no reports of drones being used for emission monitoring or odor sensing of WWTPs. In this case, the major emission problem is not methane, but odorous compounds produced during wastewater treatment, such as hydrogen sulfide (H_2S), ammonia (NH_3), mercaptans, or volatile organic compounds (VOCs) which can produce odor impact in workers and communities living nearby these facilities, even at low concentration levels [3]. Current odor assessment methodologies in WWTPs are mostly based on walkover surveys with portable H_2S detectors or via olfactometric measurements involving expensive human panels, which leads to odor measurements with poor temporal and spatial resolution. The idea of using drones to monitor odorous emissions in WWTPs is very interesting because they can measure the concentration of key odorous compounds in different locations of the plant including hard-to-reach locations, and with higher spatial resolution, less risk, and lower cost than existing methods. This information can then be used by plant operators for (i) feedback into the industrial processes, (ii) as input for atmospheric dispersion models to estimate the odor emission rate and then to predict odor impact in the plant vicinity, and (iii) to identify fugitive emissions.

The two main challenges associated with the application of drones for monitoring emissions in WWTPs are (i) the lack of reliable and lightweight sensors to detect the relevant compounds and (ii) the plume distortion produced by the downwash of the rotating propellers. While methane can be selectively detected with laser-based spectrometers amenable for drone integration, detection of H_2S, NH_3 or VOCs at the required concentration levels is yet not feasible with lightweight optical analyzers. In this case, the most straightforward approach is to use low-cost chemical sensors, such as electrochemical cells (EC) or metal oxide (MOX or MOS) sensors, which inherently have limited performance [15]. Electrochemical sensors offer decent selectivity (though not comparable to optical analysers) for compounds such as CO, SO_2, NH_3 or NO/NO_2 (among many others) and are often the technology of choice when any of these compounds is targeted [16]. MOX sensors operating in the (default) isothermal mode are not selective but are more sensitive, faster, and cheaper than electrochemical cells [17]. These features make them very popular in robotic studies addressing gas source localization and mapping tasks [18–20] where selectivity is not critical because artificial gas sources releasing a single compound (typically ethanol) are normally used.

Up to now, the use of drones fitted with low-cost chemical sensors has been mostly explored in relatively simple scenarios, such as indoor areas [19] or outdoor environments [21,22], using artificial gas sources. A few exceptions exist at the industrial and academic level. For example, Aeromon (Helsinki, Finland) has been regularly using their BH-12 multi-sensor system (based on electrochemical cells) for monitoring the emission performance of vessels and checking compliance with the new emission regulations regarding fuel sulfur content (FSC). The DR1000 "Flying Lab" from Scentroid (Whitchurch-Stouffville, ON, Canada), which uses EC and MOX sensors, has been used for monitoring the quality of fuel used for domestic heating in Poland. The recently announced Muve C360 from FLIR Systems (Wilsonville, OR, USA) is a multi-gas detector completely integrated in a DJI M210 drone for emergency responders, industrial safety, and environmental monitoring. At the research level, drones equipped with electrochemical sensors have been used for atmospheric research studies, e.g. analysing the composition of volcanic plumes [23], among other applications [6].

Despite the many advantages offered by rotorcrafts, the intense downwash generated by the propellers is a main problem for chemical sensing applications in which the drone has to fly close to point or surface emitters. In these cases, the downwash strongly distorts the gas distribution, leading to gross errors in the sensor readings. This is a well-known problem that has received lots of attention from the research community. The downwash has been simulated by numerical methods (e.g., computer fluid dynamics, CFD) and empirically characterized using smoke tracers, anemometers, and particle tracking velocimetry (PTV) [6]. These studies show that the downwash is particularly strong in the vertical axis underneath the drone where its influence can extend up to several meters (depending on the drone´s take-off weight).

The downwash is the main factor to be considered in the design of gas sampling systems for drones, or for optimizing sensor placement, especially for point-like sensors or closed-path optical analysers. Although the sensing elements can be directly exposed to the air sample, it is more convenient to place them in a sensor chamber with an aspiration system. This provides more flexibility regarding the sampling point and more control in the sample delivery. The few existing commercial systems using low-cost sensors (e.g., Aeromon BH-12, Scentroid DR1000 and FLIR Muve C360) implement a rigid horizontal sampling tube (1–2 m length) to aspirate the gas sample from outside the rotors' influence zone [23,24]. This type of boom is very convenient for monitoring elevated and channeled sources, such as chimneys or flares, but has practical inconveniences for diffusive area sources such as those encountered in WWTPs. In this case, the problem is that a drone implementing a rigid horizontal probe would have to fly very close to the ground or nearby obstacles to sample the space directly above the source, which is risky and leads to a strong mixing (dilution) of the emissions because of the downwash.

The goal of our current research is to develop a drone to monitor and map odorous emissions in WWTPs. For that, we use a commercial drone (DJI Matrice 600) fitted with a custom payload based on an array of low-cost gas sensors (electrochemical and MOX sensors) and an innovative sampling system consisting of an aspiration pump connected to a 10-m sampling tube suspended from the drone. This system allows the drone to sample the emission sources with negligible effect from the downwash and, at the same time, fly at sufficient height above the obstacles to minimize the operational risks. This paper presents the first preliminary set of experiments carried out in a real WWTP in Murcia (Spain). The objectives of these initial measurements are to (i) check if the signals recorded by the drone are consistent with the expected concentrations based on previous measurements with hand-held detectors; (ii) build rough concentration maps of the most relevant compounds to understand their spatial distribution and identify the emission hotspots; and (iii) assess if the different emission sources can be identified based on the multivariate patterns produced by the sensor array. We will discuss some of the challenges encountered during these tests, and how future developments could overcome them.

2. Materials and Methods

2.1. Test Site

Field measurements were carried out in the WWTP of Molina del Segura (Murcia, Spain), which is operated by Depuración de Aguas del Mediterráneo (DAM). The plant (Figure 1) has an extension of 35,000 m^2 and serves a population of 290,000 inhabitants. Five emission sources (settler stage A, bioreactor stage A, sludge hoppers, sludge thickener, and deodorization chimney) were suggested by the plant manager as elements with the highest emissions based on previous measurement campaigns using hand-held detectors and olfactometric campaigns (human panels). Therefore, the aerial mapping was focused on a region of ~4500 m^2 centered around these sources. An ultrasonic anemometer (Model: WindSonic, Gill Instruments, Lymington, UK) placed at 10 m a.g.l. in a clear area without nearby obstacles continuously measured wind speed and direction.

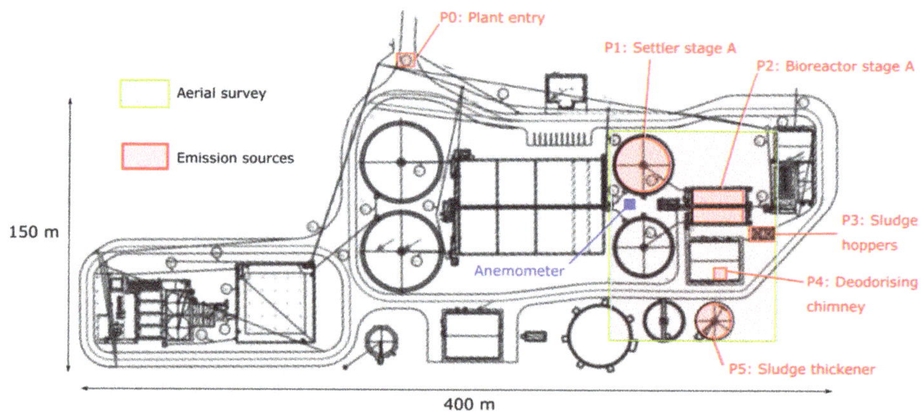

Figure 1. Map of the WWTP of Molina del Segura with the five main emission sources highlighted in red. The aerial mapping was focused on a 4500 m^2 squared region centered around these sources.

2.2. Drone and Payload

A rotary-wing drone was selected for this application due to its ability to hover, slow flight speed and vertical takeoff and landing (VTOL). These characteristics are essential for close-up monitoring of emission sources, safe navigation around the plant infrastructure, and high-resolution mapping. The selected drone was the Matrice 600 Pro (DJI International, Nanshan, Shenzhen, China) which has a high payload capacity (6 kg) and allows for a flight time between 15 min (fully loaded) and 30 min (unloaded). The drone was equipped with a custom gas sensing payload (Figure 2) composed of a custom-made electronic nose (e-nose) and a commercial multi-gas analyzer Dräger X-am 8000 (Drägerwerk AG, Lübeck, Germany).

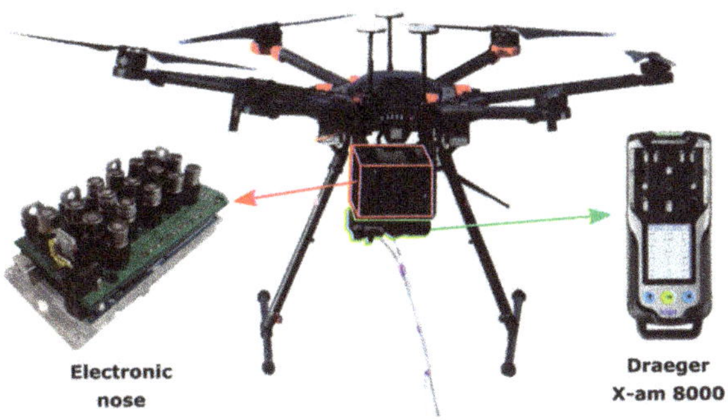

Figure 2. DJI Matrice 600 Pro drone equipped with a custom e-nose and a Dräger X-am 8000 analyzer. The inlets of both systems are connected to 10-m PTFE tubing.

Both sensor systems were attached underneath the drone using a custom mounting plate, and their inlets were connected to 10-m PTFE tubing (hanging vertically from the drone) to sample the region below the drone where the downwash has disappeared or it is greatly reduced. The required length of the tubing was determined by measuring the downwash influence with a hand-held anemometer placed below the loaded drone while it was hovering at multiple altitudes. We prefer this sampling approach over the horizontal tube because it allows the drone to fly over obstacles without risk. However, using a long sampling tube also has practical problems: (i) a delay in the measurements due to the sample transport, (ii) memory effects if some gases stick to the tubing walls, and (iii) tilt of the tube due to wind or drone motion. The delay in the measurements and the tilt of the tube lead to inaccuracies in the GPS marking of the sensor signals. To solve these issues we attached a weight of 150 g to the end of the tube as a plumb bob (to keep the tube as straight as possible during flight) and compensated the delay via software.

Regarding the e-nose architecture (Figure 3), it contains an array of 16 MOX sensors (several TGS models, Figaro Engineering Inc., Osaka, Japan) operated at various temperatures, a combo sensor for temperature, humidity and pressure, a flow sensor, GPS receiver, and long-range ZigBee 868 MHz radio communication. The specifications of the e-nose sensors are summarized in Table 1. A microcontroller reads the sensor signals and the GPS position, and sends them to the base station through the radio link at a sampling frequency of 0.2 Hz. A miniature pump delivers the gas flow to the sensing chamber at a flow rate of 1.8 L/min. Power is provided by a 7.4 V lithium polymer (LiPo) battery with 2200 mAh of capacity, allowing continuous operation for approximately 2 h. The weight of the e-nose including the battery is ~1200 g.

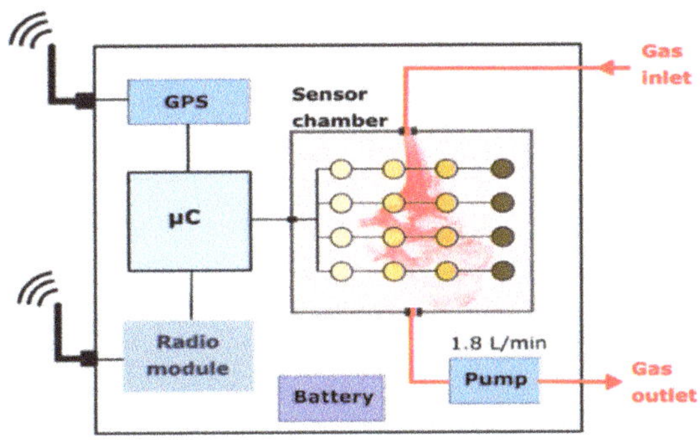

Figure 3. Internal architecture of the electronic nose.

Table 1. Sensors included in the electronic nose.

Parameter	Sensor Type	Range	Accuracy
VOCs	16× Metal oxide sensors	-	-
Temperature	MEMS	40 to +85 °C	±1 °C
Humidity	MEMS	0 to 100% RH	±3% r.h.
Pressure	MEMS	30 to 110 kPa	±0.1 kPa
Flow rate	Thermal	0 to 33 L/min	±3% m.v.

The Dräger X-am 8000 is equipped with four electrochemical sensors (for H_2S, NH_3, mercaptans and amines), a photo-ionization detector (PID) for quantifying total VOCs, an internal pump, an integrated battery, and weighs 550 g (Table 2). The sensor data are logged every second in an internal memory which can store up to 210 h of measurements.

Table 2. Sensors included in the Dräger X-am 8000 analyzer.

Parameter	Sensor Type	Range	Accuracy
H_2S	Electrochemical cell	0 to 100 ppm	±0.1 ppm
NH_3	Electrochemical cell	0 to 300 ppm	±1 ppm
Amines	Electrochemical cell	0 to 100 ppm	±1 ppm
Mercaptans	Electrochemical cell	0 to 40 ppm	±0.5 ppm
VOCs	Photo-ionization detector	0 to 2000 ppm	±0.1 ppm

2.3. Experimental Protocol

All measurements were carried out in a single day. The e-nose sensors were preheated for 24 h before the start of the measurements to stabilize the sensors' baseline. At the beginning of the experiment the drone was positioned near the entry of the plant (P0 in Figure 1), where no odor was perceivable, and measurements were taken for 7 min to determine the sensors' baseline. The drone took off from there and scanned the target area slowly at a height of approximately 12 m, keeping the inlet of the sampling tube as close as possible to the emission sources. The drone hovered for 5–7 min at each of the five emission sources (highlighted in Figure 1) to capture the variability of the gas concentration over time. The whole experiment took slightly less than 2 h to complete, which required multiple sets of drone batteries.

2.4. Data Processing and Visualization

A laptop computer with a ZigBee 868 MHz radio antenna and a custom software application developed in MATLAB R2019B (The MathWorks, Natick, MA, USA) was used as base station to receive and log in real-time the data from the e-nose (timestamp, sensor signals and GPS position). The measurement data stored in the internal memory of the Dräger X-am 8000 (timestamp and sensor signals) was downloaded into the base station at the end of the flight (no radio link available for this device). Data from both instruments were merged into a single file, using linear interpolation (MATLAB interp1) to synchronize the data to a common timestamp. Each entry of the log file is a tuple $(t, x, y, z, c_1, \ldots, c_5, s_1, \ldots, s_{16})$ where t is the timestamp, x, y, z the spatial coordinates, $c_1, \ldots, c_5$ the concentration (ppmv, parts-per-million in volume) of the five gases measured by the Dräger X-am 8000, and $s_1, \ldots, s_{16}$ the MOX sensor resistances (Ω).

For data visualization, we used MATLAB and the Google Maps Javascript API to produce a heatmap visualization of the geolocalized raw sensor data. In addition, a principal component analysis (PCA) was used to visually determine if the different emission sources could be clustered based on the sensor responses. For that, a PCA model with three principal components was applied to the e-nose signals after logarithmic transformation (to reduce the dynamic range and improve normality) and mean-centering. The PCA modelling was done also in MATLAB.

3. Results and Discussion

3.1. Weather Conditions

The weather conditions during the field measurements were favorable, with clear sky, temperature between 18 and 20 °C, and 50% relative humidity. The wind direction was predominantly north-west, with average wind speed of 10–15 km/h, and gusts of up to 50 km/h (Figure 4). The effect of wind on the sampling tube can be observed in Figure 5, which shows pictures of the drone hovering above the five emission sources. For example, while measuring at the settlers (P1) and the deodorization chimney (P4) the drone had to be positioned slightly upwind to compensate the tilt of the sampling tube. The GPS signal reception was good throughout the experiment, with more than 12 satellites in line-of-sight (LOS) with the drone.

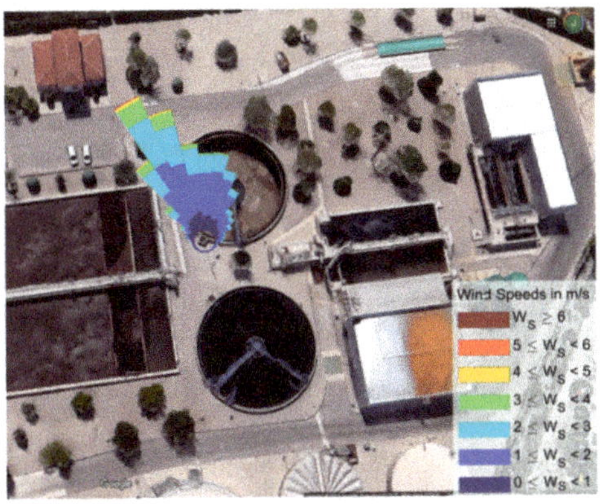

Figure 4. Wind speed and direction during the field measurements.

Figure 5. Drone hovering over the selected emission sources (P1–P5).

3.2. Gas Concentration Measurements

The raw sensor signals throughout the experiment are shown in Figure 6. The highest gas concentrations were recorded near the bioreactor stage A (P2) and the deodorization chimney (P4). The high variability of the sensor signals at the chimney is a consequence of the oscillations of the sampling tube around the chimney outlet due to the wind. The oscillations of the sampling tube were less problematic in the area sources because, since the concentration is more homogeneous, the exact location of the sampling point is not as critical as in ducted (point-like) sources. Very low concentration of all gases was measured near the settlers (P1) despite a strong malodor could be appreciated near this site. Only the response of the MOX sensors was distinguishable from the blank measurements, which may indicate that odor from this source was produced mostly by VOCs rather than by H_2S or NH_3. A peak of 100 ppm of CO_2 above the background level was measured near the sludge hoppers (P3) during sludge discharge into a truck. Finally, low concentrations were measured at the sludge thickener (P5) probably because it was covered.

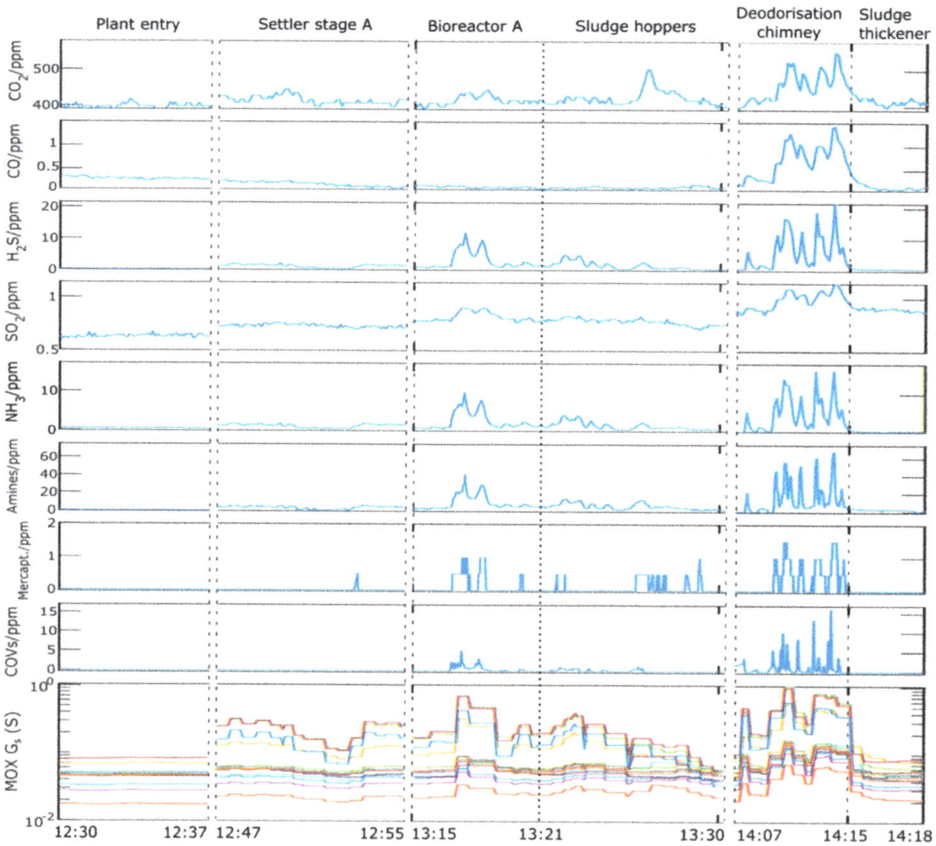

Figure 6. Raw sensor signals during the field measurements.

The measured concentrations were in line with the expected values based on previous measurement campaigns carried out at the same emission sources with a hand-held X-am 8000 detector (Table 3). It is not surprising that the measured values during a single day in very specific conditions (e.g., drone flight) differ from values obtained in other measurement campaigns carried out at a different date. This is because the pattern of emissions in a WWTP is not stationary and there is a large variability in the emissions depending on process factors (e.g., quality of influent water and flow rate) but also on environmental conditions (wind, temperature, humidity, precipitation, etc.). There are also seasonal trends. Thus, the recorded signals only represent the emissions during the time of sampling. A comprehensive characterization of the emissions, which would require a much more elaborated measurement campaign spanning several months, was out of the scope of this preliminary measurements. Similarly, a precise characterization of the uncertainty associated with the drone measurements is also subject of future experiments. The goals of this preliminary work were less ambitious, e.g., showing that drone-based measurements using the proposed sampling approach provide sensible signals.

Table 3. Comparison between drone-based measurements and those performed with a hand-held X-am 8000 detector near the same emission sources.

	Drone	Hand-Held Detector
H_2S	0–10 ppm	0–100 ppm
NH_3	0–10 ppm	0–30 ppm
Amines	0–65 ppm	0–70 ppm
Mercaptans	0–1.5 ppm	0–1 ppm
VOCs	0–15 ppm	0–14 ppm

It should be noted that while the recorded signals give a clear indication of the characteristics of emissions in the different sources, their exact values are subject to various sources of uncertainty. While low-cost sensors can provide relatively good results in the laboratory, their application in field conditions remains challenging. First of all, because the sensors react not only to the target gas but also to interfering compounds. For example, the response of an H_2S electrochemical sensor is affected by the presence of SO_2 or NH_3 because of matrix effects. Uncontrolled or unknown variations in temperature, humidity, and pressure can also affect the sensor signals, as can overheating due to direct sunlight exposure. Strong winds also affect the measurements due to the oscillations of the sampling line.

3.3. Gas Concentration Mapping

The sensor data was used to produce heatmaps indicative of the concentration of each gas. An example of an H_2S map is shown in Figure 7. As it was expected from the analysis of the raw sensor signals, the H_2S concentration hotspots are located near the bioreactor stage A (P2) and the deodorization chimney (P4). These hotspots are shifted a few meters with respect to the location of the emission sources due to the inaccuracy of the GPS position (± 3 m), the effect of wind on the gas dispersion, and the tilt of the sampling tube with respect to the vertical axis of the drone where the GPS receiver is located. This latter effect can be clearly seen in Figure 5 when the drone is sampling the chimney. In order to keep the inlet of the sampling tube centered above the chimney, the drone must be positioned a few meters upwind to compensate for the effect of wind on the tubing. Because the GPS receiver is placed on the drone and not at the inlet of the tube, the recorded position indicates the location of the drone and not the location where the gas is being sampled. This could be solved in the future by either placing the GPS receiver at the inlet of the tubing or using an on-board camera to track the position of the sampling inlet and compensate the offset via software.

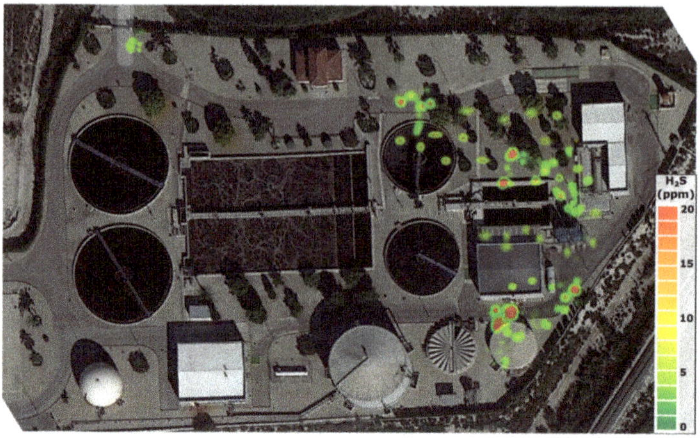

Figure 7. Map of H_2S concentration obtained from drone measurements.

3.4. Gas Source Identification

One research question of this work is whether the different emission sources could be distinguished based on the e-nose signals. A PCA score plot of the signals recorded while the drone was hovering over the sources revealed that this is indeed the case, and the different emission sources are clustered in different regions of the PCA space (Figure 8). This suggests that each source has a different gas composition, so the e-nose could be potentially used to identify which source is responsible for a certain measurement. Even the settler (P1) and sludge thickener (P5) could be differentiated from the blank measurements (P0) despite the gas concentrations measured at these sources were very close to the baseline level. This result, which may be a consequence of the low limit of detection (LOD) of MOX sensors, should be confirmed with more measurement campaigns and using external validation (blind) samples.

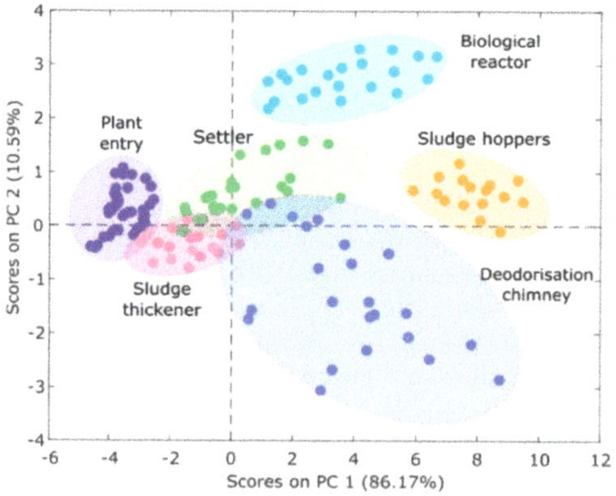

Figure 8. Principal component analysis (PCA) score plot of the e-nose signals.

4. Conclusions

This study has explored the possibility of using a small drone equipped with an array of low-cost gas sensors for real-time monitoring of odorous emissions in a WWTP. The drone was equipped with an innovative sampling system that allowed the drone to fly at a safe distance from obstacles and minimize the impact of downwash into the sensor signals. The proposed system was useful to measure gas concentrations near previously inaccessible emission sources, such as the deodorization chimney, which turned out to be the main odor source in this plant. The geolocalized sensor signals were used to build H_2S concentration maps that highlighted the location of the main emission hotspots.

During these field measurements we faced several challenges that affect the operation of the drone and the quality of the acquired data. The main challenge was the presence of strong winds which affected the drone stability, made the sampling tube oscillate considerably, and induced a high variability in the spatial distribution of the released gases. Adding a weight at the end of the sampling line improved the stability of the measurements. Flying above the obstacles was key to minimize the operational risks considering the strong and unpredictable wind gusts present in our flights. Real-time visual feedback from the sensor signals was very helpful for fine-tuning the position of the sampling inlet close to the different emission sources (especially channeled sources). Nonetheless, the geolocalization of the sensor measurements was inaccurate under strong winds because the GPS receiver and the inlet of the sampling line were not necessarily in the same vertical axis. Two possible solutions to improve this in future experiments are (i) to place the GPS receiver at the inlet of the tubing or (ii) using an on-board camera to track the position of the sampling inlet and compensate the GPS offset via software.

Another problem that we want to address in future works is the quantification of odor concentration (e.g., in standardized units such as ou/m^3 [24]) from drone-based measurements. This is much more challenging than quantification of individual gas concentrations, as the relationship between the components of a gas mixture and the perceived odour concentration is non-linear and subject to synergic and masking effects [25]. We also plan to combine the drone measurements with atmospheric dispersion models, such as CALPUFF [26], to predict the impact outside of the plant. The proposed platform could be applied in the future to other industrial sectors, such as solid waste landfills, composting plants, and animal farms.

Author Contributions: Conceptualization, J.B., M.D.E., S.D. and S.M.; Data curation, J.B.; Formal analysis, J.B.; Funding acquisition, L.P. and S.M.; Investigation, J.B., M.D.E. and S.D.; Methodology, J.B., M.D.E. and S.M.; Project administration, L.P. and S.M.; Resources, M.D.E. and S.D.; Software, J.B.; Supervision, S.M.; Validation, J.B.; Visualization, J.B.; Writing—original draft, J.B.; Writing—review & editing, J.B., M.D.E., S.D., L.P. and S.M. All authors have read and agreed to the published version of the manuscript.

Funding: This research has received funding as third party from the ATTRACT project funded by the EC under Grant Agreement 777222.

Institutional Review Board Statement: Not applicable.

Informed Consent Statement: Not applicable.

Data Availability Statement: Restrictions apply to the availability of the data presented in this study. These data can be available on private request from the corresponding author.

Acknowledgments: CERCA Programme/Generalitat de Catalunya. The Signal and Information Processing for Sensor Systems group is a consolidated Grup de Recerca de la Generalitat de Catalunya and has support from the Departament d'Universitats, Recerca i Societat de la Informació de la Generalitat de Catalunya (expedient 2017 SGR 1721). We would also like to acknowledge Luis Fernández Romero, Maria José Ibáñez, Lidia Saúco, Ana Maciá and Pilar Pradas for their support during the field campaigns. Authors of this report gratefully acknowledge the cooperation of ESAMUR (Entidad Regional de Saneamiento y Depuración de Murcia).

Conflicts of Interest: The authors declare no conflict of interest. The funders had no role in the design of the study; in the collection, analyses, or interpretation of data; in the writing of the manuscript, or in the decision to publish the results.

References

1. Agency, E.E. Spain—Industrial Pollution Profile. 2019. Available online: https://www.eea.europa.eu/themes/industry/industrial-pollution/industrial-pollution-country-profiles-2019/spain (accessed on 24 April 2020).
2. Dincer, F.; Muezzinoglu, A. Odor-causing volatile organic compounds in wastewater treatment plant units and sludge management areas. *J. Environ. Sci. Health Part A* **2008**, *43*, 1569–1574. [CrossRef] [PubMed]
3. Brinkmann, T.; Both, R.; Scalet, B.M.; Roudier, S.; Sancho, L.D. JRC Reference Report on Monitoring of Emissions to Air and Water from IED Installations. *Ind. Emiss. Dir.* **2018**. [CrossRef]
4. Directive, C. Directive 2010/75/EU of the European Parliament and of the Council. *Off. J. Eur. Union L* **2010**, *334*, 17–119.
5. Brinkmann, T.; Santonja, G.G.; Yükseler, H.; Roudier, S.; Sancho, L.D. *Best Available Techniques (BAT) Reference Document for Common Waste Water and Waste Gas Treatment/Management Systems in the Chemical Sector*; Publications Office of the European Union: Luxembourg, 2016.
6. Burgués, J.; Marco, S. Environmental chemical sensing using small drones: A review. *Sci. Total Environ.* **2020**, *748*, 141172. [CrossRef]
7. Nathan, B.J.; Golston, L.M.; O'Brien, A.S.; Ross, K.; Harrison, W.A.; Tao, L.; Lary, D.J.; Johnson, D.R.; Covington, A.N.; Clark, N.N. Near-field characterization of methane emission variability from a compressor station using a model air-craft. *Environ. Sci. Technol.* **2015**, *49*, 7896–7903. [CrossRef]
8. Ravikumar, A.P.; Sreedhara, S.; Wang, J.; Englander, J.; Roda-Stuart, D.; Bell, C.; Zimmerle, D.; Lyon, D.; Mogstad, I.; Ratner, B. Single-blind inter-comparison of methane detection technologies–results from the Stanford/EDF Mobile Monitoring Challenge. *Elem. Sci. Anth.* **2019**, *7*, 37. [CrossRef]
9. Golston, L.M.; Aubut, N.F.; Frish, M.B.; Yang, S.; Talbot, R.W.; Gretencord, C.; McSpiritt, J.; Zondlo, M.A. Natural Gas Fugitive Leak Detection Using an Unmanned Aerial Vehicle: Localization and Quantification of Emission Rate. *Atmosphere* **2018**, *9*, 333. [CrossRef]
10. Yang, S.; Talbot, R.; Frish, M.; Golston, L.; Aubut, N.; Zondlo, M.; Gretencord, C.; McSpiritt, J. Natural gas fugitive leak detec-tion using an unmanned aerial vehicle: Measurement system description and mass balance approach. *Atmosphere* **2018**, *9*, 383. [CrossRef]
11. ConocoPhillips Testing Drone Technology to Detect and Quantify Emissions. Available online: http://www.conocophillips.com/sustainability/sustainability-news/story/testing-drone-technology-to-detect-and-quantify-emissions/ (accessed on 31 March 2020).
12. BP p.l.c. BP North Sea Deploys Mars Technology in World-First Methane Monitoring Project. Available online: https://www.bp.com/en/global/corporate/news-and-insights/press-releases/bp-north-sea-deploys-mars-technology-in-world-first-methane-monitoring-project.html (accessed on 31 March 2020).
13. TOTAL Group Integrating Climate into Our Strategy. Available online: https://www.total.com/sites/default/files/atoms/files/total_rapport_climat_2019_en.pdf#page=30 (accessed on 29 March 2020).
14. Emran, B.J.; Tannant, D.D.; Najjaran, H. Low-Altitude Aerial Methane Concentration Mapping. *Remote Sens.* **2017**, *9*, 823. [CrossRef]
15. Burgués, J.; Jiménez-Soto, J.M.; Marco, S. Estimation of the limit of detection in semiconductor gas sensors through linearized calibration models. *Anal. Chim. Acta* **2018**, *1013*, 13–25. [CrossRef]
16. Mead, M.; Popoola, O.; Stewart, G.; Landshoff, P.; Calleja, M.; Hayes, M.; Baldovi, J.; McLeod, M.; Hodgson, T.; Dicks, J.; et al. The use of electrochemical sensors for monitoring urban air quality in low-cost, high-density networks. *Atmos. Environ.* **2013**, *70*, 186–203. [CrossRef]
17. Martinez, D.; Burgués, J.; Marco, S. Fast Measurements with MOX Sensors: A Least-Squares Approach to Blind Deconvolution. *Sensors* **2019**, *19*, 4029. [CrossRef]
18. Lilienthal, A.J.; Loutfi, A.; Duckett, T. Airborne Chemical Sensing with Mobile Robots. *Sensors* **2006**, *6*, 1616–1678. [CrossRef]
19. Burgués, J.; Hernández, V.; Lilienthal, A.J.; Marco, S. Smelling Nano Aerial Vehicle for Gas Source Localization and Map-ping. *Sensors* **2019**, *19*, 478. [CrossRef]
20. Burgués, J.; Hernández, V.; Lilienthal, A.J.; Marco, S. Gas distribution mapping and source localization using a 3D grid of metal oxide semiconductor sensors. *Sens. Actuators B Chem.* **2020**, *304*, 127309. [CrossRef]
21. Neumann, P.P.; Asadi, S.; Lilienthal, A.J.; Bartholmai, M.; Schiller, J.H. Autonomous gas-sensitive microdrone: Wind vector estima-tion and gas distribution mapping. *IEEE Robot. Autom. Mag.* **2012**, *19*, 50–61. [CrossRef]
22. He, X.; Bourne, J.R.; Steiner, J.A.; Mortensen, C.; Hoffman, K.C.; Dudley, C.J.; Rogers, B.; Cropek, D.M.; Leang, K.K. Autonomous chemical-sensing aerial robot for urban/suburban environmental monitoring. *IEEE Syst. J.* **2019**, *13*, 3524–3535. [CrossRef]
23. Schellenberg, B.; Richardson, T.; Watson, M.; Greatwood, C.; Clarke, R.; Thomas, R.; Wood, K.; Freer, J.; Thomas, H.; Liu, E.; et al. Remote sensing and identification of volcanic plumes using fixed-wing UAVs over Volcán de Fuego, Guatemala. *J. Field Robot.* **2019**, *36*, 1192–1211. [CrossRef]
24. Bax, C.; Sironi, S.; Capelli, L. How Can Odors Be Measured? An Overview of Methods and Their Applications. *Atmosphere* **2020**, *11*, 92. [CrossRef]
25. Capelli, L.; Sironi, S.; Del Rosso, R.; Céntola, P.; Grande, M.I. A comparative and critical evaluation of odour assessment methods on a landfill site. *Atmos. Environ.* **2008**, *42*, 7050–7058. [CrossRef]
26. Levy, J.; Spengler, J.D.; Hlinka, D.; Sullivan, D.; Moon, D. Using CALPUFF to evaluate the impacts of power plant emissions in Illinois: Model sensitivity and implications. *Atmos. Environ.* **2002**, *36*, 1063–1075. [CrossRef]

remote sensing

MDPI

Letter

Assessing the Impact of Corona-Virus-19 on Nitrogen Dioxide Levels over Southern Ontario, Canada

Debora Griffin [1,*], Chris Anthony McLinden [1,2], Jacinthe Racine [3], Michael David Moran [1], Vitali Fioletov [1], Radenko Pavlovic [3], Rabab Mashayekhi [3], Xiaoyi Zhao [1] and Henk Eskes [4]

1 Air Quality Research Division, Environment and Climate Change Canada, Toronto, ON M3H 5T4, Canada; chris.mclinden@canada.ca (C.A.M.); mike.moran@canada.ca (M.D.M.); vitali.fioletov@canada.ca (V.F.); xiaoyi.zhao@canada.ca (X.Z.)
2 Department of Physics and Engineering Physics, University of Saskatchewan, Saskatoon, SK S7N 5E2, Canada
3 Canadian Meteorological Centre Operations Division, Environment and Climate Change Canada, Dorval, QC H9P 1J3, Canada; jacinthe.racine@canada.ca (J.R.); radenko.pavlovic@canada.ca (R.P.); rabab.mashayekhi@canada.ca (R.M.)
4 Royal Netherlands Meteorological Institute (KNMI), 3731 GA De Bilt, The Netherlands; henk.eskes@knmi.nl
* Correspondence: debora.griffin@canada.ca

Received: 16 October 2020; Accepted: 11 December 2020; Published: 16 December 2020

Abstract: A lockdown was implemented in Canada mid-March 2020 to limit the spread of COVID-19. In the wake of this lockdown, declines in nitrogen dioxide (NO_2) were observed from the TROPOspheric Monitoring Instrument (TROPOMI). A method is presented to quantify how much of this decrease is due to the lockdown itself as opposed to variability in meteorology and satellite sampling. The operational air quality forecast model, GEM-MACH (Global Environmental Multi-scale - Modelling Air quality and CHemistry), was used together with TROPOMI to determine expected NO_2 columns that represents what TROPOMI would have observed for a non-COVID scenario. Applying this methodology to southern Ontario, decreases in NO_2 emissions due to the lockdown were seen, with an average 40% (roughly 10 kt[NO_2]/yr) in Toronto and Mississauga and even larger declines in the city center. Natural and satellite sampling variability accounted for as much as 20–30%, which demonstrates the importance of taking meteorology into account. A model run with reduced emissions (from 65 kt[NO_2]/yr to 40 kt[NO_2]/yr in the Greater Toronto Area) based on emission activity data during the lockdown period was found to be consistent with TROPOMI NO_2 columns.

Keywords: air pollution; TROPOMI; COVID; nitrogen oxides

1. Introduction

The outbreak of Coronavirus disease in late 2019 (COVID-19) reached Canada in early 2020, with the first Canadian COVID-related death reported in early March 2020 [1]. By mid-March, provinces were beginning to limit the size of gatherings and initiating an overall lockdown of their populations. In Ontario, the lockdown was announced on 16 March 2020. This greatly disrupted traffic patterns, with traffic density observed to decrease by roughly 50–60% by early April [2]. Travel restrictions also greatly curtailed air travel. These circumstances provided a unique and unprecedented natural experiment where emissions patterns were rapidly and drastically altered, especially in southern Ontario, home to the Greater Toronto Area (GTA), the most populous urban area in Canada [3]. The GTA consists of the City of Toronto and four surrounding regional municipalities (see Supplement Material Figure S1) and includes many limited-access highways and expressways, rail lines, and Toronto Pearson International Airport, Canada's busiest airport [4]. Its population in 2016 was over 6.4 million [3]. Ultimately, changing emissions in the GTA and the rest of southern

Ontario associated with the pandemic allow for testing and refining of emissions from different sectors, most notably those from vehicle traffic.

One pollutant that is associated with combustion processes such as vehicle traffic is nitrogen dioxide ($NO_x = NO_2 + NO$). NO_x has adverse effects on human and environmental health: it is a key ingredient in smog, as precursors to both ozone and particulate matter, and can contribute to acid deposition. NO_x concentrations strongly correlate with local emission sources due to its short lifetime of a few hours [5,6] and, because of the high and localized enhancements compared to background levels, NO_x is a good tracer of human activity near cities. For example, urban NO_x displays a strong weekly and diurnal cycle resulting from differences in traffic and manufacturing activity on weekends versus weekdays [7,8]. Observed NO_2 is not merely a function of NO_x emissions; but is also a function of the local chemical environment and meteorology. For example, it is well known that NO_2 impacts its own chemical lifetime [5]. Furthermore, meteorological parameters such as cloud cover, temperature, and wind speed and direction all have a strong effect on local NO_2 enhancements [9–11]. Given this temporal and spatial variability in NO_2, precisely where and when observations are made is also very important. Taken together, one important challenge when interpreting changes in NO_2 lies in disentangling potential changes in emissions from natural and sampling variability.

Satellite observations can help to identify NO_x emissions and their variation globally. Declines in NO_2 emissions, following the lockdown, have previously been observed by satellite instruments in China, India, Europe and North America [12–15]. In this study, observations from the European Space Agency's Sentinel-5p Tropospheric Monitoring Instrument (TROPOMI), in conjunction with forecasts from Environment and Climate Change Canada's (ECCC) operational regional air quality forecast model GEM-MACH (Global Environmental Multi-scale - Modelling Air quality and CHemistry) [16,17], are used to isolate the impact of the COVID associated lockdown on NO_2 levels in southern Ontario, Canada. In this study, we show that combining satellite observations and model output, it is possible to determine the impact of meteorology and sampling variability on the observed NO_2 column changes. The air quality model is further used to determine how possible lockdown-associated emission reductions impact the NO_2 columns, and whether those match the observed changes.

2. Materials and Methods

In the context of satellite remote sensing, one method, and the most straightforward, to assess the impact of the COVID lockdown on NO_2 is to directly compare the COVID period with a non-COVID period, perhaps using the same period from different years [13]. However, in order to completely isolate the COVID signal, this method assumes that among the two periods being compared, (i) baseline emissions do not differ, (ii) natural or seasonal variability in winds, sunlight, temperature, and other meteorological parameters are not important, (iii) differences in satellite sampling do not play any role, and (iv) any differences in the satellite retrieval algorithm are minimal. For many locations, including the Canadian domain studied here, differences in interannual NO_x emission changes should be small. However, meteorological variability can be important, and given that, sampling variability is also likely to lead to differences between the two periods. In the case of TROPOMI, different retrieval algorithms were used for spring 2019 vs. spring 2020 (v1.2 until April 2019 and thereafter v1.3, differences include the treatment of "negative" cloud fractions and the lower limit of the tropospheric air mass factor (AMF) relaxed influencing the quality flag [18]). While differences tend to be small, it is difficult at present to completely eliminate this as a possible source of difference.

With these confounding factors in mind, the method presented here is the one in which the ECCC's operational GEM-MACH air quality model forecasts are used to control for non-COVID factors such as sampling variability, meteorological variability, and other sources of variability. Furthermore, to limit potential differences in the retrieval algorithm between 2019 and 2020, the two periods considered are consecutive in 2020: a pre-COVID period and the COVID-lockdown period.

2.1. TROPOMI Observations

Observations of NO_2 from TROPOMI (2017-present [19]), an Earth-viewing spectrometer, are used here. TROPOMI has a resolution of 3.5 × 5.5 km^2 (since August 2019, before 3.5 × 7 km^2) at nadir and measures back-scattered ultraviolet/visible/solar-infrared sunlight from which NO_2 vertical column density (VCD), or the vertically-integrated NO_2 number density, can be derived. Details on the retrieval algorithm can be found elsewhere [20], but in short: a spectral fit is performed matching laboratory-measured NO_2 absorption cross-sections and other relevant parameters to these observed spectra which provide a determination of the NO_2 slant column densities (SCDs), or the number density integrated along the path of the sunlight through the atmosphere. In a second step, the stratospheric component of the SCD is determined using a chemical data assimilated system and subtracted [21]. Finally, the remaining tropospheric SCD is converted to a VCD using an AMF which quantifies the sensitivity of the satellite to a particular scene which depends on factors such as shape of the NO_2 profile, surface reflectivity, viewing geometry, and clouds. In this work, an alternative AMF is used which better accounts for the presence of snow and uses higher resolution NO_2 profile shapes to improve the effective spatial resolution [22,23]; see Supplement material for more information [24–33]. A radiative transfer model is used to calculate AMFs [34] which depend on the following factors: solar and viewing geometry, surface pressure, the presence and pressure of clouds, scene reflectivity and the vertical distribution of the NO_2 via VCD=SCD/AMF. Similar, as in the original TROPOMI AMF, the aerosols are corrected for implicitly [21]. Lastly, the TROPOMI data are filtered to use only the highest quality data (qa_value > 0.75 and the cloud cover of the pixels is at most 30%). The TROPOMI tropspheric NO_2 columns have been validated in a number of studies against ground-based, aircraft and other satellite observations [35–40]. The alternative AMFs have a smaller bias between ground-based and aircraft-borne observations over cities or near industry [23,41,42]. An evaluation of the TROPOMI NO_2 observations over the GTA in 2020 shows overall good agreement with ground-based remote-sensing PANDORA [43] measurements (see Figure S8).

2.2. GEM-MACH Air Quality Forecast Model

The Canadian operational air quality forecast model, GEM-MACH [16,17,44,45], is used in this work. GEM-MACH consists of an on-line chemical transport module that is embedded within ECCC's Global Environmental Multi-scale (GEM), weather forecast model, and is applied over a domain that covers most of North America. It includes emissions, chemistry, dispersion, and removal process representations for 41 gaseous and eight particle chemical species, and provides hourly concentrations between the surface and 0.1 Pa (on 80 hybrid vertical levels) with a 10 × 10 km^2 grid cell. The standard operational model run inputs hourly emissions fields that are prepared using the Sparse Matrix Operator Kernel Emissions (SMOKE) [46] that account for seasonal, weekly and daily variations. The performance of GEM-MACH has previously been evaluated against surface and remote-sensing measurements [16,44,47–51]. A performance evaluation of NO_2 forecasts for spring 2019 for Canada by the version of the air quality modelling system used in this study was carried out before it was implemented operationally in September 2019. As an indication of the quality of the pre-pandemic forecasts to be expected in this study, it was that found that NO_2 forecasts for Canada for that period had a mean bias of 1.4 ppbv, a correlation coefficient of 0.57, and a root mean square error of 7.8 ppbv [48]. Additionally, an evaluation with PANDORA ground-based measurements was performed over the GTA for 2020, and showed overall good agreement with the model NO_2 VCDs (see Figure S9). The current version of the emissions files used by the operational model are based on a Canadian emissions inventory compiled for the 2013 base year and a 2017 projected U.S. inventory [48]. While using year-specific NO_x emissions is ideal, suitable emission inventories are not available in a timely manner. Alternative non-operational runs were also performed for a limited time period between 15 March and 10 May 2020 with projected Canadian 2020 emissions and COVID-modified emissions for vehicle, aircraft, manufacturing, and residential sectors (see Section 3 for details). The Canadian 2020 anthropogenic emissions are based on projected national emission inventory that was generated by ECCC for policy studies [52]. The projections include expected changes in population, economic activity and energy use over a five-year period, from 2015 to 2020.

GEM-MACH output is used in this study for two purposes. The first is to provide profile shapes which are used in the calculation of revised TROPOMI AMFs as discussed above in Section 2.1, following the method proposed by Palmer et al. [34] and McLinden et al. [22]. These alternative AMFs (not the operational TROPOMI AMFs) are used to convert TROPOMI SCDs into VCDs. Thus, it is possible to carry out the direct comparison between our TROPOMI NO_2 VCDs and those obtained from the GEM-MACH (further details can be found in the supplement material). The second is to determine the time evolution of NO_2 on standard "business as usual" (BAU) input emissions that do not account for COVID impacts, which can then be contrasted with that observed by TROPOMI. In both cases, NO_2 profiles are obtained from operational forecasts, are run at 10 km spatial resolution and are launched every 12 h (and every 24 h for the special runs).

In this study, we integrate the model NO_2 profiles to obtain VCD values. The operational GEM-MACH model currently does not include NO_x sources in the free troposphere (such as lightning and aircraft at cruising altitude); as a consequence the model NO_x concentrations are near zero above the boundary layer. We obtain a more realistic free tropospheric column from GEOS-Chem [53], a 3-D model of atmospheric chemistry model (monthly averages between 18-21 UTC, from 2 km to 12 km; $0.5 \times 0.67^{\circ}$ resolution, version v8-03-01; http://www.geos-chem.org), these partial columns are on the order of 10^{14} molec/cm^2 and small compared to the partial columns in the boundary layer (see Figure S7), similar corrections have been applied in previous studies [22,23,49]. The model VCDs are then sampled (and filtered) in space and time at each TROPOMI pixel, and are filtered like the TROPOMI observations.

2.3. Determination of Expected NO_2

GEM-MACH model output is used to estimate the impact of: (1) COVID measures on NO_2 levels, (2) changes from any other possible sources of variability, including seasonal, inter-annual, or even shorter-term meteorological variability, and (3) the TROPOMI sampling variability. GEM-MACH forecasts using standard emissions inventories for both the pre-lockdown and lockdown periods are sampled at each TROPOMI pixel and overpass time.

Comparing pre-lockdown and lockdown TROPOMI observations together with pre-lockdown and lockdown GEM-MACH predictions provides an estimate of the changes in NO_x emissions purely due to the lockdown, as this method accounts for effects of meteorology, seasonality, and sampling variability. The expected TROPOMI VCDs, $V_{T,e}$, under a BAU scenario, are determined from the TROPOMI VCDs before the lockdown and are adjusted by the relative change seen in the model forecasts (GEM-MACH and free troposphere from GEOS-Chem) between the two time periods:

$$V_{T,e}(t_{covid}) = V_T(t_{pre}) \cdot \frac{V_{Model}(t_{covid})}{V_{Model}(t_{pre})}. \quad (1)$$

When averaging over time to produce spatially resolved maps, observations from 15 February to 15 March 2020 and 16 March to 8 May 2020 are used for the pre-lockdown and lockdown time periods, respectively. This end date is associated with some traffic rebound and increased emissions throughout May 2020 (see Section 3). When averaging over a larger area to produce a time series, 15-day running means are used (the satellite data need to be averaged over multiple days in order to obtain enough data over this area, approximately 50% of observations are filtered due to clouds). The expected columns for the 15-day running means are estimated as in Equation (1), where $V_{T,e}(t_{covid})$ and $V_{Model}(t_{covid})$ are the 15-day means for a specific day.

3. Results and Discussions

3.1. Spatial Averaging over Southern Ontario

Figure 1 shows the TROPOMI and operational GEM-MACH NO_2 VCDs averaged over the pre-lockdown and lockdown periods. There is excellent agreement between TROPOMI, panel (a), and GEM-MACH, panel (d), across southern Ontario for the pre-lockdown period in terms of both spatial distribution and magnitudes which provides confidence that the NO_x emissions inventory and the model itself can accurately represent the complex physics and photochemistry of the real world.

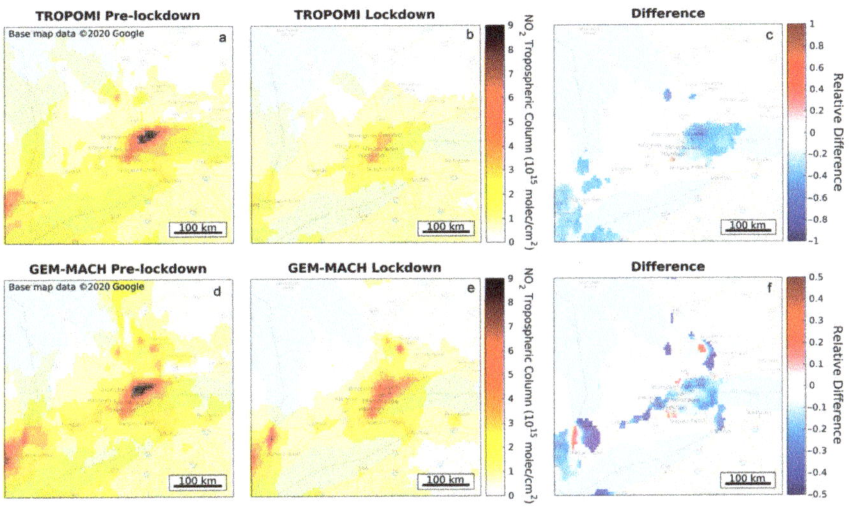

Figure 1. TROPOMI averaged VCDs over southern Ontario are shown for (**a**) a pre-lockdown (16 February–15 March 2020) and (**b**) a lockdown (16 March–8 May 2020) period. The relative differences ((lockdown-pre-lockdown)/pre-lockdown) are shown in panel (**c**) for areas that exceed 3×10^{15} molec/cm^2 in the pre-lockdown period. Panels (**d–f**) are the same but for the operational GEM-MACH model BAU NO_2 VCDs, sampled at the time and location of the TROPOMI pixels.

When comparing TROPOMI observations between the pre-lockdown and lockdown periods, panel (a)–(c), there is a large decrease in VCDs over the GTA, the Windsor-Detroit urban area (which straddles the Canada-U.S. border), and virtually the entire domain. Decreases in the urban areas can reach or exceed 50%, and in parts of the GTA the decline can even exceed 60%. However, there is also a decrease predicted by GEM-MACH, despite not accounting for COVID-related emissions reductions as shown in panels (d)–(f). This is due to a combination of a seasonal effect in which increased sunlight means a decrease in NO_x lifetime and less NO_x present as NO_2, but also expected seasonal changes in emissions (see Supplement Material Figure S2). This effect is on the order of 25% over the GTA between the two time periods, and is especially large because it occurs during the change from cold season to warm season.

Even using several weeks of TROPOMI observations, meteorological and sampling variability can impact the average. Spring 2020 was colder than 2019 and particularly cloudy over southern Ontario, leading to fewer cloud-free overpasses on which to base an average. This can have an impact on the averages, since approximately 50% of TROPOMI data are removed due to cloud cover, so that the remaining cloud-free observations are more representative of fair weather conditions. To determine the impact of the sampling variability, GEM-MACH averages are determined using all days over the entire domain, versus only those sampled as TROPOMI (qa > 0.75). For the average NO_2 between 16 March and 8 May 2020, sampling variability can lead to differences as large as 10% near cities (see Supplement material Figure S3).

As a test of the methodology to create expected TROPOMI columns for the COVID-19 period from the change in the model forecasts, the same procedure was applied to TROPOMI observations and operational GEM-MACH output from 2019. In this case, differences between expected and TROPOMI observations should be minimal, because no unusual emission reductions occurred in 2019. As can been seen in Figure 2d,e, differences are small, suggesting the method is generally reliable. Averaged over the GTA, differences are 0–2%.

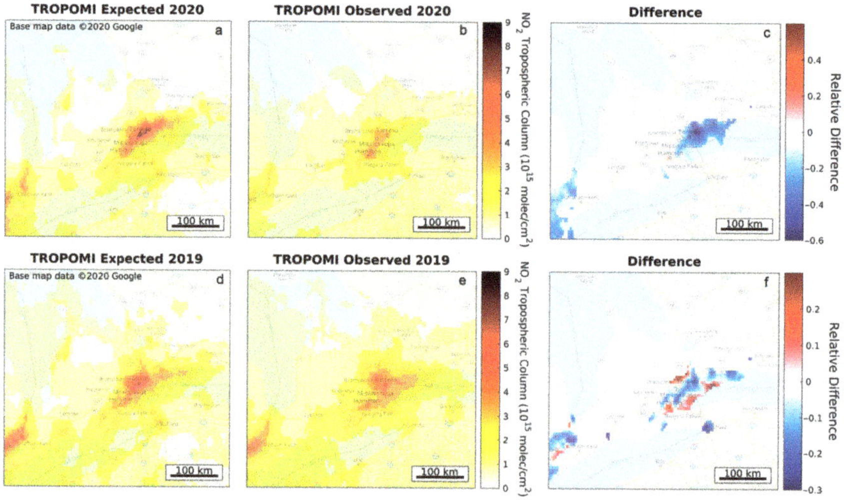

Figure 2. The figures show the expected and observed TROPOMI averaged NO2 VCDs over southern Ontario for 2020 and 2019. Expected and observed TROPOMI average VCD fields for the lockdown period (16 March–8 May 2020) are shown in panels (**a**,**b**), respectively. The same is shown in panels (**d**,**e**), but for 16 March–8 May 2019. Relative differences ((observed-expected)/expected; for areas that exceed 3×10^{15} molec/cm^2) between the TROPOMI observations and the expected columns are shown in panel (**c**,**f**) for 2020 and 2019, respectively. Note that panel (**b**) is the same as Figure 1b.

3.2. COVID-Scenario Model Run

To help evaluate the difference between expected and observed TROPOMI NO_2 columns, as shown in Figure 2, GEM-MACH is re-run using an alternative emissions scenario designed to represent COVID-19 emissions changes: (i) a 30% reduction in industrial NO_x emissions, (ii) a 60% reduction for traffic NO_x emissions, (iii) an 80% reduction in aircraft NO_x emissions (landings and takeoffs), and (iv) a 20% increase of residential fuel NO_x emissions due to people staying at home. Emissions of other air pollutants emitted by these source types (CO, VOC, NH_3, SO_2, $PM_{2.5}$, PM_{10}) are also changed by these same percentages. The change of emissions is based on the following: (i) similar emission scenarios from Europe [11], (ii) an estimate of daily driving activities which showed a reduction of 50–65% decrease [2], (iii) the reduction of airline flights which were 79% lower in April 2020 compared to April 2019 [54], and (iv) Google mobility Reports [55] showed an 20% increase spent in residential spaces and thus an increase of 20% is applied to residential emissions. Over the entire GTA, average emissions decline from 65 kt[NO_2]/yr pre-lockdown to 40 kt[NO_2]/yr lockdown (around noon; see Figures S3, S5, S6, and Table S2). Note that only Canadian emissions are adjusted in this way due to the challenge of representing the complicated mixture of city-, county-, and state-level responses to COVID-19 in the U.S., but given the short atmospheric lifetime of NO_x this is unlikely to make a big difference to NO_2 levels except close to the international border (further details can be found in the supplement on the impact of trans-border NO_2 transport, Figure S10). The results of this emissions scenario run are shown and compared to TROPOMI observations in Figure 3 (for 1 April to 8 May 2020). Good agreement is evident over much of southern Ontario. The TROPOMI observations are approximately 20–30% higher than the model output in Hamilton (an industrial city), where industry emissions might be underestimated, and parts of Mississauga, where airport or vehicle traffic emissions might be underestimated in the model run.

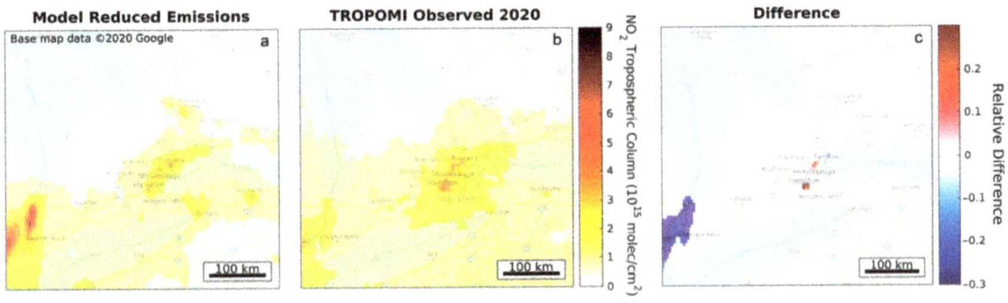

Figure 3. Model NO_2 VCDs from the reduced emissions scenario (**a**) and observed TROPOMI NO_2 VCDs (**b**) over southern Ontario averaged over the period 1 April – 8 May 2020. The relative differences ((observations-model)/model) are shown in panel (**c**) for areas that exceed 3×10^{15} molec/cm^2. Note that emissions have only been reduced in Canada; thus, large differences can be seen for the US cities near the border, especially Detroit.

3.3. Temporal Changes over Toronto

An alternative method of considering these various data sources is to average spatially and look at temporal changes. Figure 4 shows a time series of 15-day running average NO_2 over the Toronto and Mississauga area (part of the GTA with the highest emissions and population density, this area also includes Toronto Pearson Airport; see Supplement Material Figure S1). TROPOMI observations show a decline after the lockdown was announced (Figure 4a), the expected columns agree well with the TROPOMI observations during the pre-lockdown period, but, differences emerge after the lockdown begins as emissions are reduced, but the model assumes BAU emissions. The alternate model run with reduced emissions (Figure 4b) represents the decline observed by TROPOMI quite well and over the same time period, both the TROPOMI observations and the model predict a drop of roughly 40% over the GTA core (using data from 16 March to 8 May 2020) as a result of the lockdown. When the 2019 and 2020 satellite data are compared directly, however, the drop is only about half as much (20%), as the meteorology and sampling variability of the satellite are largely different in that area between 2019 and 2020. Note that the satellite data indicate that the peak of the emissions decline in Toronto and Mississauga occurred in mid-April. Throughout May 2020, the satellite measurements suggest that the NO_x emissions began to increase again gradually (Figure 4a), though they are still lower than BAU emissions. Ontario entered Phase 1 of its re-opening on 19 May 2020, when certain restrictions were lifted.

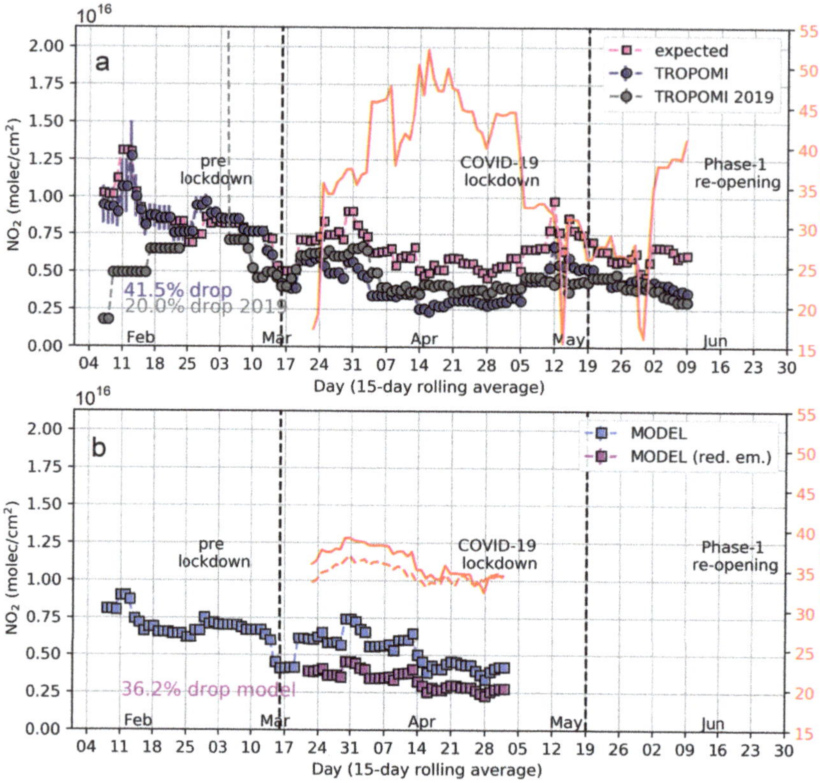

Figure 4. Timeseries of 15-day running mean of NO_2 VCDs over Toronto and Mississauga for 7 February to 9 June 2020, panel (**a**) shows the TROPOMI observations (navy), the expected columns (magenta). The timeseries of 2019 TROPOMI observations (grey) for the same period is shown as a reference. The red line indicates the percentage emission reductions based on the difference between the TROPOMI observations and expected columns. Panel (**b**) shows NO_2 columns from the model predictions sampled like TROPOMI assuming a BAU scenario with 2020 updated emissions (blue) and a 2020 COVID reduced emissions scenario (purple). The percentage decrease in model predicted VCDs (red line) is estimated from the difference between the two model runs, the red dashed line shows the drop for perfect sampling. Average emission reductions are highlighted using observations between 16 March to 8 May 2020. Approximately 200 observations are averaged for the 15-day mean, the resulting standard errors are plotted, however, the standard error is seen to be small and on the order of 10^{13}–10^{14} molec/cm^2.

4. Conclusions

We present a method to disentangle the effects of meteorology and sampling variability on the observed NO_2 changes, from the lockdown-related changes in NO_x emissions. During the period from 16 March to 8 May 2020, NO_2 columns in the center of the GTA decreased by nearly 60% compared to the previous month. About 25% of this decrease is associated with meteorological and seasonal changes independent of the COVID-19 pandemic. Even the TROPOMI sampling variability itself can impact the magnitude of the observed NO_2 columns over the course of one or two months averaging ($\sim$10%). From the TROPOMI observations and GEM-MACH air quality model results, we estimate that due to the lockdown the NO_2 columns in Toronto and Mississauga declined by over 40%. These changes vary spatially, and in certain locations columns declined by over 50%. Applying the same method to 2019 observations leads to a 0–2% decline over the GTA, which is expected as there were no emission declines in spring 2019, which gives confidence that the method is robust.

A special model run with reduced NO_x emissions of vehicle traffic, aircraft, and industry based on lockdown activity data [2,54,55] compares well with the TROPOMI observations during the lockdown and returned similar NO_2 declines in the GTA. Although, spatial patterns over cities are somewhat visible, it is hard to disentangle the emission reductions by sector with our methodology. Nevertheless, emission changes of (i) a 30% reduction in industry, (ii) a 60% reduction for traffic, (iii) an 80% reduction in aircraft landings and takeoffs, and (iv) a 20% increase in residential fuel combustion, represent the TROPOMI NO_2 observations well, at least in southern Ontario. In the GTA, NO_x emissions of 40 kt[NO_2]/yr represent the observations well, this is a drop of over 37% compared to a BAU scenario. The drop in the input emissions is almost identical to the drop determined from the model NO_2 VCDs (36%) over the same area which further indicates that the method presented works well.

This study highlights the importance of considering meteorological and sampling variability when estimating emission reductions. One needs to be cautious when simply comparing two months, since the effects of meteorological and sampling variability are not negligible when only a short series of data is averaged. We show that spring 2019 and 2020 were, with regards to the meteorology, very different years and simply looking at the difference results in about half the NO_x emission decline as compared to considering the meteorology. Further, the emission decline may vary strongly spatially, especially in cities. This can make it difficult to compare different studies unless the exact same areas are considered. The unique lockdown period associated with the 2020 COVID-19 pandemic can further be used to check and refine our existing emissions inventories for NO_x and other pollutants by looking at spatial and temporal distributions of available satellite and surface measurements for a number of different urban areas.

Supplementary Materials: The following are available online at http://www.mdpi.com/2072-4292/12/24/4112/s1, Figure S1: Boundaries of the Greater Toronto Area, Figure S2: Operational forecast model's seasonal emission changes, Figure S3: Impact of the sampling on the averaged TROPOMI columns, Figure S4: Model input NO_x emissions, Figure S5: Model input emissions in Toronto an Mississauga, Figure S6: Correlation between TROPOMI observations and model VCDs, Figure S7: Correlation between TROPOMI observations and model VCDs with and without free-tropospheric column, Figure S8: Comparison between TROPOMI and ground-based PANDORA NO_2 measurements, Figure S9: Comparison between model and ground-based PANDORA NO_2 measurements, Figure S10: Impact of US NO_2 emission changes on the GTA NO_2 concentrations, Table S1: Parameters and their reference points in the AMF look-up table, Table S2: Approximate average emissions used for the model runs in the GTA, Table S3: The statistics from the model and TROPOMI comparison.

Author Contributions: Conceptualization, D.G., C.A.M., M.D.M., and V.F.; methodology, D.G., C.A.M., M.D.M., V.F., J.R., R.M. and R.P.; software, D.G., and J.R.; validation, C.A.M., M.D.M., V.F., and X.Z.; formal analysis, D.G., J.R., and R.M.;investigation, D.G. and C.A.M.; resources, R.P., and H.E.; data curation, J.R., R.M., and H.E.; writing—original draft preparation, D.G. and C.A.M.; writing—review and editing, all authors; visualization, D.G., C.A.M., and X.Z.; supervision, C.A.M. and R.P. All authors have read and agreed to the published version of the manuscript.

Funding: This work contains modified Copernicus Sentinel data. The Sentinel 5 Precursor TROPOMI Level 2 product is developed with funding from the Netherlands Space Office (NSO) and processed with funding from the European Space. TROPOMI data can be downloaded from https://s5phub.copernicus.eu (last access: 6 June 2020).

Acknowledgments: We would like to thank you MSC-REQA employees involved in emission adjustment and modelling: Mourad Sassi, Annie Duhamel and Jessica Miville.

Conflicts of Interest: The authors declare no conflict of interest.

References

1. Statistics Canada. Provisional Death Counts and Excess Mortality, January to April 2019 and January to April 2020. Available online: https://www150.statcan.gc.ca/n1/daily-quotidien/200619/dq200619b-eng.htm (accessed on 29 September 2020).
2. Apple. COVID-19 Mobility Trends Reports. 2020. Available online: https://covid19.apple.com/mobility (accessed on 29 September 2020).
3. Statistics Canada. Population and Dwelling Count Highlight Tables, 2016 Census. 2019. Available online: https://www12.statcan.gc.ca/census-recensement/2016/dp-pd/hlt-fst/pd-pl/Table.cfm?Lang=Eng&T=801&SR=1&S=3&O=D&RPP=25&PR=0&CMA=0#tPopDwell (accessed on 29 September 2020).

4. Statistics Canada. Table: 23-10-0253-01, Air Passenger Traffic at Canadian Airports, Annual. 2020. Available online: https://doi.org/10.25318/2310025301-eng (accessed on 29 September 2020).
5. Valin, L.C.; Russell, A.R.; Cohen, R.C. Variations of OH radical in an urban plume inferred from NO_2 column measurements. *Geophys. Res. Lett.* **2013**, *40*, 1856–1860. [CrossRef]
6. Beirle, S.; Boersma, K.F.; Platt, U.; Lawrence, M.G.; Wagner, T. Megacity Emissions and Lifetimes of Nitrogen Oxides Probed from Space. *Science* **2011**, *333*, 1737–1739. [PubMed]
7. Beirle, S.; Platt, U.; Wenig, M.; Wagner, T. Weekly cycle of NO_2 by GOME measurements: A signature of anthropogenic sources. *Atmos. Chem. Phys.* **2003**, *3*, 2225–2232. [CrossRef]
8. De Foy, B.; Lu, Z.; Streets, D.G. Satellite NO_2 retrievals suggest China has exceeded its NO_x reduction goals from the twelfth Five-Year Plan. *Sci. Rep.* **2016**, *6*. [CrossRef] [PubMed]
9. Goldberg, D.L.; Anenberg, S.C.; Griffin, D.; McLinden, C.A.; Lu, Z.; Streets, D.G. Disentangling the Impact of the COVID-19 Lockdowns on Urban NO2 From Natural Variability. *Geophys. Res. Lett.* **2020**, *47*, e2020GL089269, [CrossRef]
10. Wang, P.; Chen, K.; Zhu, S.; Wang, P.; Zhang, H. Severe air pollution events not avoided by reduced anthropogenic activities during COVID-19 outbreak. *Resour. Conserv. Recycl.* **2020**, *158*, 104814. [CrossRef]
11. Menut, L.; Bessagnet, B.; Siour, G.; Mailler, S.; Pennel, R.; Cholakian, A. Impact of lockdown measures to combat Covid-19 on air quality over western Europe. *Sci. Total Environ.* **2020**, *741*, 140426. [CrossRef]
12. Zhang, R.; Zhang, Y.; Lin, H.; Feng, X.; Fu, T.M.; Wang, Y. NOx Emission Reduction and Recovery during COVID-19 in East China. *Atmosphere* **2020**, *11*, 433. [CrossRef]
13. Bauwens, M.; Compernolle, S.; Stavrakou, T.; Müller, J.F.; van Gent, J.; Eskes, H.; Levelt, P.F.; van der A, R.; Veefkind, J.P.; Vlietinck, J.; et al. Impact of coronavirus outbreak on NO2 pollution assessed using TROPOMI and OMI observations. *Geophys. Res. Lett.* **2020**, e2020GL087978, [CrossRef]
14. Shi, X.; Brasseur, G.P. The Response in Air Quality to the Reduction of Chinese Economic Activities during the COVID-19 Outbreak. *Geophys. Res. Lett.* **2020**, e2020GL088070, [CrossRef]
15. Adams, M.D. Air pollution in Ontario, Canada during the COVID-19 State of Emergency. *Sci. Total Environ.* **2020**, *742*, 140516. [CrossRef] [PubMed]
16. Moran, M.D.; Ménard, S.; Talbot, D.; Huang, P.; Makar, P.A.; Gong, W.; Landry, H.; Gravel, S.; Gong, S.; Crevier, L.P.; et al. Particulate-matter forecasting with GEM-MACH15, a new Canadian air-quality forecast model. In *Air Pollution Modelling and Its Application XX*; Springer: Dordrecht, The Netherlands, 2010.
17. Pendlebury, D.; Gravel, S.; Moran, M.D.; Lupu, A. Impact of chemical lateral boundary conditions in a regional air quality forecast model on surface ozone predictions during stratospheric intrusions. *Atmos. Environ.* **2018**, *174*, 148 – 170. [CrossRef]
18. Copernicus. Data Products: Nitrogen Dioxide. 2020. Available online: http://www.tropomi.eu/data-products/nitrogen-dioxide (accessed on 29 September 2020).
19. Veefkind, J.; Aben, I.; McMullan, K.; Forster, H.; de Vries, J.; Otter, G.; Claas, J.; Eskes, H.; de Haan, J.; Kleipool, Q.; et al. TROPOMI on the ESA Sentinel-5 Precursor: A GMES mission for global observations of the atmospheric composition for climate, air quality and ozone layer applications. *Remote Sens. Environ.* **2012**, *120*, 70–83. [CrossRef]
20. van Geffen, J.; Boersma, K.F.; Eskes, H.; Sneep, M.; ter Linden, M.; Zara, M.; Veefkind, J.P. S5P/TROPOMI NO_2 slant column retrieval: Method, stability, uncertainties, and comparisons against OMI. *Atmos. Meas. Tech. Dis.* **2019**, *2019*, 1–33. [CrossRef]
21. van Geffen, J.H.G.M.; Eskes, H.J.; Boersma, K.F.; Maasakkers, J.D.; Veefkind, J.P. TROPOMI ATBD of the Total and Tropospheric NO_2 Data Products; Issue 1.2.0 ed., 2018. S5P-KNMI-L2-0005-RP. Available online: https://sentinel.esa.int/documents/247904/2476257/Sentinel-5P-TROPOMI-ATBD-NO2-data-products (accessed on 28 September 2020).

22. McLinden, C.A.; Fioletov, V.; Boersma, K.F.; Kharol, S.K.; Krotkov, N.; Lamsal, L.; Makar, P.A.; Martin, R.V.; Veefkind, J.P.; Yang, K. Improved satellite retrievals of NO_2 and SO_2 over the Canadian oil sands and comparisons with surface measurements. *Atmos. Chem. Phys.* **2014**, *14*, 3637–3656. [CrossRef]
23. Griffin, D.; Zhao, X.; McLinden, C.A.; Boersma, F.; Bourassa, A.; Dammers, E.; Degenstein, D.; Eskes, H.; Fehr, L.; Fioletov, V.; et al. High-Resolution Mapping of Nitrogen Dioxide With TROPOMI: First Results and Validation Over the Canadian Oil Sands. *Geophys. Res. Lett.* **2019**, *46*, 1049–1060. [CrossRef]
24. Côté, J.; Gravel, S.; Méthot, A.; Patoine, A.; Roch, M.; Staniforth, A. The Operational CMC–MRB Global Environmental Multiscale (GEM) Model. Part I: Design Considerations and Formulation. *Mon. Weather Rev.* **1998**, *126*, 1373–1395. [CrossRef]
25. Girard, C.; Plante, A.; Desgagné, M.; McTaggart-Cowan, R.; Côté, J.; Charron, M.; Gravel, S.; Lee, V.; Patoine, A.; Qaddouri, A.; et al. Staggered Vertical Discretization of the Canadian Environmental Multiscale (GEM) Model Using a Coordinate of the Log-Hydrostatic-Pressure Type. *Mon. Weather Rev.* **2014**, *142*, 1183–1196. [CrossRef]
26. Houyoux, M.R.; Vukovich, J.M.; Coats, C.J., Jr.; Wheeler, N.J.M.; Kasibhatla, P.S. Emission inventory development and processing for the Seasonal Model for Regional Air Quality (SMRAQ) project. *J. Geophys. Res. Atmos.* **2000**, *105*, 9079–9090. [CrossRef]
27. Schaaf, C.B.; Gao, F.; Strahler, A.H.; Lucht, W.; Li, X.; Tsang, T.; Strugnell, N.C.; Zhang, X.; Jin, Y.; Muller, J.P.; et al. First operational BRDF, albedo nadir reflectance products from MODIS. *Remote Sens. Environ.* **2002**, *83*, 135–148. [CrossRef]
28. Makar, P.A.; Gong, W.; Milbrandt, J.; Hogrefe, C.; Zhang, Y.; Curci, G.; Žabkar, R.; Im, U.; Balzarini, A.; Baró, R.; et al. Feedbacks between air pollution and weather, Part 1: Effects on weather. *Atmos. Environ.* **2015**, *115*, 442–469. [CrossRef]
29. Makar, P.A.; Gong, W.; Hogrefe, C.; Zhang, Y.; Curci, G.; Žabkar, R.; Milbrandt, J.; Im, U.; Balzarini, A.; Baró, R.; et al. Feedbacks between air pollution and weather, part 2: Effects on chemistry. *Atmos. Environ.* **2015**, *115*, 499–526. [CrossRef]
30. Gong, W.; Makar, P.; Zhang, J.; Milbrandt, J.; Gravel, S.; Hayden, K.; Macdonald, A.; Leaitch, W. Modelling aerosol–cloud–meteorology interaction: A case study with a fully coupled air quality model (GEM-MACH). *Atmos. Environ.* **2015**, *115*, 695–715. [CrossRef]
31. Gong, W.; Beagley, S.R.; Cousineau, S.; Sassi, M.; Munoz-Alpizar, R.; Ménard, S.; Racine, J.; Zhang, J.; Chen, J.; Morrison, H.; et al. Assessing the impact of shipping emissions on air pollution in the Canadian Arctic and northern regions: Current and future modelled scenarios. *Atmos. Chem. Phys.* **2018**, *18*, 16653–16687. [CrossRef]
32. Akingunola, A.; Makar, P.A.; Zhang, J.; Darlington, A.; Li, S.M.; Gordon, M.; Moran, M.D.; Zheng, Q. A chemical transport model study of plume-rise and particle size distribution for the Athabasca oil sands. *Atmos. Chem. Phys.* **2018**, *18*, 8667–8688. [CrossRef]
33. Cooper, M.J.; Martin, R.V.; Lyapustin, A.I.; McLinden, C.A. Assessing snow extent data sets over North America to inform and improve trace gas retrievals from solar backscatter. *Atmos. Meas. Tech.* **2018**, *11*, 2983–2994. [CrossRef]
34. Palmer, P.I.; Jacob, D.J.; Chance, K.; Martin, R.V.; Spurr, R.J.D.; Kurosu, T.P.; Bey, I.; Yantosca, R.; Fiore, A.; Li, Q. Air mass factor formulation for spectroscopic measurements from satellites: Application to formaldehyde retrievals from the Global Ozone Monitoring Experiment. *J. Geophys. Res. Atmos.* **2001**, *106*, 14539–14550. [CrossRef]
35. Verhoelst, T.; Compernolle, S.; Pinardi, G.; Lambert, J.C.; Eskes, H.J.; Eichmann, K.U.; Fjæraa, A.M.; Granville, J.; Niemeijer, S.; Cede, A.; et al. Ground-based validation of the Copernicus Sentinel-5p TROPOMI NO_2 measurements with the NDACC ZSL-DOAS, MAX-DOAS and Pandonia global networks. *Atmos. Meas. Tech. Dis.* **2020**, *2020*, 1–40. [CrossRef]
36. Van Geffen, J.; Boersma, K.F.; Eskes, H.; Sneep, M.; ter Linden, M.; Zara, M.; Veefkind, J.P. S5P TROPOMI NO_2 slant column retrieval: Method, stability, uncertainties and comparisons with OMI. *Atmos. Meas. Tech.* **2020**, *13*, 1315–1335. [CrossRef]
37. Wang, P.; Piters, A.; van Geffen, J.; Tuinder, O.; Stammes, P.; Kinne, S. Shipborne MAX-DOAS measurements for validation of TROPOMI NO_2 products. *Atmos. Meas. Tech.* **2020**, *13*, 1413–1426. [CrossRef]
38. Dimitropoulou, E.; Hendrick, F.; Pinardi, G.; Friedrich, M.M.; Merlaud, A.; Tack, F.; De Longueville, H.; Fayt, C.; Hermans, C.; Laffineur, Q.; et al. Validation of TROPOMI tropospheric NO_2 columns using dual-scan MAX-DOAS measurements in Uccle, Brussels. *Atmos. Meas. Tech. Dis.* **2020**, *2020*, 1–50. [CrossRef]

39. Ialongo, I.; Virta, H.; Eskes, H.; Hovila, J.; Douros, J. Comparison of TROPOMI/Sentinel-5 Precursor NO_2 observations with ground-based measurements in Helsinki. *Atmos. Meas. Tech.* **2020**, *13*, 205–218. [CrossRef]
40. Tack, F.; Merlaud, A.; Iordache, M.D.; Pinardi, G.; Dimitropoulou, E.; Eskes, H.; Bomans, B.; Veefkind, P.; Van Roozendael, M. Assessment of the TROPOMI tropospheric NO_2 product based on airborne APEX observations. *Atmos. Meas. Tech. Dis.* **2020**, *2020*, 1–55. [CrossRef]
41. Zhao, X.; Griffin, D.; Fioletov, V.; McLinden, C.; Cede, A.; Tiefengraber, M.; Müller, M.; Bognar, K.; Strong, K.; Boersma, F.; et al. Assessment of the quality of TROPOMI high-spatial-resolution NO_2 data products in the Greater Toronto Area. *Atmos. Meas. Tech.* **2020**, *13*, 2131–2159. [CrossRef]
42. Goldberg, D.L.; Lu, Z.; Streets, D.G.; de Foy, B.; Griffin, D.; McLinden, C.A.; Lamsal, L.N.; Krotkov, N.A.; Eskes, H. Enhanced Capabilities of TROPOMI NO_2: Estimating NOX from North American Cities and Power Plants. *Environ. Sci. Technol.* **2019**, *53*, 12594–12601, [CrossRef] [PubMed]
43. Herman, J.; Cede, A.; Spinei, E.; Mount, G.; Tzortziou, M.; Abuhassan, N. NO_2 column amounts from ground-based Pandora and MFDOAS spectrometers using the direct-sun DOAS technique: Intercomparisons and application to OMI validation. *J. Geophys. Res. Atmos.* **2009**, *114*. [CrossRef]
44. Pavlovic, R.; Chen, J.; Anderson, K.; Moran, M.D.; Beaulieu, P.A.; Davignon, D.; Cousineau, S. The FireWork air quality forecast system with near-real-time biomass burning emissions: Recent developments and evaluation of performance for the 2015 North American wildfire season. *J. Air Waste Manag. Assoc.* **2016**, *66*, 819–841, [CrossRef]
45. Makar, P.A.; Staebler, R.M.; Akingunola, A.; Zhang, J.; McLinden, C.; Kharol, S.K.; Pabla, B.; Cheung, P.; Zheng, Q. The effects of forest canopy shading and turbulence on boundary layer ozone. *Nat. Commun.* **2017**, *8*. [CrossRef]
46. Coats, C.J. High-performance algorithms in the sparse matrix operator kernel emissions (SMOKE) modeling system. In Proceedings of the Ninth AMS Joint Conference on Applications of Air Pollution Meteorology with AWMA, Atlanta, GA, USA, 28 January–2 February 1996.
47. Robichaud, A.; Ménard, R.; Zaitseva, Y.; Anselmo, D. Multi-pollutant surface objective analyses and mapping of air quality health index over North America. *Air Qual. Atmos. Health* **2016**, *9*, 743–759. [CrossRef]
48. Moran, M.D.; Ménard, S. Regional Air Quality Deterministic Prediction System (RAQDPS): Update from Version 020.2 to Version 021; Canadian Centre for Meteorological and Environmental Prediction: Montreal, CA, 2019. Technical Note, p. 49. Available online: http://collaboration.cmc.ec.gc.ca/cmc/cmoi/product_guide/docs/tech_notes/technote_raqdps-021_20190703_e.pdf (accessed on 23 June 2020).
49. Zhao, X.; Griffin, D.; Fioletov, V.; McLinden, C.; Davies, J.; Ogyu, A.; Lee, S.C.; Lupu, A.; Moran, M.D.; Cede, A.; et al. Retrieval of total column and surface NO_2 from Pandora zenith-sky measurements. *Atmos. Chem. Phys.* **2019**, *19*, 10619–10642. [CrossRef]
50. Stroud, C.A.; Zaganescu, C.; Chen, J.; McLinden, C.A.; Zhang, J.; Wang, D. Toxic volatile organic air pollutants across Canada: Multi-year concentration trends, regional air quality modelling and source apportionment. *J. Atmos. Chem.* **2016**, *73*. [CrossRef]
51. Stroud, C.A.; Ren, S.; Zang, Z.; Akingunola, A.; Makar, P.; Munoz-Alpizar, R.; Leroyer, S.; Belair, S.; Sills, D.; Brook, J. Chemical analysis of surface-level ozone exceedances during the 2015 Pan American Games. *Atmosphere* **2020**, *11*, 37. [CrossRef]
52. APEI. Air Pollutant Emissions Inventory: Overview, Government of Canada. 2020. Available online: https://www.canada.ca/en/environment-climate-change/services/pollutants/air-emissions-inventory-overview.html (accessed on 28 September 2020).
53. Bey, I.; Jacob, D.J.; Yantosca, R.M.; Logan, J.A.; Field, B.D.; Fiore, A.M.; Li, Q.; Liu, H.Y.; Mickley, L.J.; Schultz, M.G. Global modeling of tropospheric chemistry with assimilated meteorology: Model description and evaluation. *J. Geophys. Res. Atmos.* **2001**, *106*, 23073–23095. [CrossRef]
54. Statistics Canada. Table 23-10-0003-01. Aircraft Movements, by Civil and Military Movements, Airports with NAV CANADA Towers, Monthly. 2020. Available online: https://doi.org/10.25318/2310000301-eng (accessed on 29 September 2020).

55. Google. COVID-19 Community Mobility Reports. 2020. Available online: https://www.google.com/covid19/mobility/ (accessed on 29 September 2020).

Publisher's Note: MDPI stays neutral with regard to jurisdictional claims in published maps and institutional affiliations.

MDPI
St. Alban-Anlage 66
4052 Basel
Switzerland
Tel. +41 61 683 77 34
Fax +41 61 302 89 18
www.mdpi.com

Remote Sensing Editorial Office
E-mail: remotesensing@mdpi.com
www.mdpi.com/journal/remotesensing

www.ingramcontent.com/pod-product-compliance
Lightning Source LLC
Chambersburg PA
CBHW061553210926
51850CB00009B/2090
* 9 7 8 3 0 3 6 5 5 8 9 3 6 *